PHOTOINDUCED
PHASE TRANSITIONS

PHOTOINDUCED
PHASE TRANSITIONS

editor

K. NASU

High Energy Accelerator Research Organization (KEK), Japan

World Scientific

NEW JERSEY • LONDON • SINGAPORE • BEIJING • SHANGHAI • HONG KONG • TAIPEI • CHENNAI

Published by

World Scientific Publishing Co. Pte. Ltd.
5 Toh Tuck Link, Singapore 596224
USA office: 27 Warren Street, Suite 401–402, Hackensack, NJ 07601
UK office: 57 Shelton Street, Covent Garden, London WC2H 9HE

British Library Cataloguing-in-Publication Data
A catalogue record for this book is available from the British Library.

PHOTOINDUCED PHASE TRANSITIONS

ISBN 981-238-763-3

Printed in Singapore by World Scientific Printers (S) Pte Ltd

Preface

There are large varieties of materials around us. In these materials a few kinds of macroscopic numbers of atoms or molecules are condensed with a definite composition ratio. Without changing this chemical composition, a material can undergo various phases from gaseous and liquid phases to crystalline ones, as temperature decreases from high to low. Even at low temperatures, a material can undergo various crystalline phases, whose lattice structure and electronic state change as temperature decreases. However, all these states are so-called equilibrium phases, in the sense that the free energy of each state takes its global minimum at each temperature.

Let us now proceed to the nonequilibrium phases generated from an equilibrium one by some external excitations or stimulations. Usually, these states are transient ones having higher energies than the equilibrium one, but, depending on the way of stimulations or excitations, we can get much more varieties of states, even if the starting equilibrium phase is the same. Surely they also relax down to the starting equilibrium phase after a period of time. However, if they are locally stable state, in the sense that their free energy takes local minima separated from the global one by substantial energy barriers, the time required for the relaxation will be long. In that case, we can complete necessary observations to determine their characteristics within this period of time. Hence, such a long-lived locally stable nonequilibrium phase is effectively the same as the equilibrium one.

Very recently a new class of insulating solids was discovered. When shone by only a few visible photons, the solids give rise to a macroscopic excited domain that has new structural, electronic and even magnetic orders quite different from the starting ground state. This phenomenon is called "photoinduced phase transition" and we can generate new long-lived locally stable macroscopic nonequilibrium phases through the excitations or stimulations by a few visible photons. The purpose of this multi-authored book is to review recent theoretical and experimental studies on this new phenomenon.

The photoinduced phase transition study has firstly started by using an organic charge transfer crystal, TTF-CA. At present, however, this study has extended to many kinds of materials, not only to other organic charge transfer crystals, but also to various metal complex crystals, Prussian blue analogues, and even to perovskite type metal oxides.

The standard technique to observe the photoinduced phase transitions is the so-called modulation spectroscopy, in which another visible photon is shone to detect a spectroscopic change between before and after the transition. Very recently, however, we succeeded in observing the photoinduced structural phase transition more directly, by using the time-resolved x-ray diffraction.

One of the most essential points of this research is to clarify "*why and how photoexcited few electrons can finally result in an excited domain with a macroscopic size*", worthy to be called a phase transition. Another important problem is "*how the resultant photoinduced nonequilibrium phase is different from the ordinary thermal induced phases*". Throughout this book, we will be fully concerned with these essential problems.

The history of the photoinduced phase transition research is quite young. It has just started and we still have various unsolved problems. There may be some cases, wherein the same experimental result is understood a little different way, depending on each researcher. For the mechanism of the photoinduced phase transition in TTF-CA, for example, a universal understanding that is commonly accepted by all relevant researchers is still not well established. In this book, we will introduce to readers this immature situation, just as it is. We believe, however, that such an up-to-date review will greatly contribute to the development of photoinduced phase transition research.

Keiichiro Nasu
Tsukuba, Japan
October 10, 2003

Contents

CHAPTER 1

THEORIES FOR PHOTOINDUCED STRUCTURAL PHASE TRANSITIONS AND THEIR DYNAMICS

Keiichiro Nasu

Solid State Theory Division
Institute of Materials Structure Science, KEK
Graduate University for Advanced Study, 1-1, Oho, Tsukuba
Ibaraki, 〒305-0801, Japan
E-mail: knasu@post.kek.jp

We review the present status of theoretical studies for photoinduced structural phase transitions and their dynamics, in connection with the two key concepts; the hidden multi-stability of the ground state and the proliferations of optically excited states. Taking the ionic (I) $\rightarrow$ neutral (N) phase transition in the organic charge transfer (CT) crystal, TTF-CA, as a typical example for this type transition, at first, we theoretically show an adiabatic path which starts from CT excitons in the I-phase, but finally reaches to a N-domain with a macroscopic size. In connection with this I-N transition, the concept of the initial condition sensitivity is also developed so as to clarify experimentally observed nonlinear characteristics of this material.

Next, using a more simplified model for the many-exciton system, we study theoretically the early time quantum dynamics of the exciton proliferation, which finally results in the formation of a domain with a large number of excitons. For this purpose, we derive a stepwise iterative equation to describe the exciton proliferation and clarify the origin of the initial condition sensitivity.

Possible differences between a photoinduced nonequilibrium phase and an equilibrium phase at high temperatures are also clarified from general and conceptional points of view, in connection with recent experiments on the photoinduced phase transition in an organo-metallic complex crystal. It will be shown that the photoinduced phase can make a new interaction appear as a broken symmetry only, even when this interaction is almost completely hidden in all the equilibrium phases, such as the ground state and other high-temperature phases. The relation between the photoinduced nonequilibrium phase and the hysteresis induced nonequilibrium one is also qualitatively discussed.

We are also concerned with a macroscopic parity violation and a ferro- (or super-para-) electricity, induced by a photogenerated electrons in the perovskite type quantum dielectric $SrTiO_3$. The photogenerated electron in the 3d band of Ti, is assumed to couple, weakly but quadratically with soft-anharmonic T_{1u} phonons and strongly but linearly to the breathing (A_{1g}) type high energy phonons. These two types of electron-phonon couplings result in two types of polarons, a "super-para-electric large polaron" with a quasi-global parity violation and an "off-center type self-trapped polaron" with only a local parity violation. This super-para-electric large polaron, being equal to an itinerant ferroelectric domain, greatly enhances both the quasi-static electric susceptibility and electronic conductivity.

Contents

1. Equilibrium Phases, Nonequilibrium Phases and Photoinduced Phase Transitions

There are large varieties of materials around us. In these materials a few kinds of macroscopic numbers of atoms or molecules are condensed with a definite composition ratio. Without changing this chemical composition, a material can undergo various phases from gaseous and liquid phases to crystalline ones, as temperature decreases from high to low. Even at low temperatures, a material can undergo various crystalline phases, whose lattice structure and electronic state change as temperature decreases. However, all these states are so-called equilibrium phases, in the sense that the free energy of each state takes its global minimum at each temperature.

Let us now proceed to the nonequilibrium phases generated from an equilibrium one by some external excitations or stimulations. Usually, these states are transient ones having higher energies than the equilibrium one, but, depending on the way of the stimulations or excitations, we can get much more varieties of states, even if the starting equilibrium phase is the same. Surely they also relax down to the starting equilibrium phase after a period of time. However, if they are locally stable states, in the sense that their free energy take local minima separated from the global one by substantial energy barriers, the time required for the relaxation will be long. In that case, we can complete necessary observations to determine their characteristics within this period of time. Hence, such a long-lived locally stable nonequilibrium phase is effectively the same as the equilibrium one.

Various amorphous crystals are well known as typical examples of these nonequilibrium phases. They are brought about by materials cooling down rapidly from their high temperature phases. The life time of the relaxation is believed to be much longer than the time scale of our daily life. Unfortunately, however, these amorphous crystals have no well-defined long range periodic crystalline or electronic order, since they are just amorphous.

Very recently, on the other hand, a new class of insulating solids, was discovered. When shone by only a few visible photons, become the solids give use to a macroscopic excited domain that has new structural and electronic orders quite different from the starting ground state (equilibrium phase). This phenomenon is called "photoinduced phase transition" and we can generate new long lived locally stable macroscopic nonequilibrium phases through the excitations or stimulations by a few visible photons. The purpose of this paper is to review recent theoretical studies on this phenomenon, in close connection with various recent experiments.

2. Relaxation of Optical Excitations, Hidden Multi-Stability and Photoinduced Phase Transitions

As is already well known, an electron in an insulating crystal induces a local lattice distortion around itself, when it is excited by a photon. A sudden change of charge distribution by this optical excitation appears, indusing a motion of the crystal lattice surrounding the excited electron, so that the whole electron-lattice system reaches a new equilibrium position within the excited state. This phenomenon is called "lattice relaxation" of an optical excitation, and the resultant state is often called "photoinduced structural change", as schematically shown in Fig. 1. This relaxation phenomenon has been studied in detail, in various kinds of insulating crystals for the last fifty years. According to the original concept of this lattice relaxation, however, it is tacitly assumed to be a microscopic phenomenon, in which only a few atoms and electrons are involved.[1]

In recent years, on the other hand, many unconventional photoactive solids were discovered, where the relaxation of optical excited states results in various collective motions involving a large number of atoms and electrons. In some cases, it results in a macroscopic excited domain with new structural and electronic orders quite different from the original ones. This situation can be called "photoinduced structural phase transition (PSPT)".[2–5]

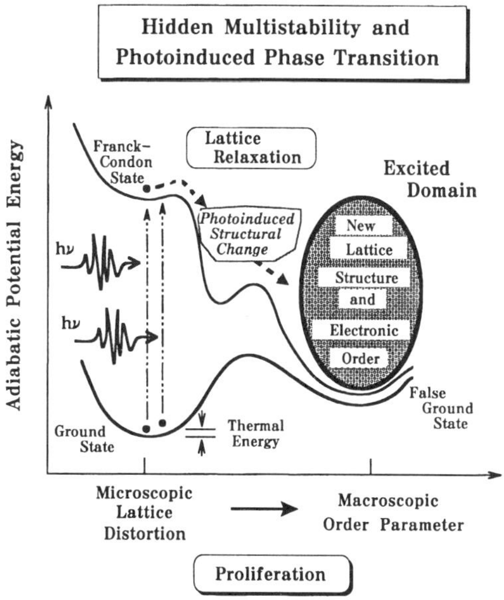

Fig. 1. Schematic and conceptional natures of photoinduced structural phase transition, hidden multi-stability and proliferation.

These problems are closely related with the hidden multi-stability intrinsic to each solid. If the ground state of a solid is pseudo-degenerate, being composed of true and false ground states with each structural and electronic orders different from others, we call it multi-stable. In this case, the photo-absorption, being initially a single-electron excitation from the true ground state, can trigger local but macroscopic instabilities. The photo-absorption can induce low-lying collective excitations during the lattice relaxation, and can finally produce a false ground state to create boundaries between the two states. Thus a local but macroscopic excited domain appears. In other words, the initially created single-electron excitation proliferates during the relaxation and grows up to be a macroscopic order, as schematically shown in Fig. 1.

The origin of the pseudo-degeneracy can be understood from the conceptual point of view related to the cohesive mechanism of each solid. As mentioned before, the solid comprises a macroscopic number of a few kinds of atoms or molecules with a definite composition ratio. However, even if the constituent atoms (molecules) are defined, the structural and electronic orders, which will be realized in the macroscopic ground state, are not always determined straightforwardly. For example, in the case of alkali halide, the ionic state and the covalent one are two well-known candidates for the ground state.[6] Moreover, if there are two predominant but mutually conflicting elements in the original Hamiltonian, there appear two candidates that inherit this conflict. Thus, it is quite ubiquitous that we have the multi-stability or the pseudo-degeneracy with the true and false ground states.

It is also very important to see the relation between photoinduced phase transitions and the ordinary ones due to thermal excitation. When the false ground state is so fortunate as to be just above the true one, being easily excited by thermal energy, we may get the ordinary phase transition and can recognize the presence of this false ground state. However, there will be various other cases where the energy of the false ground state is too high to be excited thermally. Ordinary thermodynamic measurements can be concerned only with the true ground state or small excitations thereof, and hence, they can never detect such hidden states. Even in these difficult cases, we can create the false ground state by photo-excitation and the lattice relaxation thereof, as schematically shown in Fig. 1.

Such a false ground state always disappears finally within a finite lifetime, and can never be permanent, as mentioned before. However, according to the recent progress of our laser spectroscopy techniques, an infinite life time is no more necessary for each state to be recognized as a well-defined state, provided that it can last long enough to be clearly observed by other photons to detect it. At present, the time required for this type observation is usually less than 10^{-12} second, and in some cases, even 10^{-15} second is enough.

Incidentally, let us briefly see the difference between the present photoinduced phase transition and the so-called "new material design (or search)", which is the most contemporary trend in the field of materials science. One of the most standard techniques for the new material design (or search) is to apply static external fields such as magnetic fields or pressures onto a material, which is expected to give novel or anomalous properties absent in other existing materials. The other most standard technique is to design or synthesize the material by changing its chemical composition, little by little, so that it will show rather new properties.

In the case of static external field, however, it changes all the electronic state of the material, both ground and excited states, unselectively, while photons have definite momentum, phase, helicity and energy. Hence, they can create only particular excited states, selectively and quite intensively. In contrast to the chemical design or synthesis, the photoinduced phase transition does not change the chemical composition of the material, but can realize new states. Thus the research for photoinduced phase transitions will be able to open a new multistoried concept for materials.

3. Photoinduced Ionic → Neutral Phase Transition in Organic Molecular Crystal TTF-CA

As one of the typical examples for the PSPT, here, we will be concerned with the photoinduced ionic (I) → neutral (N) transition in an organic molecular crystal Tetrathiafuluvalene-p-Chloranil and review the present stage for its experimental studies. Both Tetrathiafuluvalene (TTF) and p-Chloranil (CA) are planar organic molecules as schematically shown at the top of Fig. 2 and their crystal has a quasi 1-d chain like structure, in which these two molecules are alternately stacked along this 1-d chain axis.

In the true ground state of this crystal at absolute zero of temperature, both TTF and CA become a cation and an anion, respectively, and make a dimer with each other as shown in Fig. 2a. This is called the I-phase. On the other hand, we also have the N-phase, in which neutral TTF and CA are alternately stacked without dimerization, as shown in Fig. 2d. This is the accidentally pseudo-degenerate false ground state and at absolute zero of temperature, it is just above the ionic true ground state.

Keeping this material in the low enough temperature, but shining a strong laser light of about 0.6∼2.3 eV onto it, we can generate the N-domain even in the ionic true ground state, as shown in Fig. 2c. This change was experimentally confirmed by the change (≡ΔR) of the optical reflectivity. Figure 3 shows the spectral shape of original optical reflectivity (≡R) of the TTF-CA crystal in the I-phase at 2 K, where the exciting light is polarized perpendicular to the 1-d chain axis.[7] In this spectrum, we can see a characteristic peak structure at around 3 eV, and this peak corresponds to an intra-molecular electronic excitation of TTF^+. Hence its intensity can be used as a macroscopic indicator for the presence of cationic TTF molecules in this crystal.

In fact, when this crystal is shined by a light with an energy of 2.3 eV, we can get a relative change (≡ΔR/R) of this reflectivity, as shown in Fig. 4.[8] The intensity of the aforementioned peak is clearly seen to decrease. It means that a macroscopic number of neutral TTF molecules (N-domains) have been generated in the I-phase. This is nothing else but the PSPT. From this decrease of the peak intensity, we can determine that the domain is composed of about 200∼1000 neutral pairs and it can last for about 10^{-3} second.[8]

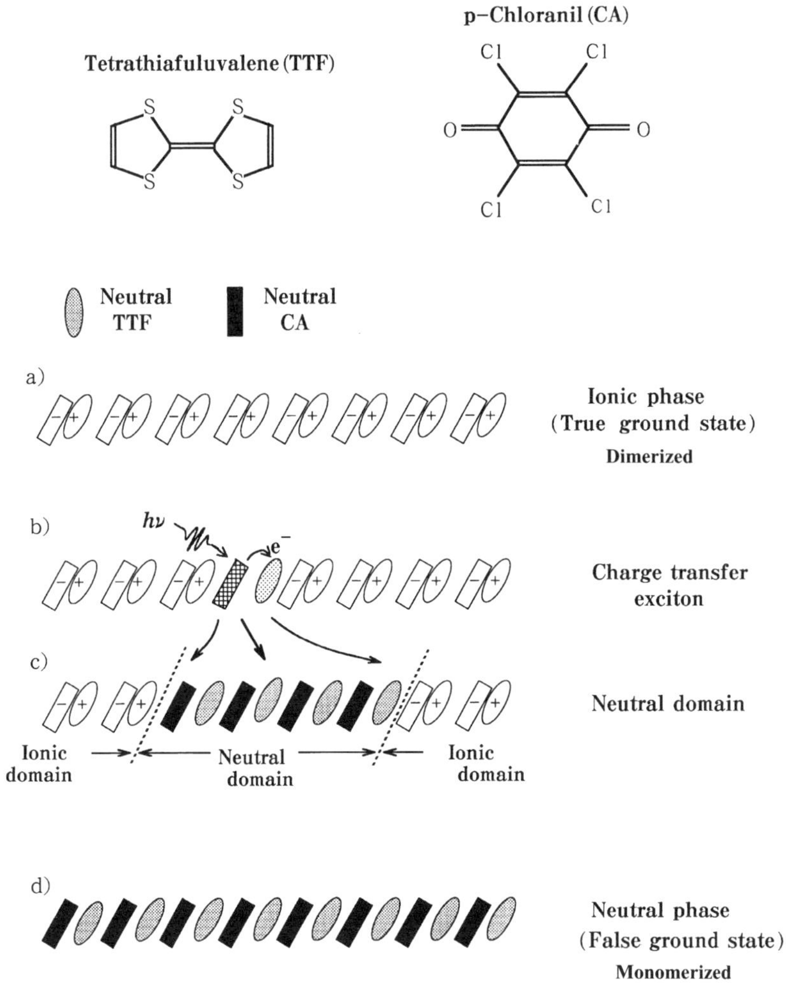

Fig. 2. Schematic nature of photoinduced ionic → neutral structural phase transition in TTF-CA crystal. a) Ionic phase, b) Charge transfer exciton, c) Neutral domain, d) Neutral phase.

4. Threshold Excitation Intensity and Initial Condition Sensitivity

For this transition, we can think of the following simple and intuitive scenario as schematically shown in Figs. 2b and 2c. That is, a single photon can make a single charge transfer (CT) excitation between neighboring molecules, which is just equal to a neutral pair generation. After that, the

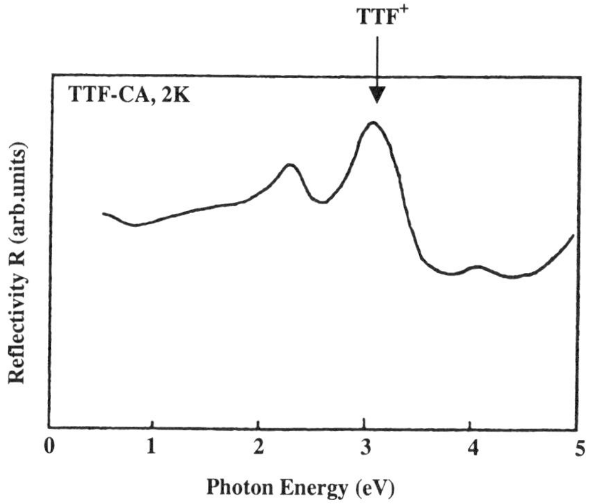

Fig. 3. Spectral shape of original optical reflectivity (≡R) of the TTF-CA crystal in the I-phase at 2 K. The exciting light is polarized perpendicular to the 1-d chain axis. Peak at around 3 eV corresponds to an intra-molecular electronic excitation of TTF^+. From Ref. 7.

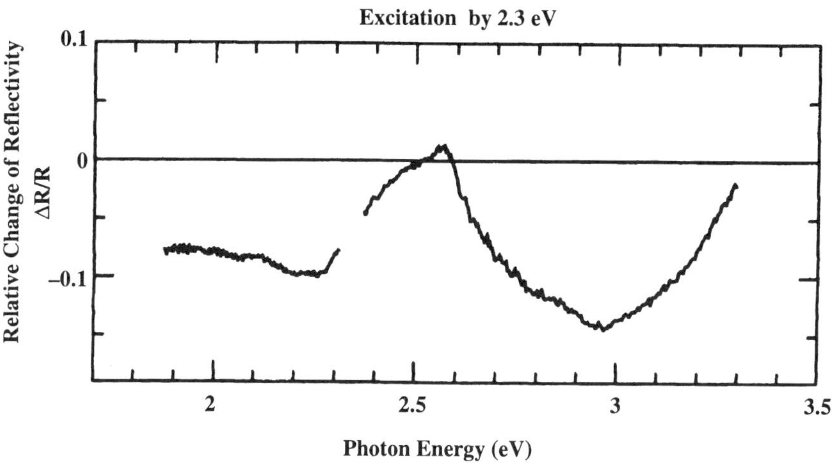

Fig. 4. Relative change (≡ΔR/R) of the reflectivity, when the TTF-CA crystal is shined by a light with an energy of 2.3 eV. From Ref. 8.

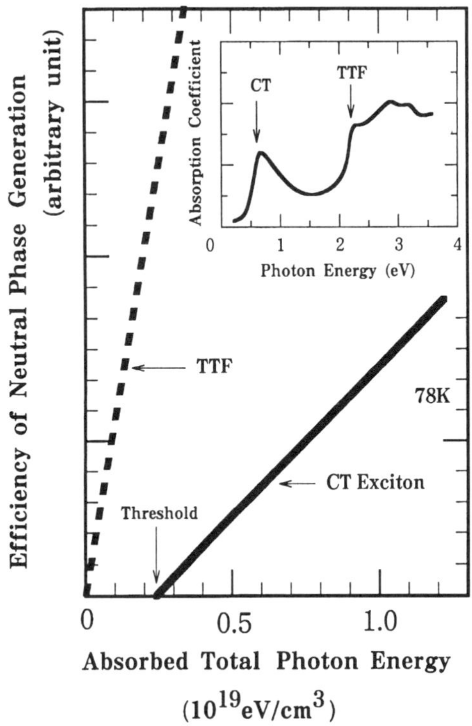

Fig. 5. The efficiency of neutral phase generation in TTF-CA as functions of total absorbed photonic energy. The horizontal axis denotes the total photon energy absorbed in the unit volume of the TTF-CA crystal. This total energy is calculated by taking the three quantities into account: the absorption coefficient, the energy of the photon and its intensity. The thick solid line is the CT excitation (0.6 eV), while the thick dashed line is the intra-molecular excitation (2.3 eV). The small inset denotes the light absorption spectrum of TTF-CA crystal. From Ref. 9.

number of this neutral pair will increase like a domino game. However, by the recent experimental studies shown in Fig. 5,[9] this simple scenario is proved wrong.

In this figure, we have shown the photo-absorption spectrum of TTF-CA, as a small inset. It has two peaks at 0.6 eV and also at 2.2 eV. The first one, being the elementary optical excitation of this crystal, corresponds to the aforementioned inter-molecular CT excitation. Among a macroscopic number of neighboring ion pairs (TTF^+ and CA^-) in Fig. 2a, only a single neighboring ion pair returns to a neutral pair (TTF and CA) by this excitation as shown in Fig. 2b, and this neutral pair also itinerates along the crystal axis, keeping all other pairs still ionic. This is nothing special but the so-called inter-molecular CT exciton, wherein a hole and an electron (a positive charge and a negative one relative to the I-phase) are bound together, so that they are always at two neighboring molecules with each other, even if their center-of-mass itinerates. The second peak at 2.2 eV corresponds to an intra-molecular electronic transition of TTF, just like the case of the aforementioned 3 eV peak. The thick solid line in Fig. 5 denotes the efficiency of the N-phase generation as a function of the exciting photon intensity, when the photon energy is fixed at this CT exciton (0.6 eV). We can clearly see that there is a threshold in the intensity, below which the macroscopic N-phase can never be generated. It means that a single CT exciton alone can never result in the macroscopic N-phase, but only through a nonlinear cooperation between several photo-excited CT excitons, the new phase can be attained. The presence of this threshold is the first note-worthy characteristic of the TTF-CA crystal.

The second note-worthy characteristic, which we can see from Fig. 5, is the difference between the dashed line and the aforementioned thick solid line. Strictly speaking, the horizontal axis of this figure does not simply denote the photon intensity itself, but denotes the total photon energy which is absorbed in the unit volume of the TTF-CA crystal. This total energy is calculated by taking the three quantities into account, the absorption coefficient, the energy of the photon and its intensity. Thus, we can compare the 0.6 eV excitation and the 2.2 eV excitation (the thick dashed line) on an equal footing. For example, if we focus on the point with a value 0.25 of the horizontal axis in Fig. 5, we can find that the efficiency becomes very high or almost zero, depending sensitively on the photon energy, although the total absorbed photonic energies are same each other. Not only these two cases, we can also excite using various other photons in between (0.6$\sim$2.2 eV), keeping the total absorbed photonic energies are same. However all these excitations give the efficiencies different with each other, and ranging between the two lines in Fig. 5.

From this fact, we can immediately conclude that it is not the ordinary thermal phase transition. In the experimental studies for photoinduced phase transitions, the first thing we have to examine is whether it is the ordinary thermal phase transition. Because, the absorbed photons may often be converted into heat in the crystal, raise up its temperature and may indirectly result in ordinary thermal phase transitions. In the case of this indirect thermal phase transition, however, its generation efficiency will depend only on the total absorbed photon energy and will never depend sensitively on the way or type of the excitation. Figure 5 shows that the generation efficiency quite sensitively depends on the type of the excitation, that is, the exciting photon energy, even if the total absorbed photonic energies are same. If we go back to Fig. 1, the aforementioned two types of excitations start from the common ground state minimum equally, and moreover, the energies of the final states of the Franck–Condon transition are almost the same.

Here, we should emphasize the so-called Franck–Condon principle. The optical transition can complete within a time of the order of 10^{-15} second, provided that its transition energy is in the visible region, while, the period of oscillation of a crystal lattice or phonon is of the order of 10^{-12} second. Thus, the configuration of the crystal lattice can never change during the optical transition, and hence, it can occur only vertically as schematically shown in Fig. 1. Consequently, possible differences between the aforementioned two transitions are only in the electronic natures of these Franck–Condon type excited states, from which the lattice relaxation and the proliferation start. However, this small difference in the initial state of the relaxation later diverges and finally determines occurrence or nonoccurrence of the photoinduced phase transition. This is the so-called "initial condition sensitivity", peculiar to the dynamics of nonlinear systems.

As is well known, the initial condition sensitivity has been studied mainly from the mathematical point of view, by taking classical nonlinear model systems with only a few degree of freedoms.[10] Setting up the initial condition, in this case, is also a purely mathematical and artificial procedure in order to solve nonlinear differential equations that describe these model systems. On the other hand, the present case is the formation of a macroscopic order in a real material and setting up the initial condition itself is also a real physical process. As mentioned before, it is set by choosing the spatiotemporal pattern of the exciting photon pulse, and hence, it is in compliance with quantum uncertainty. Thus the studies for the photoinduced phase transitions will open up new aspects of nonlinear dynamics and self-organization phenomena.

5. Adiabatic Path of Photoinduced Ionic → Neutral Phase Transition in TTF-CA Crystal

Keeping these points in mind, very recently, Hai *et al* have theoretically clarified the adiabatic relaxation path, which starts from a Franck–Condon type optical excitation in the ionic ground state of the TTF-CA crystal and terminates up to the large N-domain formation in this crystal.[11] Let us review it in detail. In order to clarify the photoinduced phenomena in TTF-CA theoretically, we have to specify our microscopic model to describe the many-electron system strongly coupling with lattice distortions (phonons) in this molecular crystal. The constituent TTF and CA molecules are, of course, originally neutral. However, when they are condensed as a crystal at low enough temperatures below 84 K, the whole system becomes the ionic and dimerized phase shown in Fig. 2(a). At temperatures higher than 84 K, it goes back to the neutral and monomeric phase as shown in Fig. 2(d) and this ordinary thermal phase transition is the first order one.[12]

As shown by Soos theoretically,[13] the change from the original neutral state to the ionic one in TTF-CA can be described as an electron transfer from the highest occupied molecular orbital (HOMO) of TTF to the lowest unoccupied molecular orbital (LUMO) of CA. The main energy gain in the ionic state relative to the neutral one is the long range Coulomb interaction between electrons and ions, in particular, the Coulomb attraction between a cation and an anion thus generated, and this energy gain increases if the inter-molecular distance decreases after the ionization, as first shown by Sakano–Toyozawa.[14]

5.1. *Model Hamiltonian*

Theory by Huai *et al*[11] is intended to take these essential points into account straightly, and based on the extended Peierls–Hubbard model for valence electrons in the HOMO of TTF molecules and the LUMO of CA molecules as schematically shown in Fig. 6. The inter-molecular Coulomb repulsion

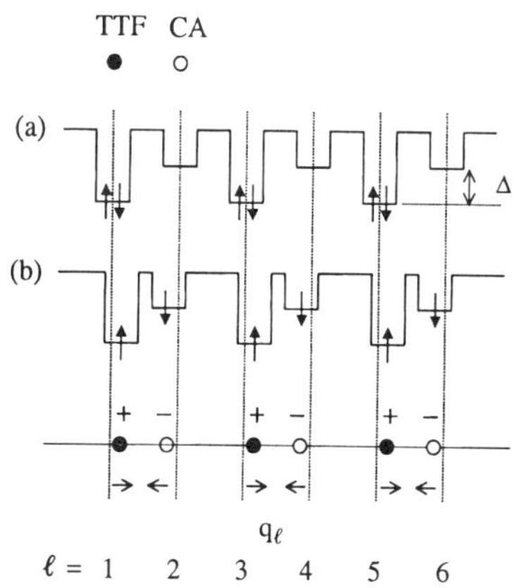

Fig. 6. Schematic nature of neutral (a) and ionic (b) phases in TTF-CA crystal. Up and down arrows denote the electrons and their spins. Δ denotes the energy difference between the HOMO and the LUMO. q_ℓ is the displacement of the molecule at that site.

in this model is assumed to depend nonlinearly on the inter-molecular distance. This model also takes weak interactions between 1-d chains into account, so that we can describe the N-domain formation in the three-dimensional TTF-CA crystal in a more realistic way. Our Hamiltonian ($\equiv$H) reads

$$\mathrm{H} = \mathrm{H_{el}} + \mathrm{H_{ph}} + \mathrm{H_{inter}}, \tag{1}$$

where $\mathrm{H_{el}}$, $\mathrm{H_{ph}}$ and $\mathrm{H_{inter}}$ denote the Hamiltonians of the electron part, the phonon part and the inter-chain interaction. The electronic part $\mathrm{H_{el}}$ is given as

$$\begin{aligned}\mathrm{H_{el}} \equiv & -\mathrm{T_i}\sum_{\ell,\sigma}\left[\left(\mathrm{C}^+_{\ell,\sigma}\mathrm{C}_{\ell+1,\sigma}\right) + (\mathrm{h.c.})\right] + \frac{\Delta}{2}\sum_{\ell}(-1)^{\ell}\mathrm{n}_{\ell} \\ & + \mathrm{U}\sum_{\ell}\mathrm{n}_{\ell,\alpha}\mathrm{n}_{\ell,\beta} - \sum_{\ell\in\mathrm{odd}}\mathrm{V}_{\ell}(\mathrm{q}_{\ell},\mathrm{q}_{\ell+1})[2-\mathrm{n}_{\ell}]\mathrm{n}_{\ell+1} \\ & - \sum_{\ell\in\mathrm{even}} V_{\ell}(q_{\ell},q_{\ell+1})[2-n_{\ell+1}]n_{\ell},\end{aligned} \tag{2}$$

$$\mathrm{n}_{\ell,\sigma} \equiv \mathrm{C}^+_{\ell,\sigma}\mathrm{C}_{\ell,\sigma}, \quad \mathrm{n}_{\ell} \equiv \sum_{\sigma}\mathrm{n}_{\ell,\sigma}, \tag{3}$$

where $\mathrm{T_i}$ is the transfer energy of an electron between a HOMO and its neighboring LUMO. $\mathrm{C}^+_{\ell,\sigma}$ is the creation operator of an electron with spin $\sigma(=\alpha,\beta)$ at ℓth lattice site, which is numbered from left to right along the 1-d chain as shown in Fig. 6, and odd sites correspond to the HOMO, while the even ones correspond to the LUMO. Δ in Eq. (2) is the energy difference between the HOMO and the LUMO, while U is the intra-molecular Coulomb repulsive energy of electrons, and is assumed common to both the HOMO and the LUMO. V_{ℓ} in the second line of Eq. (2) is the Coulomb interaction between neighboring two molecules, wherein the total charge in the TTF site ($\ell \in$ odd) is $+[2-n_{\ell}]$, while that in the CA site ($\ell \in$ even) is $-n_{\ell}$. As mentioned before, this interaction is assumed to depend nonlinearly on the change of the inter-molecular distance $(\mathrm{q}_{\ell}-\mathrm{q}_{\ell+1})$, where q_{ℓ} is the displacement of each molecule along the chain axis as shown in Fig. 6. The unit of length is the original inter-molecular distance. Thus, V_{ℓ} is given as

$$\mathrm{V}_{\ell}(\mathrm{q}_{\ell},\mathrm{q}_{\ell+1}) = \mathrm{V}_0 + \beta_1(\mathrm{q}_{\ell}-\mathrm{q}_{\ell+1}) + \beta_2(\mathrm{q}_{\ell}-\mathrm{q}_{\ell+1})^2, \tag{4}$$

where V_0 is the original value of V_{ℓ}, while β_1 and β_2 are its first and second order expanding coefficients with respect to $(\mathrm{q}_{\ell}-\mathrm{q}_{\ell+1})$.

It is well known that the electron-phonon coupling leading to the dimerization may have two origins. The first one is the present case shown in Eq. (4), being the modulation of V_ℓ, and this model was first proposed by Sakano–Toyozawa,[14] while, second one is the modulation of T_i.[15–17] Our model is not this second one. Because of the small overlap between the HOMO and the LUMO, this T_i becomes so small that its distance variation can be neglected. Moreover, as we will show later in detail, our model is rather straightforward as it makes the N-I phase transition first order, in contrast to the SSH model,[15] which makes the transition to be the second order.

The phonon part H_{ph} of Eq. (1) is given as

$$H_{ph} \equiv \sum_\ell \frac{S_1}{2}(q_\ell - q_{\ell+1})^2 + \sum_\ell \frac{S_2}{4}(q_\ell - q_{\ell+1})^4, \tag{5}$$

wherein a fourth order potential with a coefficient S_2 is introduced, as well as the ordinary second order one with the coefficient S_1, while the kinetic energy of phonons are neglected because of the adiabatic approximation.

As for the inter-chain interaction H_{inter}, we neglect it in the study of the ground state properties. This inter-chain interaction is tacitly assumed to be so small as to give almost no contribution provided that the ground state is uniform for all the chains in the crystal, no matter what is the dimerized I-phase or the monomeric neutral one. Only if a large and macroscopic domain appears in the I-phase can this inter-chain interaction be assumed to bring an appreciably energy increase. We will give its practical form later in detail.

Using this model Hamiltonian, let us now set up our simple picture for both the I- and N-phases. The N-phase is such that the HOMO of the TTF is filled up with two electrons of opposite spins (↑ and ↓), while the LUMO of the CA is vacant, as schematically shown in Fig. 6(a). Thus, the two constituent molecules are literally neutral, while, the electronic state realized in this system is just equal to the charge density wave (CDW) state of the double period. On the other hand, the I-phase is such that the HOMO and the LUMO are alternately occupied by the electrons of ↑ and ↓ spins as shown in Fig. 6(b). Within this simple picture, this state is just equal to the spin density wave (SDW) state of the double period. However, as shown in Fig. 6(b), this I-phase also has the dimerization between neighboring TTF and CA molecules.

Here we should note why Eq. (4) is nonlinear and Eq. (5) is anharmonic. If we retain only the linear term in Eq. (4), the energy gain due to the dimerization in the I-phase cancels between the left and right of each molecule, since our system is not a single ionic pair but a 1-d chain crystal. For this reason, the nonlinear term is introduced in Eq. (4). Accordingly, the quartic anharmonicity is also introduced in Eq. (5). It comes from the inter-molecular repulsion, and prevents unphysically large dimerizations.

In our theory, we have eight parameters $T_i, \Delta, U, V_0, \beta_1, \beta_2, S_1$ and S_2. The values of these parameters are determined so that, they, as a set, reproduce main experimental and theoretical results existing already prior to our theory. For T_i, we use the ab-initio calculation by Katan.[18] On the other hand, from spectroscopic studies in the visible and infrared regions, the total charge induced at the CA site of the I-phase ($\equiv\rho_I$) is determined to be $\rho_I = 0.8$.[19] By the same experiment, the total charge induced at the CA site of the N-phase ($\equiv\rho_N$) is also determined as $\rho_N = 0.3$. Unlike the simple picture shown in Fig. 6, these induced charges ρ_I and ρ_N are not natural numbers such as 1 or 0, but are fractional, because the valence electrons are quantum-mechanically itinerant, or T_i is not zero. In addition to these, the following are five well-known experimental results. By x-ray structure analysis, the dimeric displacement of each molecule in the ionic ground state is determined to be 2.5% of the original intermolecular distance.[12] Moreover, in the light absorption spectrum of the I-phase, two CT absorption peaks appear at 0.6 eV and 1.0 eV with an intensity ratio 1 : 0.5, while, in the N-phase, only a single CT peak appears at 0.6 eV.[19] Using these data, we can determine our eight parameters without serious ambiguity as $T_i = 0.17\,\text{eV}$, $V_0 = 2.716\,\text{eV}, U = 1.528\,\text{eV}$, $V_0 = 0.604\,\text{eV}$, $\beta_1 = 1.0\,\text{eV}$, $\beta_2 = 8.54\,\text{eV}$, $S_1 = 4.86\,\text{eV}$ and $S_2 = 3.4 \times 10^3\,\text{eV}$.

5.2. *Ground state properties*

Let us now see the properties of the ground state given by this Hamiltonian H, using the unrestricted Hartree–Fock approximation. Within this approximation, we can reduce the two-body terms of H_{el} into one-body terms as

$$U\, n_{\ell,\alpha} n_{\ell,\beta} \to U\big(n_{\ell,\alpha}\langle n_{\ell,\beta}\rangle + \langle n_{\ell,\alpha}\rangle n_{\ell,\beta} - \langle n_{\ell,\alpha}\rangle\langle n_{\ell,\beta}\rangle\big), \tag{6}$$

$$V_\ell n_\ell n_{\ell+1} \to V_\ell \Bigg\{ \big(n_\ell\langle n_{\ell+1}\rangle + \langle n_\ell\rangle n_{\ell+1} - \langle n_\ell\rangle\langle n_{\ell+1}\rangle\big) - \sum_\sigma \big(\langle m^+_{\ell,\sigma}\rangle m_{\ell,\sigma} + m^+_{\ell,\sigma}\langle m_{\ell,\sigma}\rangle - \langle m^+_{\ell,\sigma}\rangle\langle m_{\ell,\sigma}\rangle\big) \Bigg\}, \tag{7}$$

$$m_{\ell,\sigma} \equiv C^+_{\ell,\sigma} C_{\ell+1,\sigma}, \tag{8}$$

where $\langle \cdots \rangle$ denotes the expectation value of an operator $\cdots$. It is unknown at the present stage, but will be determined later self-consistently. Since our system is the half-filled system wherein the total number of the lattice sites ($\equiv \mathrm{N_t}$) is equal to the total number of electrons as

$$\frac{\mathrm{N_t}}{2} = \sum_{\ell} \mathrm{n}_{\ell,\alpha} = \sum_{\ell} \mathrm{n}_{\ell,\beta}, \tag{9}$$

we assume for $\langle n_{\ell,\sigma} \rangle$, $\langle m_{\ell,\sigma} \rangle$ and q_ℓ the following forms with the double period:

$$\langle \mathrm{n}_{\ell,\sigma} \rangle = \frac{1}{2} + (-1)^\ell \delta \mathrm{n}_\sigma, \quad \langle \mathrm{m}_{\ell,\sigma} \rangle = \bar{\mathrm{m}}_\sigma + (-1)^\ell \delta \mathrm{m}_\sigma, \quad \mathrm{q}_\ell = (-1)^\ell \mathrm{q}_0, \tag{10}$$

where $\delta \mathrm{n}_\sigma$, $\bar{\mathrm{m}}_\sigma$, $\delta \mathrm{m}_\sigma$ and q_0 are to be determined within the Hartree–Fock and adiabatic approximations.

The Hartree–Fock type self-consistent equation for these $\delta \mathrm{n}_\sigma$, $\bar{\mathrm{m}}_\sigma$ and $\delta \mathrm{m}_\sigma$ gives two solutions for a given q_0, and they correspond to the I-phase with $\delta \mathrm{n}_\alpha \approx -\delta \mathrm{n}_\beta$ (SDW) and to the N-phase with $\delta \mathrm{n}_\alpha \approx \delta \mathrm{n}_\beta$ (CDW). In Fig. 7, we have shown the total energy ($\equiv \mathrm{E_g}$) thus theoretically calculated as a function of q_0, and these I- and N-phases are shown to be pseudo-degenerate with each other.[11] The energy minimum of the I-phase is 0.002 eV lower that of the N-one, and they are separated by an adiabatic

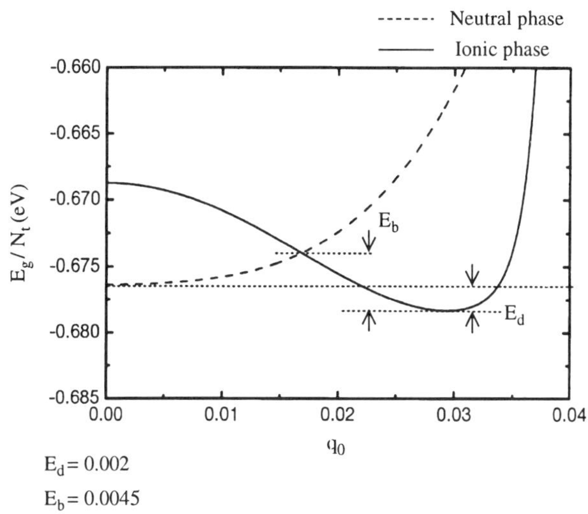

Fig. 7. Total energy per site for ionic and neutral phases as functions of q_0. From Ref. 11.

energy barrier of about 0.0045 eV. Thus the I-phase is the true ground state, while the N-phase is the false one, and both these two phases are locally stable. Furthermore, the I-phase is theoretically shown to be dimerized about 2.9% of the lattice constant, while the N-phase is not dimerized ($q_0 = 0$). Although we have been concerned only with the adiabatic energy at absolute zero temperature, the presence of the aforementioned adiabatic energy barrier means that the ordinary thermal phase transition between N-phase and I-one is first order.

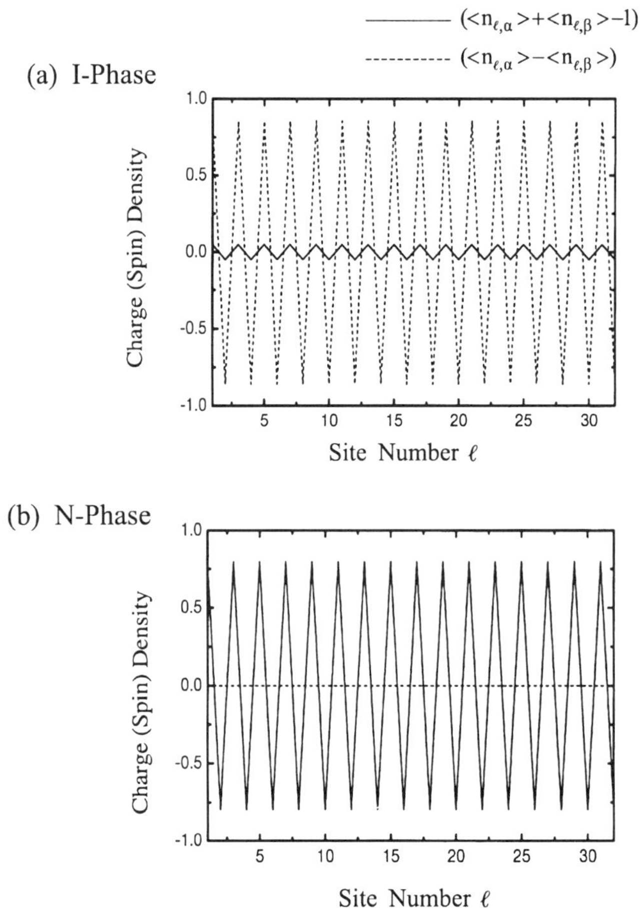

Fig. 8. Charge and spin densities of ionic (a) and neutral (b) phases as functions of ℓ. From Ref. 11.

In Fig. 8(a), we have illustrated the charge and the spin density distributions calculated for the I-phase. This phase is characterized by the strong SDW order ($\delta n_\alpha \approx -\delta n_\beta \approx 0.4$) mixed with a weak CDW type order. The calculated induce charge ρ_I is 0.95, being quite large but still fractional. Figure 8(b) demonstrates the N-phase, and there is only the CDW type order ($\delta n_\alpha \approx \delta n_\beta \approx 0.4$). The calculated induced charge ρ_N of this phase is 0.2, being comparatively small, but still significant due to the finiteness of T_i.

5.3. *Energy band structure and lowest excited state*

Figure 9(a) presents four one-electron energy bands of the I-phase, obtained as functions of a wave-vector ($\equiv$k), by using this unrestricted Hartree–Fock approximation. It has an antiferromagnetic broken symmetry in the spin space. There are two bands for each up- and down-spin electrons as distinguished by the corresponding arrows in Fig. 9(a). In the ground sate,

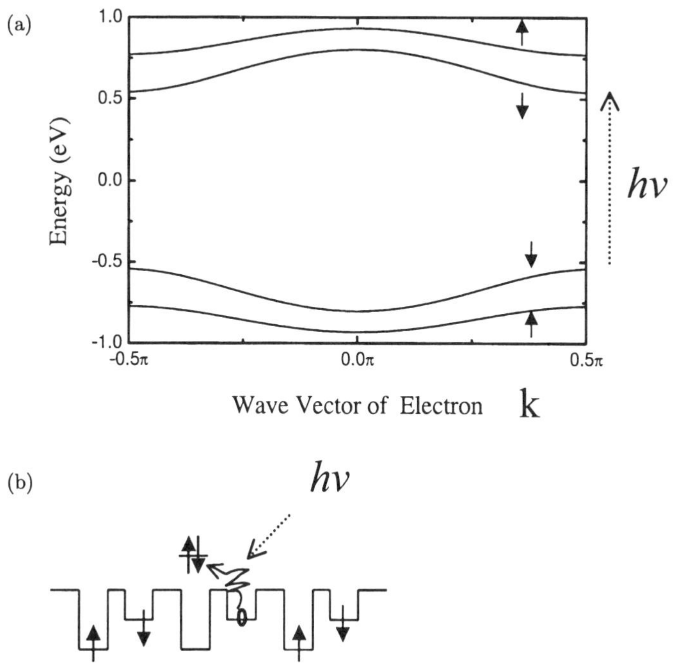

Fig. 9. (a) The energy bands in the ionic phase. The lower two bands are occupied by up and down electrons as denoted by corresponding arrows. The dashed line denotes the lowest excitation in the band picture. (b) Schematic nature of the charge transfer excitation in real space. From Ref. 11.

the lower two bands are occupied, while others are vacant. The lowest band, being occupied by up-spin electrons, is mainly composed of the LUMO of TTF, while the second lowest band, being occupied by down-spin electrons, are mainly composed of the HOMO of CA. In this system, the elementary optical excitation is such that a down-spin electron goes from the second lowest band to the third one, as shown by a dashed arrow in Fig. 9(a). This third band is mainly composed of the HOMO, which is already occupied by an up-spin electron, and hence the excitation energy partly due to U in Eq. (2).

If we change our picture from the wave-vector space to the real space, this excitation results in a doubly occupied HOMO and a vacant LUMO, while all other sites still keep original single occupancy. The electron and the hole thus created by light, attract with each other through $-V_{\ell}$, and make the aforementioned bound state called the CT exciton, as schematically shown in Fig. 9(b), and this CT exciton is the same one as already shown in Fig. 2(b). This is the lowest Franck–Condon type excited state, calculated by the theory based on the unrestricted Hartree-Fock approximation, reinforced by including the electron-hole attraction at the final state of the optical transition.[11]

Using this reinforced theory, Huai *et al* have calculated the spectral shape of the CT excitation including classical lattice fluctuations.[11] The resultant spectral shape has two peaks at 0.6 eV and at 1.0 eV in the case of the I-phase, while, in the case of N-phase, it has only a single peak at 0.6 eV. This result is consistent with the experimentally observed spectral shape of the light absorption in the energy region from 0.6 eV to 1.3 eV.[19] The theoretical origin of this change from the two peaked structure to the single peaked one was first demonstrated by Sakano–Toyozawa,[14] and the present theoretical result is identical to it, if the thermal lattice fluctuations are completely frozen. Thus, the present theory can successfully reproduce various well-known experimental and theoretical results.

5.4. *Relaxation path*

Let us now proceed to the lattice relaxation path of optical excitations. The visible photon has a long wavelength of about 1000 times of the lattice constant, and each photon can make a single CT exciton per this length. Among many excitons thus created in the whole ionic crystal, we will focus only on a single CT exciton and will clarify its lattice relaxation path. This path starts from the Franck–Condon state and terminates up to the macroscopic N-domain formation. In order to describe this path theoretically, we introduce the following domain type lattice distortion pattern q_ℓ,

$$q_\ell = (-1)^\ell q_0 \left\{1 + \Delta q \left[\tanh\left(\theta\left(|\ell| - \frac{\ell_0}{2}\right)\right) - 1\right]\right\}. \quad (11)$$

Here $(-1)^\ell\ q_0$ denotes the uniform dimeric distortion of the ionic ground state, from which we start. The second term in the curly brackets $\{\cdots\}$ denotes a local lattice displacement induced by a new excited domain. Δq is its amplitude, θ corresponds to the width of the domain boundary, and ℓ_0 is the domain size. Typical domain structures given by these parameters are demonstrated in Fig. 10. When $\Delta q < 0.5$, the lattice site ℓ inside the domain $(-\ell_0/2 \leq (\ell - 50) \leq \ell_0/2)$ has a reduced dimerization, but is still in the same phase as the original lattice outside the domain and hence it corresponds to a new I-domain. In the case of the second situation where $\Delta q = 0.5$, it is obvious that the inside lattice sites have no dimerization at all. Therefore, it corresponds to a N-domain.

Let us return to the problem of H_{inter}. The distance between the TTF and the CA molecules within a chain is about 3 Å, while the inter-chain distance is about 10 Å and hence, as mentioned before, we have neglected this interaction, assuming that it will give no serious effect. Even for the excited domain, this situation will be same, if the new displacement $(q_\ell - (-1)^\ell q_0)$ associated with this new domain is small as compared with the original displacement in the ground state, $|q_\ell - (-1)^\ell q_0| \leq q_0$. On the other hand, when the new displacement becomes very large, $|q_\ell - (-1)^\ell q_0| \gg q_0$, various

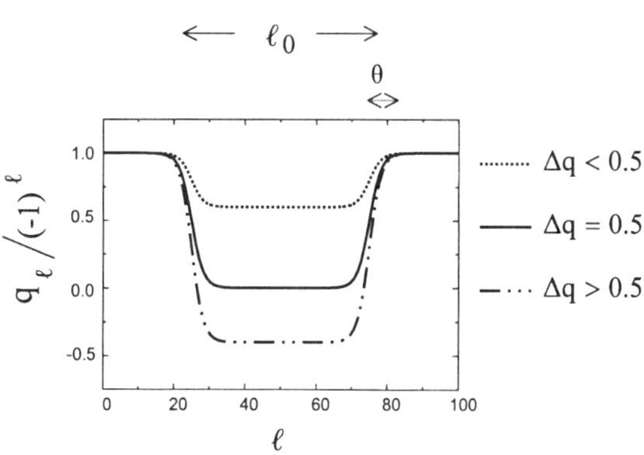

Fig. 10. The domain type lattice distortion pattern to describe the relaxation, shown as functions of ℓ. Δq is its amplitude, θ corresponds to the width of the domain boundary, and ℓ_0 is the domain size.

inter-molecular repulsions will also act, and will give a nonlinear increase of the lattice potential, just as shown in Eq. (5) and Fig. 7. Moreover, this effect will become appreciable only when the new domain expands to a macroscopic size in a single chain. To take this nonlinear inter-chain interaction into account effectively, Huai *et al* used the following form:

$$\mathrm{H_{inter}} = \sum_{\ell} \Big\{ \mathrm{K_1}(\mathrm{q}_\ell - (-1)^\ell \mathrm{q}_0)^2 + \mathrm{K_2}(\mathrm{q}_\ell - (-1)^\ell \mathrm{q}_0)^4 + K_3(q_\ell - (-1)^\ell q_0)^6 \Big\}, \tag{12}$$

where q_ℓ denotes the lattice distortion of a central chain on which we focus, while $(-1)^\ell \mathrm{q}_0$ is the representative of the displacements of environmental chains, and $\mathrm{K_i}$ $(\mathrm{i} = 1, 2, 3)$ denotes the 2ith expanding coefficient with respect to the new distortion. These environmental chains are assumed to be frozen in the ionic ground state, never being excited whatever occurs in the central chain. For practical calculations, the following values are used, $\mathrm{K_1} = 0.6949\,\mathrm{eV}$, $\mathrm{K_2} = -1.415 \times 10^3\,\mathrm{eV}$ and $\mathrm{K_3} = 9.699 \times 10^5\,\mathrm{eV}$. These values are so chosen that the interchain interaction becomes appreciable only when $|\mathrm{q}_\ell - (-1)^\ell \mathrm{q}_0| \geq 2\mathrm{q}_0$ and $\ell_0 \geq 10^2$.

Figure 11 demonstrates the adiabatic energy surface of the ground state $\mathrm{E_g}$ and that of the first excited state $(\equiv \mathrm{E_{x1}})$, thus obtained, as a function of $\Delta\mathrm{q}$ and ℓ_0; while θ is determined to minimize $\mathrm{E_{x1}}$, since we are going to clarify the relaxation path of the excited state. All the energies are referenced from the energy of the ionic true ground states $(\Delta\mathrm{q} = 0, \ell_0 = 0)$ shown in Fig. 7. In Fig. 11(a), the region with $\Delta\mathrm{q} < 0.5$ is still the I-phase, which has the SDW type order already shown in Fig. 8(a).

Nevertheless, if we go up from this ionic true ground state $(\Delta\mathrm{q} = 0, \ell_0 = 0)$ to the plateau region $(\Delta\mathrm{q} \cong 0.5$ and $\ell_0 \geq 40)$ of Fig. 11(a), the N-domain appears. In Fig. 12, we have shown charge- and spin-density distributions of this N-domain by solid lines, taking the case $\Delta\mathrm{q} = 0.5$ and $\ell_0 = 50$ as a typical example. From Fig. 12, we can see that the original SDW type order inside of the domain have now disappeared, and a CDW type order newly appears instead, although the outside of the domain still remains unchanged. According to our calculation,[11] this domain boundary between the I-phase and N-one has both a charge and a spin.

In this plateau region of Fig. 11(a), there are various shallow local minima, one of which is shown in an enlarged inset of the right-hand side. The

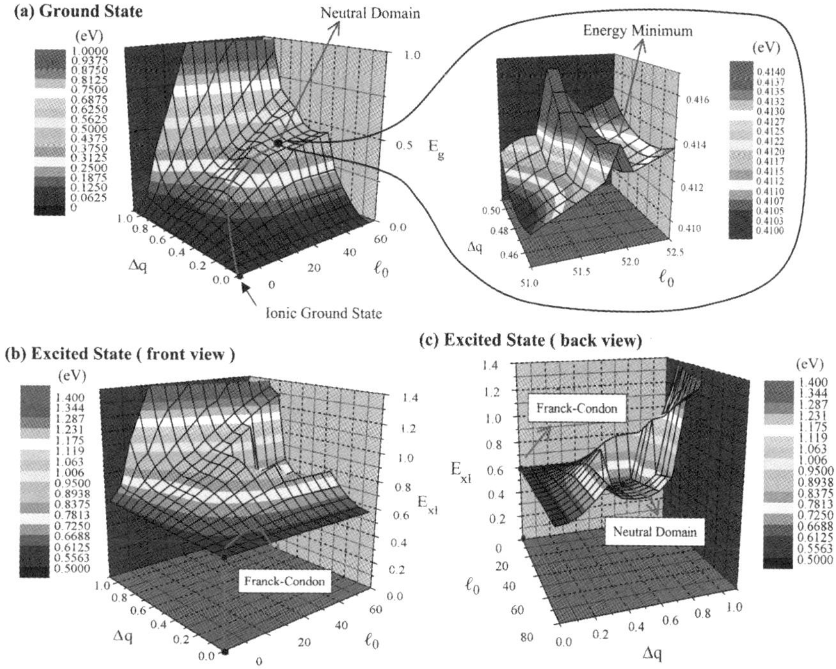

Fig. 11. The calculated adiabatic energy surfaces in a two dimensional space spanned by Δq and ℓ_0. (a) The ground state E_g. The gray line denotes the steepest ascending line from the ionic true ground state to the neutral domain. The enlarged inset denotes one of local energy minima. (b) The front view of the lowest excited state E_{x1}. (c) The back view of the lowest excited state E_{x1}. All the energies are referenced from the ionic true ground state. From Ref. 11.

presence of these local minima is mainly due to the discreteness of the lattice, and makes the N-domain stable and long-lived. As Δq and ℓ_0 increase further ($\Delta q \geq 0.5, \ell_0 \geq 50$) from this plateau region, E_g also increases further. This increase is mainly due to H_{inter} given by Eq. (12), and partly due to the intrinsic energy difference between the I-phase and the N-one shown in Fig. 7.

Figures 11(b) and (c) show the front and back views of the adiabatic energy surface of the lowest excited state E_{x1}. At around the origin ($\Delta q = 0$, $\ell_0 = 0$), or at the Franck–Condon state, a local minimum appears, and it is due to the CT exciton. We will see this Franck–Condon region later in detail. There is another local minimum at around $\Delta q = 0.5$ and $\ell_0 = 45$, and it corresponds to the excited state of the N-domain. In Fig. 12, we have also

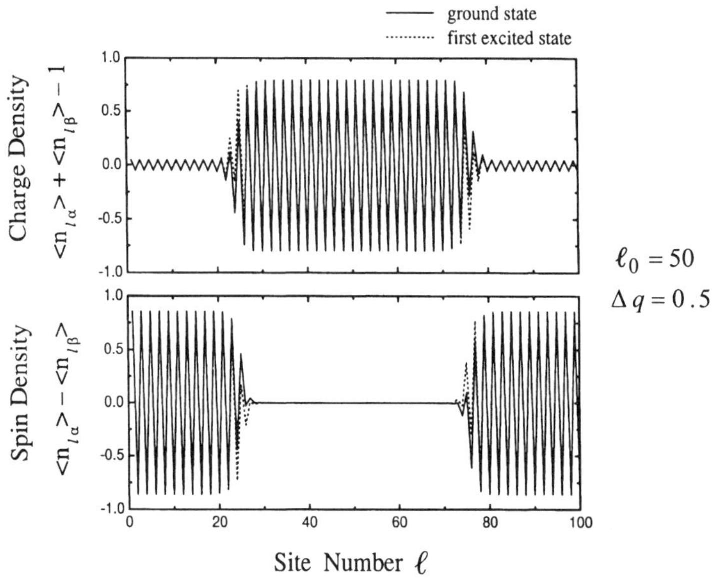

Fig. 12. Charge- and spin-density distributions of the N-domain in the I-phase. Solid lines (ground state), dashed lines (lowest excited state). $\Delta q = 0.5$, $\ell_0 = 50$. The original SDW type order inside of the domain has disappeared and a CDW type order has appeared instead. From Ref. 11.

shown its charge- and spin-density distributions by dashed lines. We can see the same CDW and SDW type orders as that of the ground state, except the domain boundary. As Δ_q and ℓ_0 increase further ($\Delta q \geq 0.5, \ell_0 \geq 50$), E_{x1} also increases further. This increase is again mainly due to H_{inter}, and partly due to the intrinsic energy difference between the I-phase and the N-one. Thus, H_{inter} is essential to make the N-domain locally stable.

In Fig. 13, we have shown both E_g and E_{x1} as a function of ℓ_0, along the steepest ascending line from the true ionic ground state to the N-domain. This line, denoted by gray in Fig. 11(a), is determined to be always orthogonal to the equi-potential line. In the adiabatic energy curve of E_{x1}, we can see two local minima mentioned before. The first one at around $\ell_0 = 0$ corresponds to the CT exciton as mentioned before. While the second one at $\ell_0 = 45$ corresponds to the N-domain, which is a little above the CT exciton. Moreover, these two minima are clearly separated by a high adiabatic potential barrier with each other. As shown in Figs. 2(c) and 12, we have two boundaries between the I-phase and N-one. The energy increases to create these two boundaries makes a small size N-domain ($\ell_0 < 40$) too unstable in the excited state. It can be stable only when its size exceeds a critical value ($\ell_0 \approx 40$), just like the formation of a gaseous bubble in an overheated liquid. We should also note that it is the characteristic only of the excited state E_{x1}. In the ground energy curve E_g shown in Fig. 13, we do not have such a high barrier, but only low barriers as shown in the enlarged inset.

Thus, we can conclude that the single CT exciton alone can never result in the straightforward formation of the macroscopic N-domain. Only when our system is excited by a large excess electronic or vibrational (phonon) energies than the single exciton state, can it overcome this high barrier and result in the macroscopic N-domain formation, as shown in Fig. 13. In other words, only when we have such a large excess energy at the very beginning of the relaxation, can this excess energy be converted into the exciton proliferation and can induce other various nonlinear processes during the relaxation.

In this connection, let us focus our attention only on the adiabatic potential energy surface of the Franck–Condon region in Fig. 13. It is shown in

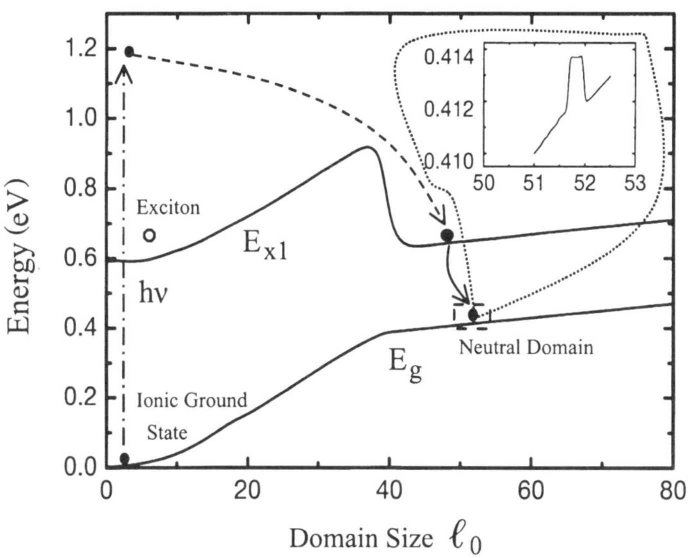

Fig. 13. The adiabatic energies of the ground state E_g and the lowest excited state E_{x1} along the steepest ascending line shown by the gray line in Fig. 11. The dashed line denotes the expected relaxation path to generate the neutral domain. The enlarged inset denotes one of local minima. From Ref. 11.

Fig. 14 in detail. We can see that the exact local energy minimum associated with the CT exciton is, not just at the origin ($\ell_0 = 0$), but at around the point with $\ell_0 \approx 7$ and $\Delta q \neq 0$. This is but the ordinary self-localization of the CT exciton.[1] Since our system has a high intrinsic nonlinearity, however, even this single CT exciton can somewhat proliferate during this self-localization process, and grow up to be a small cluster composed of a few CT excitons. Nevertheless, because of the high energy barrier mentioned before, this growth is limited only within the microscopic scale. After this limited lattice relaxation within the Franck–Condon region, the resultant cluster will disappear nonradiatively as schematically shown in Fig. 14. This theoretical result is consistent with the recent experiments.[9,20] This small scale proliferation accompanied by self-localization is already well known in various other quasi one-dimensional insulators.[2]

Finally, it should be noted that the process described above is only the early stage or the nucleation process of the macroscopic photoinduced phase transition. The neutral domain thus created will stay at the

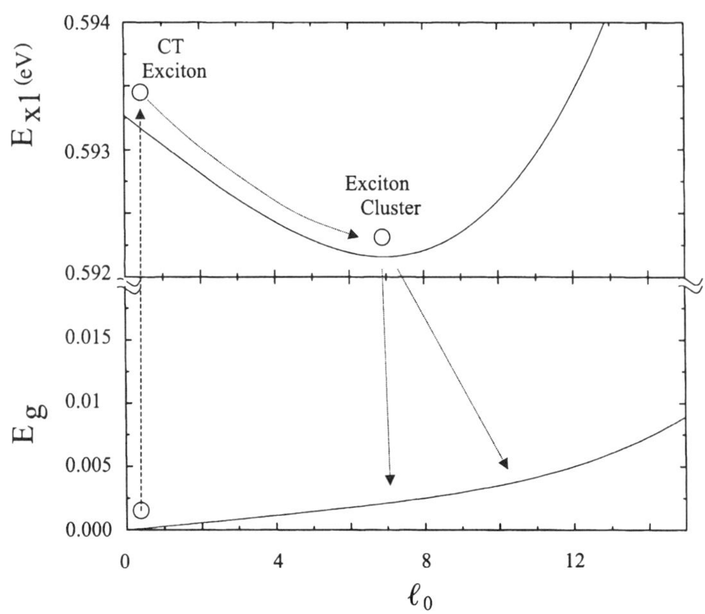

Fig. 14. The adiabatic energies of the ground state E_g and the lowest excited state E_{x1} around the Franck-Condon region. The local energy minimum at $\ell_0 \approx 7$ is a small CT exciton cluster. From Ref. 11.

aforementioned shallow local energy minimum of E_g for a longer time than the one-oscillation period of the phonon. Hence, during this stay, it can move diffusively within a chain and can merge with other N-domains created by other photons, and results in a global phase transition.

5.5. *High energy optical excited states in SDW state*

As mentioned above, the high energy optically excited states are responsible for the photoinduced phase transition, rather than the single CT exciton. Hence, we have to clarify the nature of these highly excited states of the SDW state. In the Franck–Condon state, as mentioned before, quantum and dynamical natures of the electron-phonon interaction does not work, since the lattice is fixed at the ground state configuration. As is well known, this electron-phonon interaction is a retarded one, which can start working quantum-mechanically about 10^{-12} second after the optical excitation. The inter-electron Coulomb interaction, being the instantaneous force, will dominate the nature of highly excited states, since the starting ground state is the SDW state. In order to clarify this effect, however, the unrestricted Hartree–Fock approximation is too crude and we have to take into account effects of quantum fluctuations of itinerant electrons due to the Coulomb interaction. For this reason, we clarify higher optical excited states of the half-filled SDW state, using a more simple one-dimensional extended Hubbard Hamiltonian ($\equiv H_{ex}$). It is given as

$$\begin{aligned} H_{ex} = -T_i \sum_{\ell,\sigma} \left[\left(C^+_{\ell,\sigma} C_{\ell+1,\sigma} \right) + (\text{h.c.}) \right] \\ + U \sum_{\ell} n_{\ell,\alpha} n_{\ell,\beta} + V_0 \sum_{\ell} n_\ell n_{\ell+1}. \end{aligned} \tag{13}$$

When $U > 2V_0$, this Hamiltonian gives the SDW state with the antiferromagnetic order, just like Fig. 6(b), although we have neither Δ nor the dimeric electron-phonon interaction. While, when $U < 2V_0$, we have the CDW state just like Fig. 6(a). Moreover, when U exceeds $2V_0$ only a little, we get such a situation similar to Fig. 7 that this SDW state is just below the CDW one. Thus, as far as the Franck–Condon states are concerned, we can well reproduce the situation realized in the TTF-CA crystal, using this simple model. The benefit of this simple model is that we can calculate higher excited states, using the non-Grassmann path-integral theory,[21] which takes the coulombic inter-electron quantum correlations straightforwardly. By this theory, various optical spectra have been calculated under the bistable condition,[22,23]

$$(\mathrm{U} - 2\mathrm{V}_0) \ll \mathrm{U}. \tag{14}$$

From these results, we can conclude that high energy optical excited states with energies a few times greater than the CT exciton are the random excitations of several CT excitons coupled with many magnons or spin-excitations. As is well known, in the SDW state, we have the Goldstone mode called magnon, whose energy is gapless and has a linear dispersion. When the SDW state has a small dimerization, a small energy gap opens in this dispersion. However, the above characteristic of high-energy excited states is not affected.

One can think of such a possibility that, a macroscopic CDW domain (N-domain) can be directly excited at once, just after the Franck–Condon transition from the SDW ground state by using only a few photons. However, it is impossible since these two states are "macroscopically orthogonal" with each other. It is theoretically obvious that a dipole operator of a photon can change the state of only one electron, before and after the transition. An electron and a hole thus created by this dipole transition can also change the states of many other surrounding electrons through the Coulomb interaction. However, this is but random scattering, being far from making a well-defined macroscopic spatial order. The macroscopic spatial order can be established only after this highly excited state is cooled down to some local minimum, by dissipating its excess energy one by one into the phonon system, which acts as a heart reservoir, just as shown in Fig. 13. This step-wise nature, being not at once, is the essence of the macroscopic order formation, resulting in the PSPT.

6. Early Time Nonlinear Nonequilibrium Quantum Dynamics and Initial Condition Sensitivity

In the previous sections, we have been concerned only with the adiabatic natures of the TTF-CA, and so, let us now proceed to the early time nonlinear nonequilibrium quantum dynamics of the PSPT. To theoretically describe this dynamics, however, the practical situation realized by Eq. (1) is too much complicated. Hence, we have to simplify our problem, extracting only its essences from the viewpoint of the PSPT. These five essential points are summarized as.

1. When the exciting photon is resonated to the CT exciton ($\sim$0.6 eV), these excitons are created homogeneously in the 1-d crystal and hence, the mean inter-exciton distance becomes rather long as compared with the lattice constant. Just after the intra-molecular excitation ($\sim$2.2 eV), an electron-hole pair is created in a same molecule, with the energy a few times greater than the CT exciton. This high-energy pair is electronically unstable even at the Franck–Condon state and will immediately decompose into several CT excitons, through the inter-molecular Auger decays or other coulombic many-body scatterings. These several CT excitons are expected to be very close to each other around the original molecule. Therefore, the difference in the initial condition between the CT excitation and the intra-molecular one is reduced only to the difference of the initial distance between the adjacent photogenerated CT excitons. We also have many other excitations in between ($0.6 \sim 2$ eV), and corresponding spatial arrangements of the CT excitons will be different each other, even if the total absorbed photonic energies are same. To describe this difference, we have schematically shown in Fig. 15, typical three cases of such spatial arrangements, that is, sparse (Fig. 15(a)), close (Fig. 15(b)) and moderately distant (Fig. 15(c)) excitations. Using these three typical arrangements of excitons, we will clarify the initial condition sensitivity.
2. There must be hidden but intrinsic multi-stability. A false ground state has to be just above the true one and the energy difference between them per unit volume should be larger than the thermal energy, but

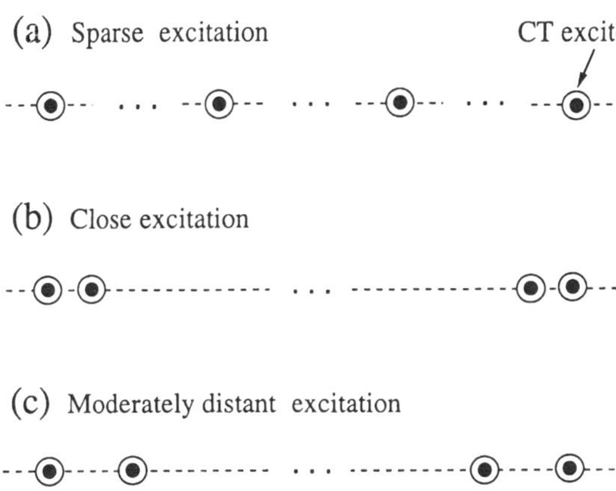

Fig. 15. Schematic nature of CT exciton excitations with various distances. (a) Sparse, (b) close and (c) moderately distant excitations.

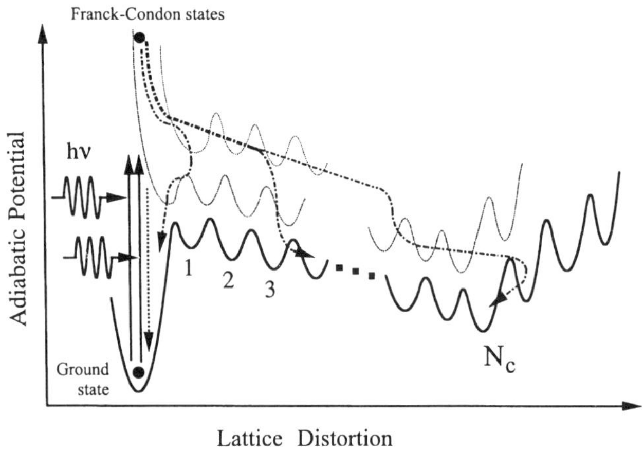

Fig. 16. Schematic nature of the dynamics of exciton relaxation and the domain formation. The solid curve and thin dotted curves denote adiabatic potential surfaces of the ground and excited states. The global minimum of the lowest curve denotes the ground state. The other local minima $1, 2, \ldots$ of this curve denote the lowest energy states with $1, 2, \ldots$ excitons, respectively. The local minimum N_c has the lowest energy among these local minima $1, 2, \ldots$. The upward solid arrows denote the initial photoexcitations. The dashed-and-dotted arrows denote the main relaxation paths. The downward dotted arrow denotes radiative decay of excitons.

should be much smaller than the visible photon energy as schematically shown in Figs. 1, 7, 13 and 16.

3. There must be a nonlinear mechanism, through which the excitons can proliferate. This mechanism should be efficient enough to overcome various radiative and nonradiative decay channels of the CT excitons. These decay channels will act in every intermediate stages of the relaxation, to hinder the proliferation as schematically shown in Fig. 16.
4. At the final state of the Franck–Condon transition, a large excess energy should be given to the excitons, as schematically shown in Fig. 13 and Fig. 16. By using these excess energies, the initially created excitons can proliferate under the energy conservation law.
5. The resultant domain should be a local minimum in the adiabatic potential surface of the ground state, so that this domain can have a sufficiently long life time and is worthy to be called the PSPT. Thus, the initially created excitons have to go through various intersections of the potential surfaces diabatically, so that it can finally reach the local energy minimum, not in the excited state potential surface, but in the ground state one, as shown schematically in Figs. 13 and 16. In this sense, the so-called adiabatic approximation is not applicable to describe the real lime dynamics of the PSPT.

6.1. *Model Hamiltonian for quantum dynamics*

In order to describe the above points from a unified theoretical point of view, Mizouchi *et al* have used a more simplified model than Eq. (1), and have clarified the early time quantum dynamics of the PSPT. They have studied the following many-exciton system coupling strongly with Einstein phonons in a quasi one-dimensional crystal.[24,25] Its Hamiltonian ($\equiv \mathrm{H_{me}}$) is given as

$$\begin{aligned}\mathrm{H_{me}} &= \mathrm{E_e}\sum_{\ell}\mathrm{B}_\ell^+\mathrm{B}_\ell - \mathrm{T_e}\sum_{\langle\ell,\ell'\rangle}[\mathrm{B}_\ell^+\mathrm{B}_{\ell'} + (\mathrm{h.c.})] \\ &\quad + \omega\sum_{\ell}\mathrm{b}_\ell^+\mathrm{b}_\ell - \omega\mathrm{S}^{1/2}\sum_{\ell}\mathrm{B}_\ell^+\mathrm{B}_\ell(\mathrm{b}_\ell^+ + \mathrm{b}_\ell) - \mathrm{V_e}\sum_{\langle\ell,\ell'\rangle}\mathrm{B}_\ell^+\mathrm{B}_\ell\mathrm{B}_{\ell'}^+\mathrm{B}_{\ell'} \\ &\quad + \sum_{\ell,\ell'(\neq\ell)}\mathrm{G}(|\ell-\ell'|)\mathrm{B}_\ell^+\mathrm{B}_\ell(\mathrm{B}_{\ell'}^+ + \mathrm{B}_{\ell'}) + \mathrm{H_{ic}},\end{aligned} \tag{15}$$

where B_ℓ^+ is a creation operator of an exciton (a boson) at a site ℓ with an energy $\mathrm{E_e}$, and $\mathrm{T_e}$ is the transfer energy between ℓ to its neighboring site ℓ', and the bracket $\langle\ell,\ell'\rangle$ denotes this neighboring relation. b_ℓ^+ is the creation operator of the Einstein phonon at site ℓ with a frequency ω, and S is the exciton-phonon coupling constant. $\mathrm{V_e}$ denotes a phenomenological attraction between two excitons at neighboring two sites. In the present theory, we will not specify its microscopic origin. It can be an effective attraction due to the van der Waals force, other coulombic interactions, or some intersite exciton-phonon couplings. The occupancy of a single site by more than one exciton is excluded from the beginning. $\mathrm{G}(\ell-\ell')$ in the sixth term represents the third-order anharmonic inter-exciton coupling. Through this anharmonicity, an exciton at site ℓ' is created or annihilated by another exciton at site ℓ. Such a nonlinearity results from the long range Coulomb interaction between electrons and holes constituting excitons. The Coulomb interaction itself is originally quartic with respect to these Fermion operators, being quite nonlinear from the beginning. Hence, to take into account this nonlinearity or anharmonicity is quite natural. However, through this nonlinearity the excitons can proliferate, as shown later in detail. Furthermore, since this coulombic nonlinearity is an instantaneous force with no retardation effect, it can cause the proliferation readily.

The last term $\mathrm{H_{ic}}$ of Eq. (15) denotes the interchain interaction

$$\mathrm{H_{ic}} = \sum_{\mathrm{N}}\mathrm{h(N)}\sum_{\mathrm{i}}|\mathrm{N,i}\rangle\langle\mathrm{N,i}|, \quad \mathrm{N} \equiv \sum_{\ell}\mathrm{B}_\ell^+\mathrm{B}_\ell, \tag{16}$$

where N denotes the total number of excitons, and i specifies each quantum state ($\equiv|\mathrm{N,i}\rangle$) within the N exciton states. Thus, $\sum_{\mathrm{i}}|\mathrm{N,i}\rangle\langle\mathrm{N,i}|$ is a projection operator into the N exciton state, while h(N) denotes an effective potential given from the other chains to this N-exciton system, when it is created in the central chain on which we focus. This potential h(N) is assumed to depend only on the total number N, irrespective of various spatial arrangements of N excitons. This functional form is also assumed to be quite nonlinear in the sense that it is almost zero when N is smaller than its critical value $N_c(\sim 10)$, while it becomes nonlinearly appreciable when N exceeds N_c. For such h(N), we have taken the following form as

$$\begin{gathered} h(N) = \omega \sum_{j=0}^{4} a_j N^j, \quad a_4 = 3.27 \times 10^{-4}, \quad a_3 = -7.08 \times 10^{-3}, \\ a_2 = 4.98 \times 10^{-2}, \quad a_1 = 2.03 \times 10^{-2}, \quad a_0 = -6.33 \times 10^{-2}. \end{gathered} \tag{17}$$

The role of this interaction is almost the same as that of Eq. (12), and it makes the domain with N_c excitons locally stable, as schematically shown in Fig. 16.

It should be noted that the present exciton-phonon system is our relevant system and is also assumed to couple linearly with the photon field and acoustic phonon modes, which act as a heat reservoir during the relaxation, although they are not written here explicitly.

6.2. *Relaxation and occurrence or nonoccurrence of the PSPT*

Using Fig. 16, let us now schematically explain the characteristics of the relaxation processes, which lead to the occurrence or nonoccurrence of the PSPT. The solid curve and thin dotted curves denote adiabatic potential surfaces of the ground and excited states. The global minimum of the lowest curve denotes the ground state. The other local minima $1, 2, \ldots$ of this curve denote the lowest energy states with $1, 2, \ldots$ excitons, respectively. The local minimum N_c has the lowest energy among these local minima $1, 2, \ldots$. The upward solid arrows denote the initial photoexcitations. The dashed-and-dotted arrows denote the main relaxation paths. The downward dotted arrow denotes radiative decay of excitons. These nonradiative and radiative decays are brought about through the interaction between our relevant system and the acoustic phonon and photon fields mentioned before. We also have assumed a multi-stable situation where the lowest energy of (N + 1)-exciton states is energetically close to that of N-exciton state. This condition is realized when $(E_e - S)$ and V_e are well balanced in Eq. (15) as, $(E_e - S) \approx V_e$.

In Fig. 16, the excitons are created from the ground state by photoabsorption (the leftmost upward solid arrows). In the next, the exciton relaxes along the following two paths. One path is the vibrational (or lattice) relaxation, where the system changes along an adiabatic potential curve of the excited state (the dashed-and-dotted curves in Fig. 16). The photoexcited state along this path has a large excess electronic and vibronic (phonon) energies. The other path is the direct radiative decay (the downward thick dotted arrow in Fig. 16). Here, the former path is main one because vibrational relaxation is faster than radiative decay. As seen from this figure, many adiabatic potential surfaces come close to each other at many points. Around such points, the diabatic transition occurs. If the relaxation proceeds toward the right through such processes, the proliferation becomes successful.

As mentioned in Sec. 4, various relaxations can start from almost the same Franck–Condon states, as far as the initially absorbed total photonic energies are the same. However, these starts are a little different from each other, according to the way or type of the photoexcitation, because their electronic structures are mutually different. Then, during the relaxation, the difference between these adjacent starts diverges. Some of them quickly return to the ground state nonradiatively, while others proceed toward the right in Fig. 16 and stays in multi-exciton states for a long time. Furthermore, among the latter cases, some of them can reach quickly to the destined lowest energy state with N_c excitons. While others, because of the tunneling, reach rather slowly the destined N_c-exciton state, although they relax down quickly only as far as to the lowest potential curve. This is nothing else but the initial condition sensitivity.[10]

6.3. *Life time prolongation*

Here we should note the prolongation of the life time of these various transient states, which appear during the relaxation. As mentioned above, the PSPT is a transient phenomenon, and the system surely goes back to the ground state finally. However, the state generated by the PSPT can become sufficiently stable and it takes quite a long time to go back to the ground state. This is because the overlap integral between the multi-exciton state and the ground state with no exciton, becomes smaller and smaller, as the proliferation proceeds from left to right in Fig. 16. Hence the transition between them becomes very difficult, even if the photon and the acoustic phonon spontaneously try to induce it through their quantum fluctuations. Thus, the more the decay time elongates, the more the resultant domain becomes large and stable, just as the hen-and-egg relation.

6.4. *Iterative method for exciton proliferation*

In order to investigate the time evolution of the aforementioned exciton-phonon system, we are now at the stage to derive the master equation under the Markov approximation for the photon field and the reservoir. However, there is a serious numerical difficulty in the direct calculation of this time evolution in the large size systems, because the quantum mechanical treatment of Einstein phonons leads to too large dimensional calculations. Therefore, in order to overcome this numerical difficulty, we derive an iterative equation for the exciton proliferation. The basic idea is schematically shown in Fig. 17.

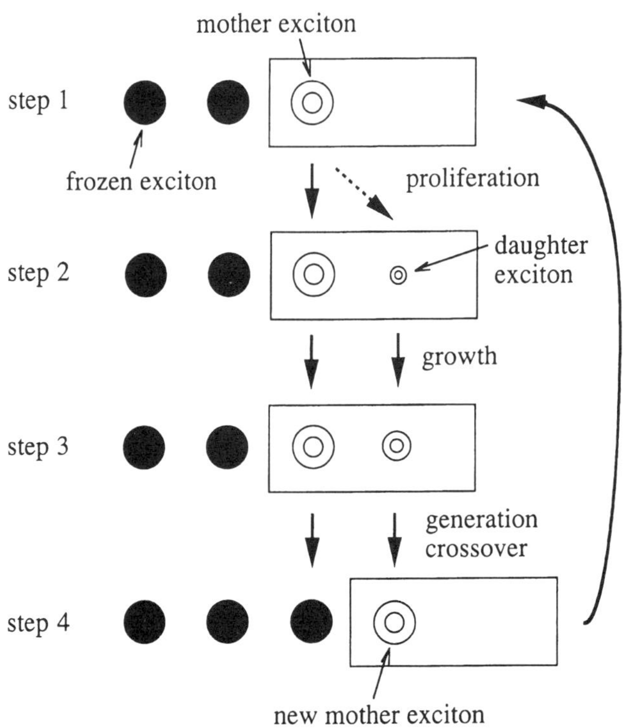

Fig. 17. Schematic nature of the iterative equation, and the generation crossover.

We focus only on the most front of the expanding photoinduced domain and the contributions from the other excitons, not in this front, are approximated by a mean field. As proliferation procccds, the position of this front also moves. At the first step in Fig. 17, the front of the domain is assumed to be inside the box. This box is our relevant system, within which we calculate excitons, Einstein phonons and their interactions, full-quantum-mechanically. We take the size of the box as 4 lattice sites around the front. Inside of this box, we prepare an exciton with an excess energy, and it is represented by a double circle. In the following, we call it "mother exciton". We also prepare several excitons outside of our relevant system (the box), and they are represented by black circles. They are localized at each lattice site, and their energies are fixed to the lowest vibronic ones. We call them "frozen excitons".

Under the existence of these frozen excitons, the Hamiltonian H_{me} of our exciton-phonon system is approximated by H'_{me}, which is defined only within the Hilbert space of the front region as

$$\begin{aligned}
H'_{me} \equiv{}& E_e \sum_{\ell} B_\ell^+ B_\ell - T_e \sum_{\langle \ell,\ell' \rangle} [B_\ell^+ B_{\ell'} + (\text{h.c.})] \\
&+ \omega \sum_{\ell} b_\ell^+ b_\ell - \omega S^{1/2} \sum_{\ell} B_\ell^+ B_\ell (b_\ell^+ + b_\ell) \\
&- V_e \sum_{\langle \ell,\ell' \rangle} B_\ell^+ B_\ell B_{\ell'}^+ B_{\ell'} + \sum_{\ell,\ell'(\neq \ell)} G(|\ell - \ell'|) B_\ell^+ B_\ell (B_{\ell'}^+ + B_{\ell'}) \\
&+ \sum_{\ell_f} N_{\ell_f} [E_e - \omega S] - V_e \sum_{\langle \ell_f,\ell'_f \rangle} N_{\ell_f} N_{\ell'_f} \\
&- V_e \sum_{\langle \ell_f,\ell \rangle} N_{\ell_f} B_\ell^+ B_\ell + \sum_{\ell_f,\ell} G(|\ell_f - \ell|) N_{\ell_f} (B_\ell^+ + B_\ell) \\
&+ \sum_{N} h(N + N_f) \sum_{i} |N,i\rangle\rangle\langle\langle N,i|, \qquad (18)
\end{aligned}$$

$$N \equiv \sum_{\ell} B_\ell^+ B_\ell, \quad N_f \equiv \sum_{\ell_f N_{\ell_f}} . \qquad (19)$$

Here, ℓ and ℓ' now denote the sites only within the front(box). N_{ℓ_f} denotes the average density of the frozen exciton at a site ℓ_f outside of the front, and hence, this N_{ℓ_f} is a *c*-number. N now denotes the total number of excitons only within the front, and $|N,i\rangle$ denotes ith eigenstate with N excitons in this front. The first six terms of Eq. (18) denote the Hamiltonian within the front, while the seventh and eighth terms denote the energy of frozen excitons outside of the front. The ninth and tenth terms denote the interactions between the front and its outside. The last term is the interchain interaction, in which the presence of the frozen exciton is already taken into account.

Using Eq. (18) and the master equation method, we can now describe the dynamics of the exciton proliferation in this front. As seen in the sixth term of Eq. (18), the mother exciton can proliferate through the original $G(\ell - \ell')$ by using her excess energy, and can make her "daughter exciton" (step 2 in Fig. 17). However, we can see that even the frozen excitons can also help this proliferation as an external mean field, through the tenth term of Eq. (18).

Thus, as time goes by, the daughter exciton grows up and the total number of excitons increases (step 3 in Fig. 17). When the total number of excitons in the front increases by 1 from the initial number, we regard the daughter as being grown up to be an adult, and approximate that the following generation crossover occurs (step 4 in Fig. 17). The mother exciton is replaced by a frozen one and the daughter exciton is taken as a new mother. Here, as seen from Eq. (18), it should be noted that excitons except frozen ones can move between lattice sites in the box. Thus, the exciton density has a non-integer value at each lattice site. Therefore, in the aforementioned crossover, we regard the excitons whose densities are the largest and the second largest in the box, as the mother exciton and the daughter one, respectively.

Furthermore, we assume the new mother inherits the excess energy, after this generation crossover. We determine this excess energy of the new mother so that the total energy in the system is conserved before and after this generation crossover. However, this excess energy becomes smaller and smaller than the starting one, because of the dissipation or relaxation.

By this crossover, the front moves, and accordingly the box moves. In fact, in the case of Fig. 17, the front moves one lattice site towards the right. Then, we focus again only on this new front, and iterate the aforementioned procedure, until the excess energy is exhausted through the interaction between our relevant system and the aforementioned heat reservoir. During this iteration, the size of the box around the front is kept unchanged as shown Fig. 17. Therefore, without enlarging our computer memory, we can numerically calculate the time evolution dynamics of a large number of excitons in the large system.

Hereafter, we call one iteration (from the step 1 to the step 3 in Fig. 17) "one generation". We should also note that this iterative procedure is justified, when T_e is small as compared with $S\omega$ in Eq. (15).

6.5. *Numerical results*

Let us now specify the values of the parameters in Eq. (15). Our purpose here is not a comprehensive study for Eq. (15), but one of typical situations realized by a set of parameter values, which are in compliance with the aforementioned five points, and makes the PSPT quite successful. As one of such examples, the following values are taken; $\omega = 0.1\,\text{eV}(\hbar = 1)$, $E_e/\omega = 9.5$, $T_e/\omega = 1.0$, $S = 8$, $V_e/\omega = 1.7$, $G(1)/\omega = 0.2$, $G(2)/\omega = 0.1$, $G(3)/\omega = 0.067$, $G(4)/\omega = 0.05$, $G(5)/\omega = 0.04$ and zero for other G's. N_c is taken to be 10. The radiative damping rate of the exciton is assumed to be 10^{-9} second at the Franck–Condon state and afterward decreases in proportion to the third power of the transition energy, while the damping of the Einstein phonon due to its coupling with the acoustic phonon reservoir is assumed to be 20% of the phonon energy ω. The large S corresponds to the large excess energy at the Franck–Condon state, while finite ω/T_e makes aforementioned diabatic transitions possible.

Here, we start from the initial condition that there are two localized excitons, just as shown in Fig. 15. One of them is an exciton created by the Franck–Condon excitation from the ground state. This is a mother exciton. While, the other exciton is taken as a frozen one, for the reason mentioned before. In the following, the distance between these two excitons at the initial state is represented by d_0. Results are shown in Fig. 18, and the time evolution dynamics become different according to this d_0. The net exciton proliferation occurs only when $d_0 = 2$, 3 and 4. When $d_0 = 1$ and ∞, the number of excitons increases a little from 2 only at early time, however, the net proliferation does not occur finally. Therefore, the initial distance between excitons should be moderate in order to get net proliferation. In the too-far distance case ($d_0 = \infty$), the nonlinearity among excitons does not work and it leads to no net proliferation. On the other hand, in the too close excitation case ($d_0 = 1$), too strong nonlinearity works, and it leads to the exciton annihilation rather than its proliferation. Therefore, the critical nonlinearity is necessary to realize net proliferation. Here, the case with $d_0 = \infty$ corresponds to the CT excitation below the threshold mentioned in Sec. 4. In the intra-molecular excitation case, d_0 takes various values within the force range of the aforementioned intermolecular Auger decay. Among them, only the successful cases can finally survive

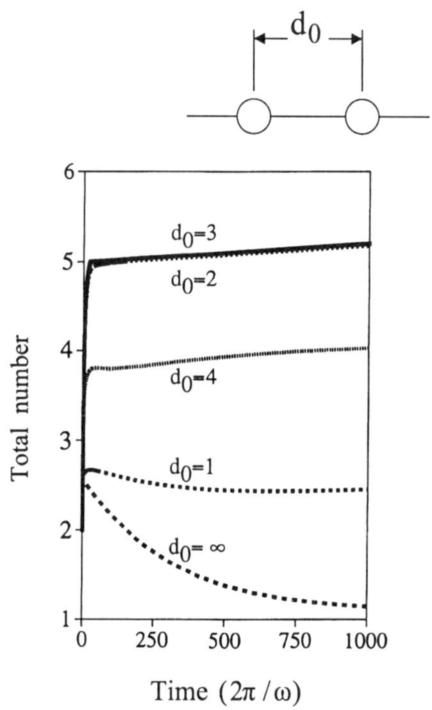

Fig. 18. The total number of exciton as functions of time. d_0 is the distance between the initial two excitons. From Ref. 25.

(d_0 = 2, 3 and 4). Furthermore, even when the net proliferation occurs, the time evolution behavior of each proliferation is not same but chaotic, according to the value of d_0. For example, when $d_0 = 4$, the proliferation occurs more slowly than the other proliferating cases (d_0 = 2 and 3). These results show the initial condition sensitivity still exists even when the proliferation is successful.

In Fig. 19, we show a spatial arrangement of excitons at the beginning of each generation (the step 1 in Fig. 17), taking the case with $d_0 = 3$. Here, G_e denotes the generation, and the time ($\equiv t$) is measured from the beginning of the first generation. The black circle and the double one denote a frozen exciton and a mother one, respectively. In particular, the white circle with a black one denotes the mother exciton created by the Franck–Condon excitation from the ground state. The values under each arrangement denote the exciton densities in the front, just before the generation crossover (the step 3 in Fig. 17). The values of frozen exciton densities are almost equal to 1, and hence they are not written explicitly.

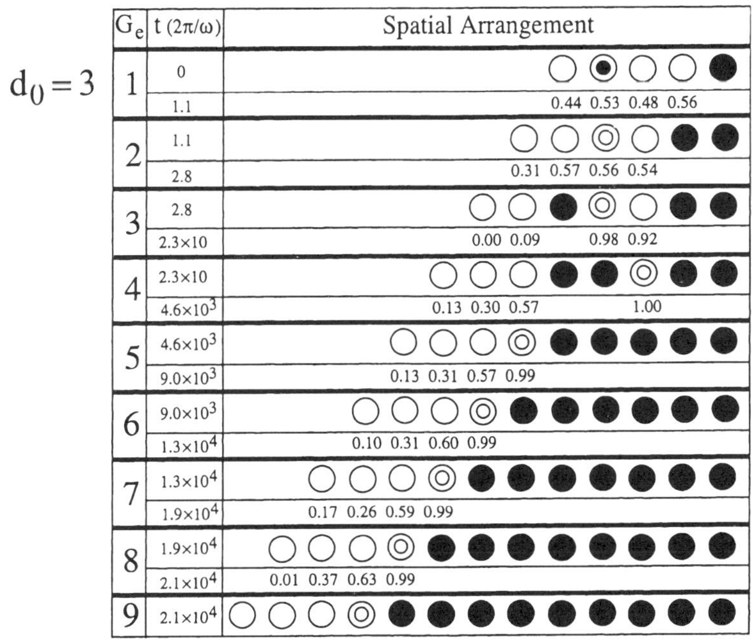

Fig. 19. A spatial arrangement of excitons at the beginning of each generation (the step 1 in Fig. 17). $d_0 = 3$. G_e and t denote the generation and the time measured from the beginning of the first generation, respectively. The black circle and the double one denote a frozen exciton and a mother one, respectively. The white circle with a black one denotes the mother exciton created by the Franck-Condon excitation from the ground state. The values under each arrangement denote the exciton densities in the front, just before the generation crossover (the step 3 in Fig. 17). The values of frozen exciton densities are almost equal to 1, and hence they are not written. From Ref. 25.

From Fig. 19, we can see that the two mutually unconnected clusters of excitons are formed during the first four generations ($G_e = 1$ to 4). At $G_e = 5$, however, these two clusters have merged with each other, and this merged one grows longer and longer, until the total number of excitons reaches N_c. Thus, we see that the dynamics of the exciton proliferation are not monotonic but fluctuating, since our system is quantum-mechanical.

7. Difference between Photoinduced Nonequilibrium Phase and High Temperature Equilibrium Phase

According to these progresses of experimental and theoretical studies on photoinduced phase transitions, a new but quite basic question has now emerged. That is, how the photoinduced phase is practically different from the high temperature equilibrium one. As mentioned occasionally, the photoinduced phase is a nonequilibrium phase, brought about through the multi-stability, or pseudo-degeneracy of the ground state. However, when this multi-stable situation is realized in our system, a state similar to the false ground state, is also inferred to appear as an equilibrium phase at high temperatures, that is, through the ordinary thermal phase transition.

In fact, both this photoinduced phase at low temperature and the equilibrium one at high temperature are observed as shown in Sec. 3 and 5, and these two phases are quite similar with each other. Thus the aforementioned issue becomes quite serious and significant.

7.1. *Photoinduced phase transition in organo-metallic complex crystal*

However, recent experiments on the photoinduced phenomenon in the organo-metallic complex crystal [Fe(2 − pic)$_3$]Cl$_2$ · Et OH (2 − pic = 2-amino-methyl-pyridine), have opened a new prospect in the study for the aforementioned difference. In this crystal, as shown in Fig. 20, an Fe^{2+} ion and its neighboring six nitrogen atoms of three 2-pic molecules, are making a complex, which approximately has an O_h symmetry. As schematically shown by dashed lines of this figure, this metal complex is bonded with other three neighboring ones through the hydrogens, and these hydrogen bonds result in inter-complex interactions. An Fe^{2+} ion has six electrons in its 3d orbitals (t_{2g}, e_g), and these electrons are in a crystal field, whose symmetry is almost O_h as mentioned before. At absolute zero temperature, these 6 electrons, being in the three t_{2g} orbitals, become diamagnetic

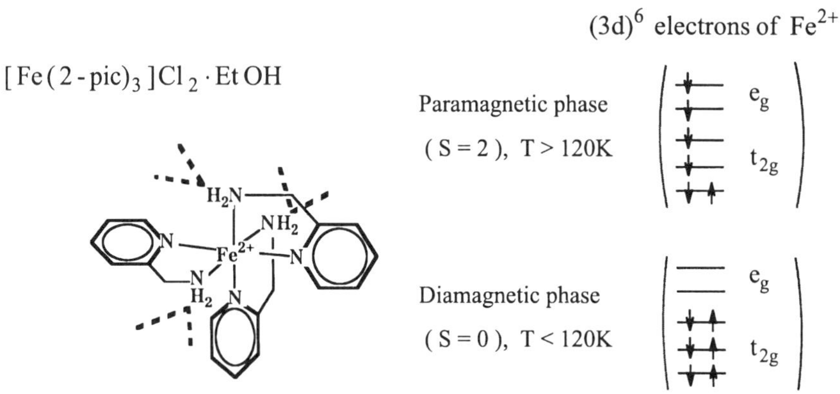

Fig. 20. The structure and electronic states of [Fe(2 − pic)$_3$]Cl$_2$ · EtOH.

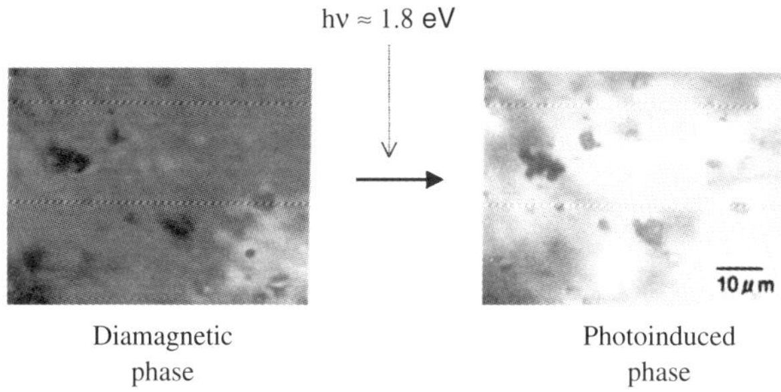

Fig. 21. The photoinduced phase transition in $[Fe(2-pic)_3]Cl_2 \cdot EtOH$. From Ref. 26.

(S = 0) as shown in Fig. 20. This diamagnetic phase has a strong light absorption band at around 2 eV, and the color of this crystal is deep red. While, at about 120 K, a first order phase transition occurs from this diamagnetic phase to a paramagnetic one (S = 2), as shown in Fig. 20, and the color of the crystal changes from deep red to yellow.

On the other hand, Ogawa *et al*[26] have recently discovered the photoinduced phase transition in this crystal. Shining 1.8 eV light on to the low temperature diamagnetic phase of this crystal, they have succeeded to generate a macroscopic paramagnetic domain. By this photoinduced phase transition, as shown in Fig. 21, the color of the crystal changes from deep red to yellow, which is quite similar to the yellow of the high temperature paramagnetic phase. They also have found the bi-directional nature of this photoinduced diamagnetic $\leftrightarrow$ paramagnetic transition, and the threshold type behavior just like the case of TTF-CA mentioned in Sec. 4.

In connection with this discovery, very recently, Tayagaki *et al*[27] have also succeeded in observing Raman scattering spectra of these three phases, the low temperature diamagnetic phase, the high temperature paramagnetic phase, and the photoinduced paramagnetic phase at low temperature. The resultant three Raman spectra are shown in Fig. 22. We can see clearly that the photoinduced phase is different from the other two phases, especially in the shaded region, although other spectral regions are similar to that of the high temperature paramagnetic phase. It should be noted that this difference between the three phases is a macroscopic difference, being not an ordinary photoinduced structural change which, very often occurs only in a microscopic region of optical excited crystals. Tayagaki *et al* have assigned this difference in the shaded region to come from a new parity violation of the aforementioned O_h symmetry around the 3d orbital.[27]

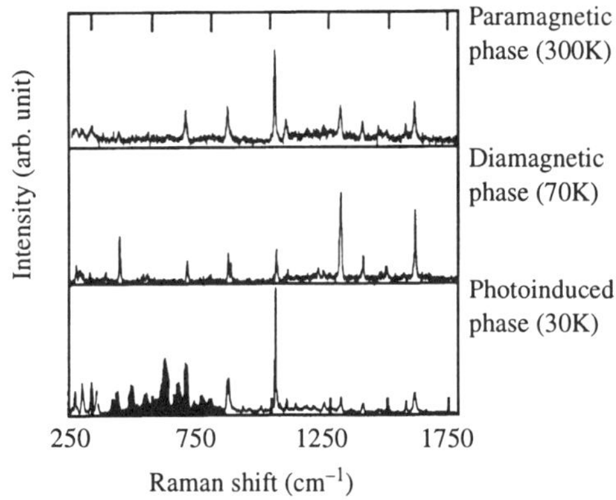

Fig. 22. The Raman spectra of $[Fe(2-pic)_3]Cl_2 \cdot EtOH$, in the paramagnetic phase (300 K), the diamagnetic phase (70 K) and the photo-induced phase (30 K).The shaded area of the photoinduced phase denotes the eccentric difference from other two spectra. From Ref. 27.

According to the present status of our experimental study on this organo-metallic complex crystal, however, a new interaction which originates this new parity violation cannot be clarified sufficiently because this crystal is really complex, as we can easily infer from Fig. 20.

7.2. *Broken symmetry only in photoinduced nonequilibrium phase*

It is quite clear that this new broken symmetry (or new parity violation) does not occur in both equilibrium phases (the low temperature diamagnetic phase and the high temperature paramagnetic one), but occurs only in the photoinduced nonequilibrium phase at low temperature. This difference between the equilibrium phase and the nonequilibrium one, if once well-established conceptually, will greatly affect our studies for photoinduced phase transitions in various other materials. Because, it means that, we can discover a new interaction and its resultant broken symmetry through the photoinduced nonequilibrium phase at low temperatures, even if this new interaction is not clearly observed in any equilibrium phases, such as the true ground state or high temperature equilibrium phases.

Here, we should note our conventional way to recognize a new interaction to exist and operate in a material. According to our ordinary way of recognition, such an interaction (or its resultant broken symmetry) is often neglected or regarded not to exist at all from the beginning, if it is not clearly observed in any equilibrium phase. The forementioned experimental result,[27] however, clearly tells us that it is a prejudice which we have to overcome.

7.3. *Two-dimensional extended Peierls–Hubbard model*

In order to make this point clear, Haui *et al*[5,28] studied theoretically a model system which is more standard and simpler than the organo-metallic case. Its purpose is to describe theoretically the situation where a new interaction ($\equiv H_n$) appears explicitly only in the low temperature nonequilibrium phase as a broken symmetry, under the condition that it is almost completely hidden in any other equilibrium phase at low and high temperatures. To address such an issue theoretically, they introduced a square lattice composed of N_t sites and N_t electrons, which are strongly coupling with site-localized phonons. Then this two- dimensional (2-D) many-electron system has the following Hamiltonian ($\equiv H_{PS}$), which we often call the Peierls–Hubbard model:

$$H_{PS} = -T_D \sum_{\langle \ell,\ell' \rangle,\sigma} C^{+}_{\ell,\sigma} C_{\ell',\sigma} - S_D \sum_{\ell} Q_\ell (n_\ell - 1) + \sum_{\ell} V_{ah}(Q_\ell) - \frac{1}{2M} \sum_{\ell} \frac{\partial^2}{\partial Q_\ell^2} + H_n, \quad (20)$$

$$H_n = U_D \sum_{\ell} n_{\ell,\alpha} n_{\ell,\beta}, \quad (21)$$

where T_D denotes the transfer energy of an electron between two neighboring lattice sites, and ℓ is a 2-D vector with the Cartesian components ℓ_x and ℓ_y. S_D denotes the coupling constant between electrons and the site-localized phonon mode whose dimensionless coordinate is Q_ℓ. $V_{ah}(Q_\ell)$ denotes a potential energy of this site-localized phonon mode, while M is the effective mass of this mode, and its dimension is $(\text{energy})^{-1}$. This potential $V_{ah}(Q_\ell)$ is assumed to be highly anharmomic as

$$V_{ah}(Q_\ell) = (aQ_\ell^2 - bQ_\ell^4 + cQ_\ell^6), \quad (22)$$

where a, b, c are 2i-th expanding coefficients of $V_{ah}(Q_\ell)$ with respect to Q_ℓ $(i = 1, 2, 3)$. Such an anharmonicity is necessary to make various thermally induced and photoinduced phases stable, and also to make phase transitions first order ones. The last term H_n is the aforementioned new interaction on which we focus. In the present case, for simplicity, we assume that this H_n is a weak on-site Coulomb interaction as shown in Eq. (21).

It is expected to bring an SDW type broken symmetry, and competes with a CDW type order which surely comes from the strong electron-phonon coupling S_D in this 2-D half-filled many-electron system.

7.4. *Mean field theory and phase diagram*

Within the mean-field theory and the adiabatic approximation ($M \to \infty$), H_{PS} is approximated by a Hartree–Fock Hamiltonian ($\equiv H_{HF}$) which is given as

$$H_{HF} = -T_D \sum_{\langle \ell, \ell' \rangle, \sigma} C^{+}_{\ell,\sigma} C_{\ell',\sigma} - S_D \sum_{\ell} Q_\ell (n_\ell - 1) + \sum_{\ell} V_{ah}(Q_\ell)$$
$$+ U_D \sum_{\ell,\sigma} \left(\langle n_{\ell,-\sigma} \rangle n_{\ell,\sigma} - \langle n_{\ell,-\sigma} \rangle \langle n_{\ell}, \sigma \rangle / 2 \right), \tag{23}$$

where in $\langle n_{\ell,\sigma} \rangle$ denotes the thermal average of $n_{\ell,\sigma}$, and it should be determined self-consistently. As for Q_ℓ, we assume that the lattice distortion has a staggered order in both two directions of the square lattice as

$$Q_\ell = (-1)^{\ell_x + \ell_y} Q_0. \tag{24}$$

Here Q_0 is the amplitude of the CDW or the Peierls type distortion. Correspondingly, $\langle n_{\ell,\sigma} \rangle$ is also assumed as

$$\langle n_{\ell,\sigma} \rangle = \frac{1}{2} + (-1)^{\ell_x + \ell_y} \delta n_\sigma, \tag{25}$$

where δn_σ again denotes the amplitude of the density wave of electrons with $\sigma (= \alpha, \beta)$ spin, and this wave also has the double period for both two directions of the square lattice. When the strong coupling and the weak correlation condition ($S_D > T_D \geq U_D$) is satisfied, the ground state of this system is expected to be a CDW type insulator, in which two electrons with up and down spins make a strongly bound singlet pair, and this pair occupies a single site every two sites, along both the x and y axes of this 2-D lattice. While, its high temperature phase will be the ordinary paramagnetic metallic state, although it has a very weak CDW type order, since this model has the complete nesting.

As one of the typical examples to describe this situation, we take the following parameter values,

$$T_D = 0.3\,\text{eV}, \quad S_D = 1.0\,\text{eV}, U_D = 0.3\,\text{eV}, \quad a = 4.445\,\text{eV},$$
$$b = 2.642 \times 10^3\,\text{eV}, \quad c = 5.487 \times 10^3\,\text{eV}.$$

In the present study, periodic boundary condition is imposed on a 16×16 square lattice whose total number of sites is $N_t = 256$. Figure 23 shows

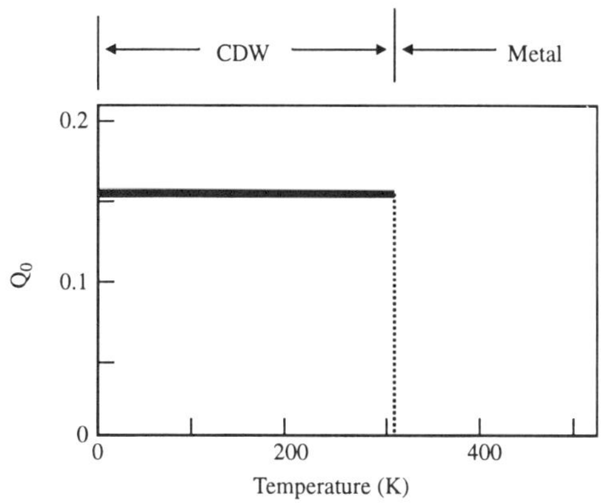

Fig. 23. The phase diagram, and Q_0 as a function of temperature T. From Ref. 28.

the phase diagram obtained by this mean field theory, and Q_0 is plotted as a function of temperature. The abrupt change of Q_0 at a critical temperature 305 K ($\equiv T_C$) indicates that below T_C, the whole system is in the CDW type insulating state in which up- and down-spin electrons make a strongly bound singlet pair ($\delta n_\alpha = \delta n_\beta > 0,\ Q_0 \neq 0$), above T_C, this system changes into a paramagnetic metallic state ($\delta n_\alpha = \delta n_\beta = 0,\ Q_0 = 0$). This phase transition is the first order, and it just comes from the anharmonicity of V_{ah}, in contrast the case of harmonic phonons, which result in the ordinary second order CDW-metal transition. Thus, the anharmonicity is necessary to separate various equilibrium and nonequilibrium phases by high adiabatic energy barriers and to make them locally stable.

Figure 24(a) shows us the free energy (per site) at zero temperature in the two-dimensional space spanned by the lattice distortion amplitude Q_0 and the SDW type order parameter ($\delta n_\alpha - \delta n_\beta$). There are two types of minima: the global one at finite lattice distortion $Q_0 = 0.154$ and a shallow one without lattice distortion. They are both locally stable and separated by a barrier of 0.01 eV.

As already mentioned in the last section, the electron-phonon coupling results in the Peierls distortion $Q_0 \neq 0$ and the CDW type electronic order ($\delta n_\alpha = \delta n_\beta > 0$), while the on-site Coulomb interaction yields the SDW type electronic order ($\delta n_\alpha = -\delta n_\beta > 0$). Based on the present set of parameters, the CDW order apparently overwhelms the SDW order in the ground state, due to the very strong electron-phonon coupling ($S_D > U_D$). However, this SDW order will also be able to appear as a locally stable false ground state or a nonequilibrium phase, as long as U_D is finite. From Fig. 24(a),

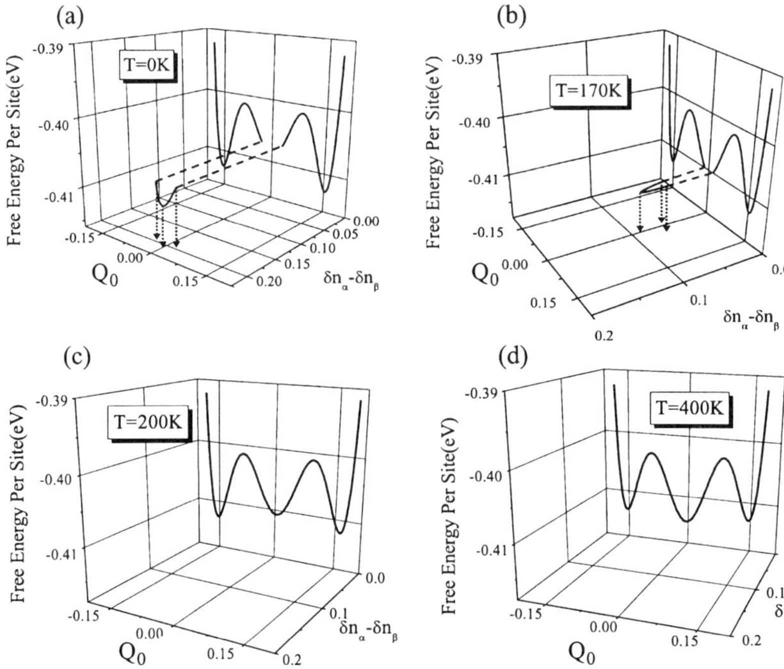

Fig. 24. The free energy (per site) in the two-dimensional space spanned by the lattice distortion amplitude Q_0 and the SDW type order parameter $(\delta n_\alpha - \delta n_\beta)$. (a) T = 0 K, (b) T = 170 K, (c) T = 200 K, (d) T = 400 K. From Ref. 28.

we can see that the global minimum, being the true ground state, has a finite lattice distortion with no SDW order. However, we can also see that there is another local minimum, being a false ground state, which has a finite SDW order without lattice distortion. Therefore it is quite clear that the SDW order appears in the false ground state, although it is hidden in the true ground state.

Let us now turn to the cases of finite temperatures. Figures 24(b), (c) and (d) show us the free energy (per site) at the sequence of temperatures 170 K, 200 K and 400 K, respectively. The SDW order remains a locally stable nonequilibrium state at T = 170 K, as shown in Fig. 24(b), although this order has significantly decreased from that of zero temperature. At a much higher temperature T = 200 K shown in Fig. 24(c), we can see the two free energy minima corresponding to the metallic state ($Q_0 = 0$) and the CDW one ($Q_0 \approx 0.15$), and both of them have no SDW order. It also should be noted that the difference of the free energy between these two states is decreasing due to the increase of temperature. Then, as shown in

Fig. 24(d), the free energy minimum at $Q_0 = 0$ becomes lower than that of $Q_0 = 0.154$, so that the global free energy minimum now locates at $Q_0 = 0$. That is, the first order thermal phase transition occurs from the CDW state to the metallic one just as shown in Fig. 23.

Thus, we have seen that the SDW order is always hidden, although our original Hamiltonian shown in Eq. (21) surely has a finite U_D. At zero temperature, the SDW order is completely killed in the true CDW ground state, because, in this state, the electron-phonon coupling makes up- and down-spin electrons to form a strongly bound singlet pair within a single lattice site. The high temperature phase is a paramagnetic metallic one, but this high temperature itself also destroys the SDW order. Therefore the Coulomb interaction and its resultant SDW order are almost completely hidden in all the thermal equilibrium phases.

In the present study, the lattice is always assumed to have a staggered order in all possible CDW type phases, and lattice fluctuations are excluded from the beginning. Hence, we may expect that thermal fluctuations below T_C will cause some signs of the SDW order. In this connection, Huai *et al* also have calculated the SDW order $(\delta n_\alpha - \delta n_\beta)$ of the false ground state at each temperature, and found that this order decreases rapidly when temperature is above 100 K, and goes to zero at about 183 K, which is well blow T_C. Thus even if the thermal lattice fluctuations are taken into account, we will get only a blurred sign of the SDW order.

7.5. *Photoinduced phase*

Since this SDW order exists only in the false ground state at very low temperature, it may be detected by using photo-excitation. If this system is shinned by a strong laser light at low enough temperature, a CT excitation occurs. This excitation is such one that, the up- and down-spin electrons, making a singlet pair in a lattice site, will be separated into two neighboring lattice sites as an antiferromagnetic pair. This antiferromagnetic pair, once generated by light, will proliferate and will grow to be a nonequilibrium SDW domain.

To study such a lattice relaxation of CT excitation, we can introduce the following disk type lattice distortion pattern Q_ℓ:

$$Q_\ell = (-1)^{\ell_x+\ell_y} Q_0 \left[1 + \Delta Q \left\{ \tanh\left[\theta\left(\sqrt{\ell_x^2+\ell_y^2} - \ell_D/2\right)\right] - 1\right\}\right]. \quad (26)$$

Here $(-1)^{\ell_x+\ell_y} Q_0$ denotes the aforementioned Peierls distortion in the true ground state. The second term in the square brackets $[\cdots]$ denotes a local lattice displacement induced by a new excited domain. ΔQ is its amplitude, θ again corresponds to the width of the domain boundary and ℓ_D is the diameter of the domain, which in the present case, is assumed to be a disk type in the square lattice. When $\Delta Q = 0.5$, it is obvious that the lattice inside the domain has no distortion at all, while we have already seen in Fig. 24(a) that the SDW false ground state also has no lattice distortion. Hence, the SDW order is expected to appear in this domain too, if we choose ℓ_D and θ properly.

In the present study, we will not investigate the lattice relaxation process in detail, but just want to demonstrate a possible relaxation path from the true ground state to the expected SDW domain. Hence, we fix the domain diameter ℓ_D at $\ell_D = 10$, and describe the adiabatic energy of the ground state E_g only as a function of ΔQ, while θ is determined to minimize the domain energy at $T = 0$. Figure 25 shows the adiabatic energy curve of E_g, which is referenced from the true ground state. If we start from the origin $\Delta Q = 0, E_g$ rapidly increases up to 0.9 eV, and after that it turns to go down to another minimum near $\Delta Q = 0.5$. Thus, this minimum is locally stable and has the SDW order. To elucidate the nature of this domain, we present its lattice distortion pattern Q_ℓ and spin density pattern in Fig. 26. The diameter of each dot in Fig. 26(a) is proportional to $|Q_\ell/Q_0|$, while the diameter of each dot in Fig. 26(b) is proportional to $|\langle n_{\ell,\alpha}\rangle - \langle n_{\ell,\beta}\rangle|$, at each lattice site. The value of $|\langle n_{\ell,\alpha}\rangle - \langle n_{\ell,\beta}\rangle|$ at the center of the domain is almost

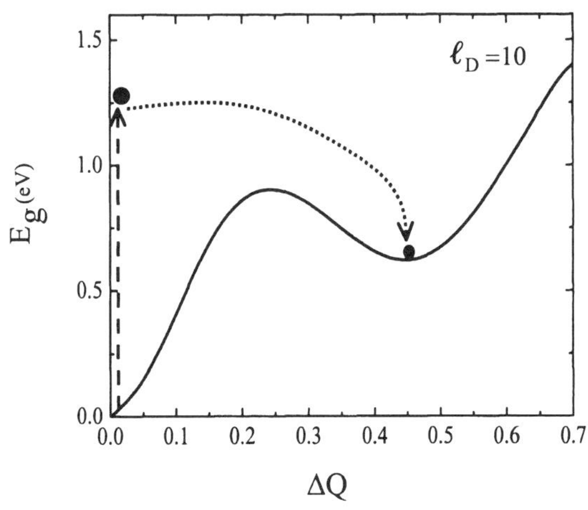

Fig. 25. The adiabatic energy of the ground state E_g as a function of ΔQ. Energy is referenced from the true ground state. $\ell_D = 10$. From Ref. 28.

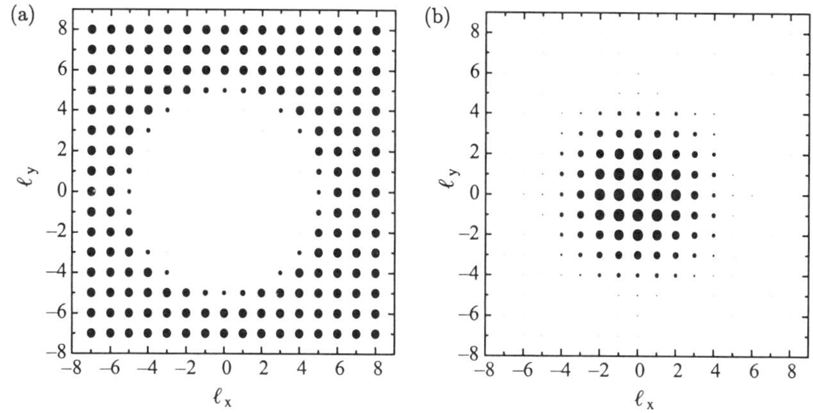

Fig. 26. (a) The lattice distortion Q_ℓ of the domain. The diameter of dots are proportional to $|Q_\ell/Q_0|$. (b) The spin density of the domain $\langle n_{\ell,\alpha}\rangle - \langle n_{\ell,\beta}\rangle$. The diameter of dots are proportional $|\langle n_{\ell,\alpha}\rangle - \langle n_{\ell,\beta}\rangle|$ at each lattice site. The value of $|\langle n_{\ell,\alpha}\rangle - \langle n_{\ell,\beta}\rangle|$ at the center of the domain is almost equal to that of Fig. 24(a). From Ref. 28.

equal to that of Fig. 24(a). It is now clear that this local domain exhibits the SDW order just as we have expected and it can be generated by the photoexcitation at low enough temperature. By the dotted and dashed lines in Fig. 25, we have schematically shown a possible relaxation path of the CT excitation from the CDW ground state to this SDW domain. Although this photoinduced phase may not have a long lifetime, there is no fundamental difficulty to detect it by using modern spectroscopy techniques.

Thus, through the photoinduced nonequilibrium phase, we can recognize the presence of an interaction H_n and its resultant broken symmetry, even if it is almost completely hidden in any equilibrium phase from absolute zero temperature to high ones.

Let us now consider this conclusion from a more general point of view. Real solids are so complex that although they are made of only a few kinds of atoms and molecules, their structures show a great number of variations under different chemical and physical environments, and are stabilized by delicately balanced interactions. It is far from rare that an interaction, being only a little weaker than its counterparts, can really appear under very limited conditions. In many cases, it just disappears from our scope of study, simply because its strength is subsidiary to other more dominant counterparts. Therefore it is vitally important to elucidate such a new interaction and its resultant broken symmetry by a photoinduced nonequilibrium phase.

7.6. *Hysteresis and nonadiabatic effect*

Let us qualitatively discuss the effect of hysteresis in connection with nonadibabatic effects coming from the fourth term of Eq. (20). The CDW-metal transition shown in Fig. 23 is the first order phase transition, and hence we have to expect the following hysteresis effects around T_c. Slowly warming the whole system from low temperature to a high one across T_c, we can make the CDW state remain even above T_c as a locally stable nonequilibrium state, whose free energy is higher than the globally stable (equilibrium) metallic state. This situation is also the same for the cooling process from above T_c and the difference between the warming process and the cooling one is usually called hysteresis loop. The temperature region in which this hysteresis phenomenon occurs is mainly determined by the nonadiabatic effects coming from the fourth term of Eq. (20). When the energy difference between the nonequilibrium state and the globally stable equilibrium one becomes large, a nonadiabatic (or diabatic) transition will surely occur from this nonequilibrium state to this equilibrium one, provided that M is large but finite.

In this context, the aforementioned hysteresis phenomena can be understood as the following situation, that these two states are almost degenerated from each other and hence, the life time of the aforementioned decay transition becomes quite long as compared with the time scale of our daily life. Thus the nonequilibrium state due to the hysteresis effect can appear only in the vicinity of T_c.

On the other hand, the photoinduced nonequilibrium state mentioned before can appear even outside of this hysteresis loop, although its life time is often shorter than the time scale of our daily life. As mentioned before, according to the recent progress of our laser spectroscopy techniques, an infinite life time is not necessary for each state to be recognized as a well-defined state, provided that it can last long enough to be clearly observed by other photons to detect it.

8. Photoinduced Macroscopic Parity Violation and Ferroelectricity

In previous sections, we have been concerned with various photoinduced structural phase transitions, which, after the complicated nonlinear lattice relaxation, finally result in false ground states. These states are, of course, no more luminescent, as schematically shown in Fig. 1. However, from recent experimental studies on $SrTiO_3$, we discovered a new type photoinduced structural phase transition, which occurs only within the optical excited state, keeping this state still luminescent. In these materials, a photogenerated conduction electron results in a ferro- (or super-para-) electric momain with a macroscopic parity violation.

8.1. *Large polaron, self-trapped polaron, linear and quadratic couplings*

An electron, excited from a valence band to a conduction one by a photon in an insulating solid, often forms a quasi-particle called polaron, being composed of the original electron and phonon cloud around itself. This polaron effect, coming from the interaction of the electron with phonons or lattice vibrations, has been a matter of considerable interest for these past 40 years and many experimental and theoretical studies have already been devoted. It is one of most basic themata related with various fields of solid state physics, such as optics and electronic conductivities in many kinds of semiconductors and insulators.[29–31]

At present, it is well known, that this polaron can be clearly classified into two types, the large polaron and the self-trapped one, provided that the electron-phonon (e-p) interaction is short ranged in ordinary three-dimensional crystals.[30] These two different types are brought about through the competition between the quantum itineracy of electron and the strength of e-p coupling.

When the e-p coupling is weak as compared with the itineracy, or the energy-band width of this conducting electron, we can get the large (or free) polaron, extending over a wide region of the crystal. In this case, the phonon cloud or the lattice distortion cloud around the electron, has a large radius as compared with the lattice constant of this crystal; while, the thickness of this cloud is rather thin, because of the weakness of this e-p coupling. These large polarons, once formed in a crystal by photoexcitations, can greatly contribute to increase the photoconductivity or the electronic conductivity. On the other hand, when the e-p coupling is strong as compared with the energy-band width, the electron self-induces a potential well of local lattice distortion only around a single lattice site of the crystal and is trapped in it. This is the self-trapped (or small) polaron, which never contributes to ordinary electronic conductivity.

It is also well established that these two states, the large and self-trapped polaron states, are clearly separated by an energy barrier in the adiabatic potential surface, just like the first order phase transitions. Hence, the large polaron state can remain as a locally stable state in the adiabatic potential surface, even when the e-p coupling is strong enough to make the self-trapped state globally stable.[30] However, in this competition between the large polaron and self-trapped one, the original e-p coupling is tacitly assumed to be linear and short ranged. That is, a local electron density at a certain position only linearly couples with a phonon field of the same position through a contact interaction. Consequently, if the original crystal lattice has a space inversion symmetry, only the phonon mode with even parity can contribute to this coupling and possible contributions coming from odd phonon modes are excluded.

For this reason, in this section, we will see possible contributions coming from odd phonon modes under the condition that our starting crystal lattice has a space inversion symmetry.[32] In this case, the conduction-band electron couples, not linearly, but quadratically with odd mode phonons. The effects of this quadratic coupling have not been seriously taken into account, or often been neglected, because, this quadratic coupling can be easily merged into the original restoring force of this mode, if this force is very strong or hard. However, if this odd mode is quite soft and anharmonic, the quadratic e-p coupling, switched on by the photogeneration of an electron, will result in parity violating instabilities and ferroelectric phase transitions, which are absent in the cases of even modes.

8.2. *Recent optical experiments on $SrTiO_3$ and soft-anharmonic* $\mathbf{T}_{1u}$ *mode*

This problem is closely related with recent optical experiments on the perovskite type $SrTiO_3$.[33–35] As shown in Fig. 27, the structural unit of this material is an O^{2-} octahedron with a Ti^{4+} ion in its center.[36] The six apices of this octahedron are connected with each other, three dimensionally. The top of the valence band of this material is composed mainly of the 2p orbital of O, and the bottom of the conduction band is mainly composed of the 3d orbital of Ti.[37] In between, there is a wide indirect energy gap of about 3.2 eV.[37,38]

Irradiating this wide gap material by ultraviolet light, Katsu *et al*,[33] Takesada *et al*,[34] and Hasegawa *et al*,[35] very recently found that the electronic conductivity and the quasi-static electric susceptibility are macroscopically enhanced. From the hole coefficient measurements by Yasunaga,[39] this increase of the conductivity has already been known to come, not from the holes, but from the electrons. Hence, a pair of the 3d electron and the 2p hole generated by the light, is shown to separate into a mutually independent electron and hole, with no exciton effect in between.

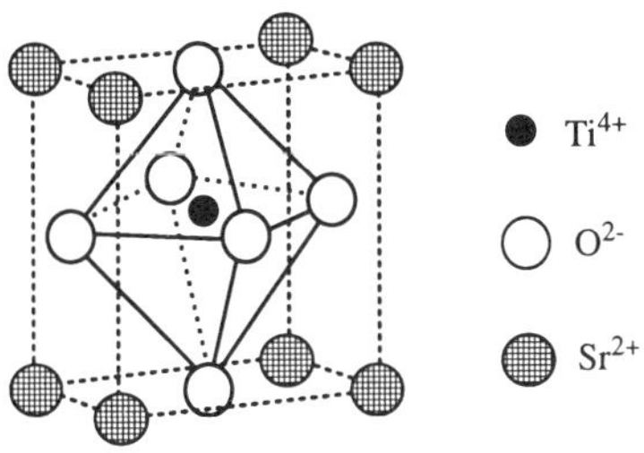

Fig. 27. Schematic structure of $SrTiO_3$.

After this separation, only the electron remains as a mobile carrier, while the hole is assumed to be trapped and localized by some reason, which is unknown. On the other hand, the quasi-static electric susceptibility, or the real part of the quasi-static dielectric constant ($\equiv\varepsilon_1$), is shown to increases about 10^3, between before and after this irradiation.[34,35]

An extraordinarily long lived luminescence, arising from this irradiation, is also found by Hasegawa *et al*[38] at 2.4 eV, and the electron and hole are, thus, shown to recombine and disappear; while this luminescence has a large Stokes shift of about 0.8 eV. It tells us that the electron or (and) hole strongly couples with the phonon mode, which corresponds mainly to the breathing (A_{1g}) type motion of O^{2-}s around Ti^{4+}, and has an energy of about 20 meV.[38] What is very interesting is that the aforementioned enhancement of ε_1 disappears as this luminescence terminates. It clearly means that this enhancement is directly related to the presence of electron or (and) hole.[35]

On the other hand, long before these optical studies, many elaborate works have already been systematically devoted to the ground state properties of this material with no photogenerated electron (hole).[40–45] They are mainly concerned with low temperature properties of ε_1, in relation to soft and anharmonic natures of T_{1u} (TO) mode, which mainly corresponds to an off-center type displacement motion of a Ti^{4+} ion, from the central position of the O^{2-} octahedron. In Fig. 28, we have schematically shown the pattern of this T_{1u} mode together with the aforementioned A_{1g} mode. Qualitatively speaking, the Ti^{4+} ion is only loosely confined within this octahedron, and hence, together with the quantum effect of its motion, we can expect to get various parity violating instabilities and ferroelectric phase transitions.

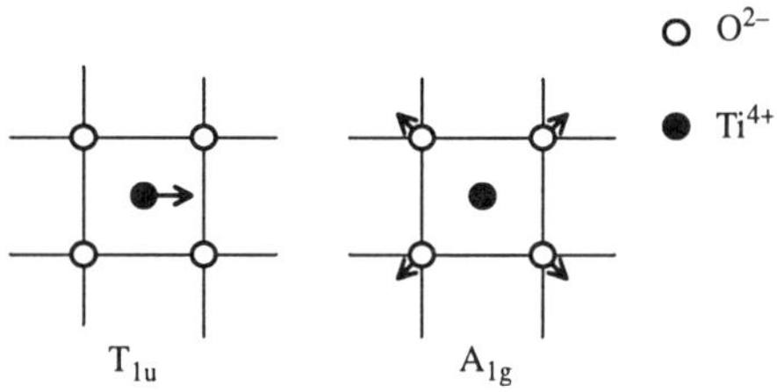

Fig. 28. Schematic nature of T_{1u} and A_{1g} modes.

According to Muller *et al*,[41] this T_{1u} mode has almost no, or even a negative restoring force, if it is described only in terms of this T_{1u} mode coordinate. However, it has an effective positive frequency of about 1 meV, which comes from a quite specific anharmonicity, called "bi-quadratic mode-mode coupling" between this T_{1u} mode and other acoustic phonon modes.[41,42] In many cases of soft-anharmonic phonons, we are often tempted to simply imagine an ordinary single-mode double-well type anharmonicity, composed of a positive quartic potential and a negative quadratic one, only in terms of this site localized T_{1u} phonon coordinate. However, by Muller *et al*,[41] this single-mode quartic anharmonicity is shown not be realized in this material at low temperatures.

By Raman scattering measurements, this soft T_{1u} phonon is also shown to couple not linearly but quadratically with electronic excitations,[46] although its coupling strength seems to be one order smaller as compared with the linear coupling of the breathing (A_{1g}) mode. In connection with this photoinduced quadratic e-ph coupling, we have to supply a deficiency of our aforementioned knowledge for the anharmonicity of this soft mode. Even if this e-p coupling is not very strong, the amplitude of the displacement motion of Ti^{4+} ion induced in the photoexcited states is usually much greater than the motion in the ground state at low temperatures. Hence, after the photoexcitaion, an unphysically large lattice displacement is inferred to occur in this soft mode coordinate, if this mode has almost no restoring force. However, as long as our present system is a real solid crystal, an inter-ionic hard-core repulsion between O and Ti ions surely exists in this crystal. This hard-core repulsion finally works to suppress the occurrence of such an unphysically large displacement in the photoexcited states, although this repulsion is hidden in the ground state at low temperatures. Thus, in the cases of photoexcited states at low temperatures, we have to take it for granted, that a positive sextic anharmonicity, or more higher-order ones coming from this repulsion surely exist, although the aforementioned quartic one is not realized.

8.3. *Possible scenario*

Let us now proceed to a possible scenario that describes the aforementioned photoinduced enhancement of the conductivity and ε_1. By the photonexcitation, an conduction electron is generated in the 3d orbital of Ti, and this orbital is making an itinerant energy band with a width of about 2 eV, through its hybridization to the 2p orbital of O.[37] This electron linearly and strongly couples with the breathing (A_{1g}) mode of O around the Ti, and hence, as explained before, we have both the metastable large polaron state and the globally stable self-trapped state. Since the large polaron state is energetically higher that the self-trapped one, the photogenerated electron reaches this state first.[21] As explained before, this state can contribute to the electronic conductivity.

On the other hand, this large polaron state quadratically couples with the T_{1u} phonons, and the sign of this quadratic coupling constant is "negative" for the reason we see later in detail. Thus, the quadratic coupling makes this soft mode more "soft". This further softening due to the e-p coupling occurs at a very large number of lattice sites, which are included in this large polaron radius. Since this mode is dipole active, this further softening consequently contributes to the enhancement of quasi-static ε_1.

8.4. *Model Hamiltonian*

Keeping this scenario in mind, let us start from the following model Hamiltonian ($\equiv H_f$) that describes the electrons in the 3d conduction band of $SrTiO_3$, coupling linearly with the breathing mode, and quadratically with the T_{1u} mode, as, ($\hbar = 1$)

$$H_f = -T_f \sum_{\ell,\ell',\sigma} \{c^{+}_{\ell,\sigma} c_{\ell',\sigma} + \text{h.c.}\} - S_b\omega_b \sum_{\ell} n_\ell A_\ell + \frac{\omega_b}{2} \sum_{\ell} \left(-\frac{\partial^2}{\partial A_\ell} + A_\ell^2 \right)$$
$$- \frac{S_d\omega_d}{2} \sum_{\ell,i} n_\ell D^2_{\ell,i} + \frac{\omega_d}{2} \sum_{\ell,i} \left(-\frac{\partial^2}{m\partial D^2_{\ell,i}} + \frac{D^6_{\ell,i}}{3} \right) + U_f \sum_{\ell} n_{\ell,\alpha} n_{\ell,\beta}. \tag{27}$$

Here, $c^{+}_{\ell,\sigma}$ is the creation operator of an electron at a lattice site ℓ with spin σ ($=\alpha,\beta$) in a simple cubic crystal, and T_f is the transfer energy between neighboring two lattice sites ℓ and ℓ'. S_b is the dimensionless constant of linear coupling between this electron and the site localized breathing (A_{1g}) mode, whose energy is ω_b and dimensionless coordinate is A_ℓ. On the other hand, S_d (> 0) is the dimensionless constant of the quadratic coupling between the electron and the site localized T_{1u} mode, whose energy is ω_d and dimensionless coordinate in the direction i ($=$ x, y, z) is $D_{\ell,i}$; while U_f denotes the intra-site (intra-orbital) Coulomb repulsion. The dispersions of phonons are neglected, and only the long wave characteristics of each mode are taken into account.

The origin of a short range quadratic coupling between a site localized T_{1u} mode and an electron around it, is already well known to be an off-center effect.[43,47–50] It comes from the local mixing between the occupied atomic orbital of the electron, and unoccupied ones which are energetically higher than this occupied one but have opposite parities. This mixing is brought about by a linear change of the crystal field coming from the off-center type T_{1u} displacement of a central atom. Within the second order perturbation theory, the resultant energy change ($\equiv \Delta E$) of the occupied state is given as

$$\Delta E = -D^2 \sum_j \frac{\langle o|V|j\rangle\langle j|V|o\rangle}{E_{jo}}, \tag{28}$$

where, $|o\rangle$ and $|j\rangle$ denote occupied and unoccupied atomic orbitals, whose parities are opposite each other, and E_{jo} is the energy difference between them. V is an electronic operator with T_{1u} symmetry, and denotes the linear change of the crystal field due to the T_{1u} type displacement D, whose site and direction indices ℓ and i are omitted. We can see that ΔE is always negative as long as $|o\rangle$ is the highest occupied atomic orbital. Thus, we can get a negative quadratic interaction and it is a driving force for the central atom to push off its original position. We should note that this "off-center" effect is nothing but the local parity violation, resulting in a local and microscopic ferroelectricity.

In the present case, the Wannier function of the conduction band electron is mainly composed of the 3d orbital of Ti, hybridized with the 2p orbitals of O. This Wannier function mixes with the 4p orbital of Ti and the 3s orbital of O due to the T_{1u} type off-center displacement of the central Ti ion. In addition to these typical atomic orbitals, there are many other unoccupied higher electronic states, that can locally mix with the Wannier function. Thus, we can get a negative quadratic coupling as described by the fourth term of Eq. (27), wherein S_d only phenomenologically represents the effects of all aforementioned local mixings. The typical energy difference E_{jo} is about 10 eV, and hence this off-center effect is usually neglected. As mentioned before, if the original restoring force of this mode is very strong or hard, this quadratic coupling can be easily merged into the original hard force. However, in the cases of soft-anharmonic modes, the quadratic coupling becomes very important in connection with parity violating instabilities, even though it is not strong.

It should be noted that, our purpose in the present section, is not to estimate this coupling energy S_d quantitatively. Our purpose is to clarify possible parity violations coming from this quadratic e-p coupling, under the condition that this effect is one order smaller than both the itineracy of the electron and linear e-p coupling.

As for the anharmonicity of the fifth term of Eq. (27), we have taken a sextic one, for the reason mentioned before in detail, while "m" in this fifth term denotes the dimensionless effective mass of this site localized T_{1u} mode. This mass will be determined so that the lowest vibronic excitation energy of this soft-anharmonic Hamiltonian becomes equal to ω_d, which is obtained by the experiment.[42]

We now rewrite H_f into the following dimensionless form as

$$\begin{aligned} h \equiv H_f/\omega_b = & -t \sum_{\ell,\ell',\sigma} \{c^{+}_{\ell,\sigma} c_{\ell',\sigma} + \text{h.c.}\} + u \sum_{\ell} n_{\ell,\alpha} n_{\ell,\beta} \\ & - s_b \sum_{\ell} n_\ell a_\ell + \frac{s_b}{2} \sum_{\ell} a_\ell^2 - \frac{1}{2 s_b} \sum_{\ell} \frac{\partial^2}{\partial a_\ell^2} \\ & - \frac{\gamma s_d}{2} \sum_{\ell,i} n_\ell d_{\ell,i}^2 + \frac{\gamma s_d}{6} \sum_{\ell,i} d_{\ell,i}^6 - \frac{\gamma}{2\,m\, s_d^{1/3}} \sum_{\ell,i} \frac{\partial^2}{\partial\, d_{\ell,i}^2}, \end{aligned} \tag{29}$$

where

$$\begin{aligned} & t \equiv T_f/\omega_b, \quad u \equiv U_f/\omega_b, \quad s_b \equiv (S_b)^2, \quad \gamma \equiv \omega_d/\omega_b, \\ & s_d \equiv (S_d)^{3/2}, \quad a_\ell \equiv s_b^{-1/2} A_\ell, \quad d_{\ell,i} \equiv s_d^{-1/6} D_{\ell,i}. \end{aligned} \tag{30}$$

Assuming t, u, s_b, s_d are all greater than unity

$$(t, u, s_b, s_d) \gg 1, \tag{31}$$

we can define an adiabatic Hamiltonian $\mathrm{h_{ad}}$ as

$$\mathrm{h} \to \mathrm{h_{ad}} \equiv -\mathrm{t}\sum_{\ell,\ell',\sigma}\{\mathrm{c}^{+}_{\ell,\sigma}\,\mathrm{c}_{\ell',\sigma} + \mathrm{h.c.}\} + \mathrm{u}\sum_{\ell}\mathrm{n}_{\ell,\alpha}\mathrm{n}_{\ell,\beta} - s_{\mathrm{b}}\sum_{\ell}\mathrm{n}_{\ell}\mathrm{a}_{\ell}$$
$$+ \frac{s_{\mathrm{b}}}{2}\sum_{\ell}\mathrm{a}^2_{\ell} - \frac{\gamma s_{\mathrm{d}}}{2}\sum_{\ell,\mathrm{i}}\mathrm{n}_{\ell}\mathrm{d}^2_{\ell,\mathrm{i}} + \frac{\gamma s_{\mathrm{d}}}{6}\sum_{\ell,\mathrm{i}}\mathrm{d}^6_{\ell,\mathrm{i}}. \tag{32}$$

8.5. *Variational method for polaron*

Let us now proceed to a variational calculation for a polaron. For this single polaron state $|\mathrm{p}\rangle$ we use a trial function ($\equiv\varphi(\ell)$) as,

$$|\mathrm{p}\rangle \equiv \sum_{\ell}\varphi(\ell)\mathrm{c}^{+}_{\ell,\alpha}|0\rangle, \quad \sum_{\ell}|\varphi(\ell)|^2 = 1, \tag{33}$$

where $|0\rangle$ is the true electron vacuum. After taking the expectation value of $\mathrm{h_{ad}}$ with respect to this $|\mathrm{p}\rangle$, we get

$$\langle\mathrm{h_{ad}}\rangle = \left\langle -\mathrm{t}\sum_{\ell,\ell',\sigma}\{\cdots\}\right\rangle - \mathrm{s_b}\sum_{\ell}\langle\mathrm{n}_{\ell,\alpha}\rangle\mathrm{a}_{\ell} + \frac{\mathrm{s_b}}{2}\sum_{\ell}\mathrm{a}^2_{\ell}$$
$$- \frac{\gamma s_{\mathrm{d}}}{2}\sum_{\ell,i}\langle n_{\ell,\alpha}\rangle d^2_{\ell,i} + \frac{\gamma s_{\mathrm{d}}}{6}\sum_{\ell,\mathrm{i}}\mathrm{d}^6_{\ell,\mathrm{i}}, \quad \langle\cdots\cdots\rangle \equiv \langle\mathrm{p}|\cdots\cdots|\mathrm{p}\rangle. \tag{34}$$

Using the Hellman–Feynman theorem, we obtain

$$\frac{\partial\langle\mathrm{h_{ad}}\rangle}{\partial\mathrm{b}_{\ell}} = 0, \ \to \langle\mathrm{n}_{\ell,\alpha}\rangle = \mathrm{a}_{\ell}, \quad \frac{\partial\langle\mathrm{h_{ad}}\rangle}{\partial\mathrm{d}_{\ell,\mathrm{i}}} = 0, \ \to \langle\mathrm{n}_{\ell,\alpha}\rangle^{1/4} = \mathrm{d}_{\ell,\mathrm{i}}. \tag{35}$$

Substituting Eq. (15) into Eq. (14), we finally get

$$\langle\mathrm{h_{ad}}\rangle = \left\langle -\mathrm{t}\sum_{\ell,\ell',\sigma}\{\ldots\}\right\rangle - \frac{\mathrm{s_b}}{2}\sum_{\ell}\langle\mathrm{n}_{\ell,\alpha}\rangle^2 - \frac{\gamma s_{\mathrm{d}}}{3}\sum_{\ell,\mathrm{i}}\langle\mathrm{n}_{\ell,\alpha}\rangle^{3/2}. \tag{36}$$

8.6. *Continuum approximation*

Let us take the following Gaussian type trial function with a reciprocal localization length ($\equiv\Delta_{\mathrm{p}}$) as

$$\varphi(\ell) \propto \exp\left[-\frac{\Delta^2_p(\ell\cdot\ell)}{2}\right], \quad \ell = (\ell_{\mathrm{x}}, \ell_{\mathrm{y}}, \ell_{\mathrm{z}}), \tag{37}$$

where ℓ_{x}, ℓ_{y} and ℓ_{z} are the Cartesian components of ℓ. Using the continuum approximation, we regard ℓ_{i} ($\mathrm{i} = \mathrm{x}, \mathrm{y}, \mathrm{z}$) as a continuous variable from $-\infty$ to ∞. In this case, the first term of Eq. (36) is reduced as

$$\left\langle -t\sum_{\ell,\ell',\sigma}\{\cdots\}\right\rangle \to \int_{-\infty}^{\infty} d\ell_x \int_{-\infty}^{\infty} d\ell_y \int_{-\infty}^{\infty} d\ell_z\, \varphi(\ell) \times \left[-t\left(\frac{\partial^2}{\partial \ell_x^2}+\frac{\partial^2}{\partial \ell_y^2}+\frac{\partial^2}{\partial \ell_z^2}\right)\right]\varphi(\ell), \tag{38}$$

wherein all the one-body energies are referenced from the bottom of the conduction band. The other two terms in Eq. (36) can be easily calculated within this continuum approximation, and we finally get as

$$\langle h_{ad}\rangle = \frac{3t\Delta_p^2}{2} - \frac{s_b\Delta_p^3}{2^{5/2}\pi^{3/2}} - \frac{2^{5/2}(\gamma s_d)\Delta_p^{3/2}}{3^{3/2}\pi^{3/4}}. \tag{39}$$

Before we go in detail about the numerical results calculated by using Eq. (39), let us qualitatively examine energetics of $\langle h_{ad}\rangle$. It is shown in Fig. 29 as a function of Δ_p. When $(\gamma s_d) = 0$, the first term of $\langle h_{ad}\rangle$, being the kinetic energy, increases in proportion to Δ_p^2, while, the second term, being the energy gain due to the self-trapping by the breathing (A_{1g}) mode, decreases in proportion to $-\Delta_p^3$, as shown in Fig. 29(a). If these two competing quantities are of the same order, we get two minima of $\langle h_{ad}\rangle$. The minimum at $\Delta_p = 0$ in Fig. 29(a), denotes the well-known large (or free) polaron, while the other minimum at large Δ_p, being not written explicitly, is the self-trapped (small) polaron. It should be noted that all these states have even parities irrespective of Δ_p. As mentioned before, these two states are separated by an adiabatic energy barrier, and hence, the large polaron state can be locally stable, even when the self-trapped polaron state is globally stable.[30]

When $(\gamma s_d) \neq 0$, however, the energy gain due to the T_{1u} type parity violation is proportional to $-\Delta_p^{3/2}$, being most dominant in the small limit of Δ_p as shown in Fig. 29(b). Hence the previous minimum at $\Delta_p = 0$ now moves to a small but finite Δ_p. Thus, we can get a very important conclusion within the adiabatic approximation. An infinitesimal (γs_d) is enough to change the even-parity large polaron into the parity violating one, although the energy gain due to this symmetry breaking is quite small. This parity violation occurs at a large number of lattice sites included in the large polaron radius, and hence it is a quasi-global parity violation, as schematically shown in Fig. 30. While, the self-trapped state, being not written explicitly in Fig. 29(b), also has a parity violation, but it is quite local. Hence, we can call it the off-center type self-trapped state, wherein only the central atom will move off the original position, as shown in Fig. 30.

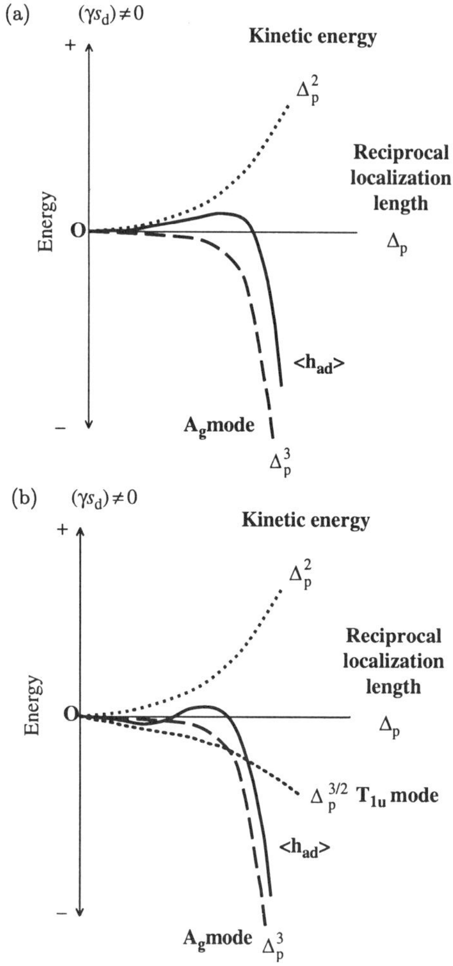

Fig. 29. Schematic energetics of the self-trapping. (a) Without T_{1u} mode. (b) With T_{1u} mode.

This off-center effect will be rather slight, since the lattice distortion of this state is mainly dominated by the strong coupling of the breathing mode.

It should also be noted that this parity violation of the T_{1u} mode is tacitly accompanied by the parity violation of the Wannier function itself, as mentioned in Eq. (28). Originally, this function has T_{2g} or E_g symmetry of the O_h point group, since it is the 3d orbital of Ti. However, as the T_{1u} type off-center displacement occurs, it mixes with unoccupied T_{1u} type

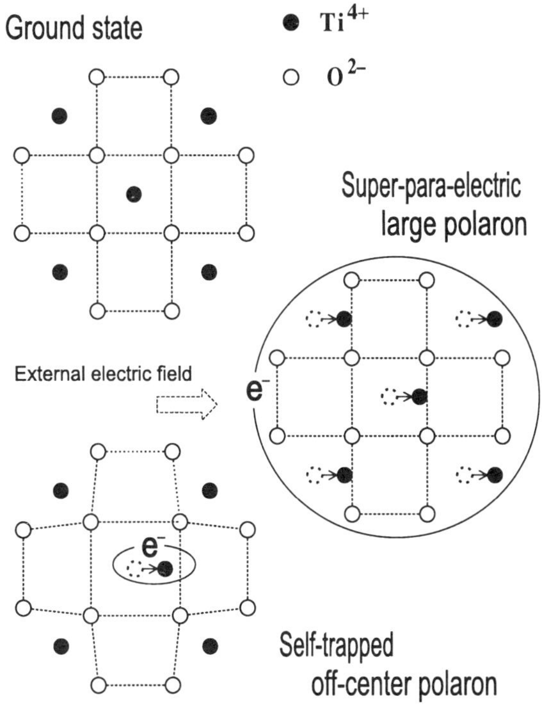

Fig. 30. Schematic natures of the ground state, the super-pare-electric large polaron, and the off-center type self-trapped polaron.

atomic orbitals, resulting in a finite dipole, and the parity violation in the electronic level too, although this change is hidden.

It is also very essential that the phase of this local parity violation is quite random and has no inter-site coherence, since phonons are site-localized ones, and the quadratic coupling is also independent of this phase. However, a weak external electric field may be enough to make it spatially coherent, as schematically shown in Fig. 30. In this sense, our new large polaron is super-pare-electric (SPE) one, which has a quasi-global parity violation. It is effectively same as a ferroelectric domain with an electron in it.

Let us now examine the nature of the anharmonicity used in this theory. For the reason mentioned above in detail, the sextic anharmonicity is used in Eq. (27). In this sextic case, an infinitesimal $\gamma s_{\rm d}$ is enough for the parity violation to occur. However, even if we change this sextic anharmonicity to other ones higher than quintic as

$$\frac{\omega_d}{2}\sum_{\ell,i}\frac{D_{\ell,i}^6}{3}\to\frac{\omega_d}{2}\sum_{\ell,i}\frac{|D_{\ell,i}^{2M}|}{M},\quad M>2.5, \tag{40}$$

we can also easily prove the occurrence of the same adiabatic instability for an infinitesimal γs_d. In the case of the aforementioned single-mode quartic double-well, the instability already occurs in the ground state itself and hence, the photogenerated electron may not induce further instabilities.

8.7. *Numerical results*

Keeping these general characteristics in mind, let us see results of further numerical calculations. According the quantum chemical theory by using a small cluster model for various perovskite type metal-oxides, the binding energy of the self-trapped poralon is shown to be few times of 0.1 eV,[51] which is much smaller that the aforementioned conduction band width (2 eV) or the energy gap (3.2 eV). Hence, we can say that in this type compound family, the self-trapping is marginal in the sense that the itineracy and the e-p coupling are well balanced. We, hereafter, will be concerned only with this marginal case.

In order to describe this marginal situation in the context of the present continuum approximation, we, at first, have to determine the meaningful maximum of Δ_p. For this maximum, we use $\Delta_p \le 2$. The present continuum model has no meaning for too large Δ_ps, and this limit $\Delta_p = 2$ is determined, since $|\varphi(\ell)|^2$ at $\ell = (1,0,0)$ becomes almost zero for larger Δ_ps than 2. Using this maximum value $\Delta_p = 2$, and neglecting the effects of the T_{1u} mode in Eq. (39), we get the marginal condition for s_b as,

$$6t \le \frac{\sqrt{2}s_b}{\pi^{3/2}}. \tag{41}$$

Since, $12\,T_f = 2$ eV, t becomes $t = 8.33$. In this case, the marginal s_b is given as $s_b \ge 200$. Under this condition the self-trapped state becomes the globally stable one, while from the following equation for Δ_p,

$$\frac{\partial\langle h_{ad}\rangle}{\partial\Delta_p}=0,\to\frac{s_b\Delta_p^{3/2}}{2^{5/2}\pi^{3/2}}-t\Delta_p^{1/2}+\left(\frac{2}{3}\right)^{3/2}\frac{(\gamma s_d)}{\pi^{3/4}}=0, \tag{42}$$

we can drive the condition to make the SPE large polaron locally stable. If this equation for Δ_p gives two roots in the region $0 < \Delta_p < 2$, the large polaron becomes locally stable in the adiabatic potential surface, just as schematically shown in Fig. 29(b). Regarding this Eq. (42) as a cubic equation with respect to $\Delta_p^{1/2}$, we get this condition

$$(\gamma s_d)^2 < \frac{(\sqrt{2}\pi t)^3}{s_b}. \tag{43}$$

When $t = 8.33$ and $s_b = 200$, it becomes, $\gamma s_d < 15.9$.

As mentioned before, our purpose in the present section is not to estimate this quadratic coupling constant γs_d quantitatively. Our purpose is to clarify possible parity violating instabilities coming from this quadratic e-p coupling, under the condition that this coupling is one order smaller that both the itineracy and the linear e-p coupling. In Fig. 31, we have shown one of such typical cases. It is obtained by the following set of parameters, $\omega_b = 20\,\text{meV}$, $t = 8.33$, $s_b = 200$, $\gamma = 0.05$, $\gamma s_d = 15$. The resultant SPE large polaron extends over about 1000 lattice sits, and its binding energy is far smaller than ω_b. While, the self-trapped state has a binding energy of about $20\omega_b$, although it is not written explicitly in Fig. 31. Throughout this review paper, we will not be concerned with quantitative calculations for the electronic conductivity. However, we can say, that this SPE large

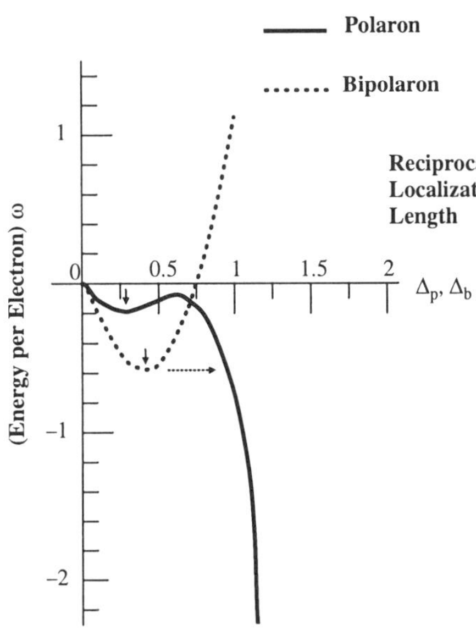

Fig. 31. (Energy per electron)/ω_b of poaron and bipolaron, as functions of Δ_p and Δ_b. $t = 8.33$, $s_b = 200$, $\gamma = 0.05$, $\gamma s_d = 15$, $u/s_b = 1.75$. From Ref. 32.

polaron (or this charged ferroelectric domain) is far mobile than the self-trapped state, and can surely contribute to the increase of the conductivity.

8.8. *Singlet bipolaron*

Let us now proceed to the next problem, whether two neighboring photo-generated electrons can make a stable singlet bipolaron, as compared with two independent polarons. For a singlet bipolaron state ($\equiv|\text{bp}\rangle$), we use a trial function ($\equiv\varphi_{\text{b}}(\ell)$) as

$$|\text{bp}\rangle \equiv \sum_{\ell,\ell'} \varphi_{\text{b}}(\ell)\varphi_{\text{b}}(\ell')a^{+}_{\ell,\alpha}a^{+}_{\ell',\beta}|0\rangle, \quad \sum_{\ell}|\varphi_{\text{b}}(\ell)|^2 = 1,$$

$$\varphi_{\text{b}}(\ell) \propto \exp\{-\Delta_{\text{b}}^2(\ell\cdot\ell)/2\}, \tag{44}$$

where Δ_{b} is the reciprocal localization length of this bipolaron. By the same method used before, we can get the energy per electron as

$$\frac{\langle\text{bp}|\text{h}_{\text{ad}}|\text{bp}\rangle}{2} = \frac{3\,\text{t}\Delta_{\text{b}}^2}{2} - \left(s_{\text{b}} - \frac{\text{u}}{2}\right)\frac{\Delta_{\text{b}}^3}{2^{3/2}\pi^{3/2}} - \frac{8(\gamma s_{\text{d}})\Delta_{\text{b}}^{3/2}}{3^{3/2}\pi^{3/4}}. \tag{45}$$

Since this system is based on the 3d orbital of Ti, the inter-electron Coulomb repulsion is stronger than both the itineracy and linear e-p coupling, and hence the self-trapped bipolaron seems to be unstable. However, the large bipolaron will be stable enough, since its mean inter-electron distance is very large. As one of typical examples for such cases, we take $\text{u}/s_{\text{b}} = 1.75$, keeping other parameters same as before, and can get the result shown in Fig. 31.

We can see that the SPE large bipolaron is stable relative to the two separated SPE polarons, while the self trapped bipolaron is unstable. From Fig. 31, we can also see that the SPE large bipolaron can fissure into two isolated self-trapped polarons, through a quantum tunneling process, as schematically shown by a dashed allow. This may be a rather slow process, since the localization radius of the polarons has to change largely and almost discontinuously between initial and final states of this tunneling.[52]

8.9. *Further softening and enhanced ε_1*

Let us now proceed to the aforementioned further softening of the $\text{T}_{1\text{u}}$ mode and the enhancement of ε_1. At first, we will reconfirm the original soft-anharmonic nature of the $\text{T}_{1\text{u}}$ mode at the ground state. It is given only by the fifth term of Eq. (27) without the electron. By using a variational method, we determine the lowest vibronic state ($\equiv|+\rangle$) and its energy

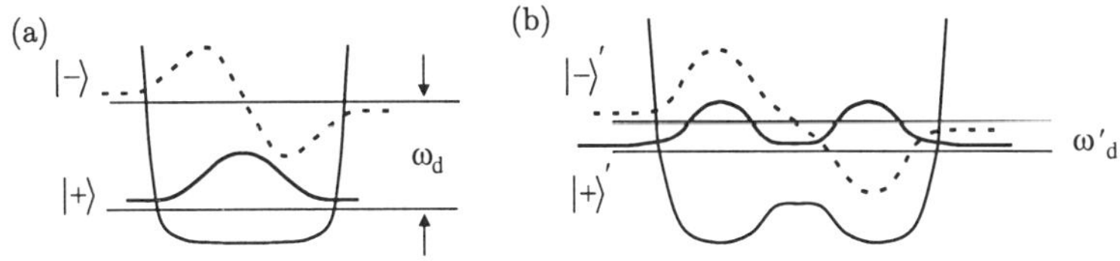

Fig. 32. The lowest two vibronic states of the T_{1u} mode. (a) At the electronic ground state. (b) At the central lattice site of the super-pare-electric large polaron. $\omega_b = 20\,\mathrm{meV}, t = 8.33, s_b = 200, \gamma = 0.05, \gamma s_d = 15, u/s_b = 1.75$.

($\equiv E_+$) at site ℓ and in the direction i, as well as those ($\equiv|-\rangle$, $\equiv E_-$) of the second lowest state. Hence, we use the following trial functions:

$$|+\rangle \propto \exp\{-(\kappa_+ D)^2/2\}, \quad |-\rangle \propto D\exp\{-(\kappa_- D)^2/2\}, \tag{46}$$

where $\kappa_\pm$ is the variational parameter of ($\equiv|\pm\rangle$), and the indices ℓ and i of D are omitted. After the variational calculation, we get $\kappa_+ = 1.2253$, $E_+/\omega_d = 0.3693$, $\kappa_- = 1.3622$, $E_-/\omega_d = 1.3693$, and these energies are referenced from the minimum point of the sextic potential $D = 0$. Since we have used $m = 1.3351$, the lowest excitation becomes $(E_- - E_+), /\omega_d = 1$, as shown in Fig. 32(a).

Let us roughly estimate the further softening due to the aforementioned negative quadratic e-p interaction. By this interaction, the effective potential at each lattice site will become a double well type one as schematically shown in Fig. 32(b). Strictly speaking, the T_{1u} type motions of Ti ions are now not mutually independent, but are correlated with each other since they are coupling to the same electron which is not localized. In order to avoid this complexity, we focus only on a single mode at a typical lattice site, for example, the x direction of the central site of the bipolaron, and fix all the coordinates of other modes at the values determined by the previous adiabatic and variational method, Eqs. (44) and (45). Then, we can get an effective Hamiltonian ($\equiv H_{ef}$) only for this mode as

$$H_{ef} = -S_d\omega_d|\varphi_b(0)|^2 D^2 + \frac{\omega_d}{2}\left(-\frac{\partial^2}{m\partial D^2} + \frac{D^6}{3}\right). \tag{47}$$

It gives a new double well potential with two minima at $D = \pm D_0$

$$D_0 \equiv (2S_d|\varphi_b(0)|^2)^{1/4}. \tag{48}$$

The electronic state $\varphi_b(0)$ may depend also on D of this mode. However, this electron is extended over about 1000 lattice sites, and hence we can neglect it. Using this H_e, we can determine the lowest vibronic excitation energy by using the following variational method.

$$|\pm\rangle' \propto \{\exp[-\kappa'_\pm(D - D_0)^2/2] \pm \exp[-\kappa'_\pm(D + D_0)^2/2]\}, \quad (49)$$

where $|\pm\rangle'$ denotes the lowest (second lowest) state with a variational parameter ($\equiv\kappa'_\pm$) and an energy ($\equiv E'_\pm$). Using the result shown in Fig. 31, we get $\kappa'_+ = 1.760, E'_+/\omega_d = 0.977$, $\kappa'_- = 2.010$, $E'_-/\omega_d = 1.232$ and these energies are referenced from the minimum point $D = D_0$ in Fig. 32(b). Thus, the lowest excitation energy ($\equiv\omega'$) is given as $\omega' = (E'_- - E'_+) = 0.255\,\omega_d$ and we finally got the further softening. We can say that, after the photoexcitation, the original vibronic wave function shown in Fig. 32(a), bifurcates as shown in Fig. 32(b). The lowest excitation energy decreases into about 1/4 of the original, because of this bifurcation.

From this further softening, we can roughly estimate the enhancement of ε_1. The matrix element of the dipole transition in this T_{1u} mode will not be changed so much in spite of this bifurcation. Under this condition the relative change of the quasi static dielectric constant is estimated as

$$\frac{(\text{relative change of } \varepsilon_1)}{(\text{polaron density})} = \frac{(\text{polaron volume})}{(\omega'_d/\omega_d)}, \quad (50)$$

wherein the polaron density itself varies depending on the exciting light intensity of each experiment, while, the polaron volume can be roughly estimated from the values of Δ_p and Δ_b shown in Fig. 31. Thus, by the present theory, we can conclude that the enhancement factor is just the right hand side of Eq. (50). It is concluded to be of the order of 10^4.

As mentioned before, this is also a kind of photoinduced structural phase transition, which occurs in the optical excited state and we can finally get a charged and conductive ferroelectric domain.

9. Further Problems in Photoinduced Phase Transitions

Thus we have reviewed the present status of theoretical and experimental studies on the photoinduced phase transiton phenomena. The history of this research is quite young, and we have various further problems, which are important but are not referred in preceding sections.

As for the theoretical models for the photoinduced phase transitions, here, we have been concerned only with the so-called itinerant models, in which the electrons or the excitons are itinerant from site to site. In this connection, Koshino, Ogawa and Sakai[53,54] have developed rather unique theories based on a site-localized two level system (ground and excited states), interacting with other site ones through a classical spring constant. They call it "domino theory", since in this model, a site-localized electronic excitation can proliferate through the spring constant just like the domino game. This theory could successfully describe the photoinduced collective phenomena of the previous organo-metallic complex crystal.

In the previous sections, we have been concerned only with the PSPT, which involves some change of the lattice structural. However, there are various other cases, such as photoinduced magnetic phase transitions[55] and the photoinduced superfluid transitions of high density excitons,[56–58] in which, phase transitions or order formations occur only in the electronic degree of freedoms, without the structural change of the lattice. Thus, through the photoinduced phase transition research, we will be able to discover various new phases, with and without structural change.

Acknowledgements

The author is very grateful to Professors K. Tanimura, S. Koshihara, K. Tanaka, H. Okamoto and Doctor T. Ishikawa for presenting their experimental results prior to publication. This work was supported by NAREGI Nanoscience Project, Ministry of Education, Culture, Sports, Science and Technology, Japan.

References

1. M. Ueta, H. Kanzaki, K. Kobayashi, Y. Toyozawa and E. Hanamura, *Excitonic Processes in Solids* (Springer-Verlag, Berlin, 1986), p. 203.
2. K. Nasu, *Relaxations of Excited States and Photoinduced Structural Phase Transitions* (Springer-Verlag, Berlin, 1997).
3. K. Nasu, *Photoinduced Phase Transitions and Their Precursor Phenomena,* Phase Transitions **74** (Gordon Breach, Oxford, 2001).
4. K. Nasu, P. Huai and H. Mizouchi, J. P. CM. **13** R693 (2001).
5. K. Nasu, K. Tanimura, T. Ogawa and A. Nakamura, *Photoinduced Phase Transitions, Their Dynamics and Precursor Phenomena,* Phase Transitions **75B** (Gordon Breach, Oxford, 2002).
6. C. Kittle, *Introduction to Solid State Physics,* Seventh edition (John Wiley, New York, 1996), p. 53.
7. Y. Tokura, T. Koda, T. Mitani and G. Saito, *Solid State Commun.* **43** 757 (1982).
8. S. Koshihara, H. Takahashi, H. Sakai, Y. Tokura and T. Luty, *J. Phys. Chem.* **B103** 2592 (1999).
9. T. Suzuki, T. Sakamaki, K. Tanimura, S. Koshihara and Y. Tokura, *Phys. Rev.* **B60** 6191 (1999).

10. J. Thompson and H. Stewart, *Nonlinear Dynamics and Chaos* (John Wiley, New York, 2002), p. 3.
11. P. Huai, H. Zheng and K. Nasu, *J. Phys. Soc. Jpn.* **69** 1788 (2000).
12. M. Cointe, M. Cailleau, H. Cailleau, B. Toudic, L. Toupet, G. Heger, F. Moussa, P. Schweiss, K. Kraft and N. Karl, *Phys. Rev.* **B51** 3374 (1995).
13. Z. Soos, *Annual Review of Phys. Chem.* **25** 121 (1974).
14. T. Sakano and Y. Toyozawa, *J. Phys. Soc. Jpn.* **65** 671 (1996); *Relaxations of Excited States and Photoinduced Structural Phase Transitions*, ed. K. Nasu (Berlin, Springer-Verlag, 1997), p. 160.
15. W. Su, J. Schrieffer and A. Heeger, *Phys. Rev.* **B22** 2099 (1980).
16. N. Nagaosa, *J. Phys. Soc. Jpn.* **55** 2754 (1986).
17. H. Hayashi and K. Nasu, *Phys. Rev.* **B32** 5295 (1985).
18. C. Katan, C. Koenig and P. Blochl, *Solid State Commun.* **102** 589 (1997).
19. C. Jacobsen and J. Torrance, *J. Chem. Phys.* **78** 112 (1983).
20. S. Iwai, S. Tanaka, K. Fujinuma, H. Kishida, H. Okamoto and Y. Tokura, *Phys. Rev. Lett.* **88** 057402 (2002).
21. K. Nasu, *Electron Conduction in Oxides*, ed. N. Tsuda (Springer-Verlag, Berlin,2000), p. 107.
22. N. Tomita and K. Nasu, *Phys. Rev.* **B63** 085107 (2001).
23. N. Tomita and K. Nasu, *Phys. Rev.* **B64** 125118 (2001).
24. H. Mizouchi and K. Nasu, *J. Phys. Soc. Jpn.* **69** 1543 (2000).
25. H. Mizouchi and K. Nasu, *J. Phys. Soc. Jpn.* **70** 2175 (2001).
26. Y. Ogawa, S. Koshihara, K. Koshino, T. Ogawa., C. Urano and H. Takagi, *Rhys. Rev. Lett.* **84** 3181 (2000).
27. T. Tayagaki and K. Tanaka, *Phys. Rev. Lett.* **86** 2886 (2001).
28. P. Huai and K. Nasu, *J. Phys. Soc. Jpn.* **71** 1182 (2002).
29. Y. Toyozawa, *Polarons and Excitons*, ed. C. Kuper (Oliver and Boyd, Edinburgh, 1963), p. 211.
30. Y. Shinozuka and Y. Toyozawa, *J. Phys. Soc. Jpn.* **46** 505 (1989).
31. G. Mahan, *Many Particle Physics* (Kluwer Academic, New York, 2000).
32. K. Nasu, *Phys. Rev.* **B67** 174111 (2003).
33. H. Katsu, H. Tanaka and T. Kawai, *Jpn. J. Appl. Phys.* **39** 2657 (2000).
34. M. Takesada, T. Yagi, M. Itoh and S. Koshihara, *J. Phys. Soc. Jpn.* **72** 37 (2003).
35. T. Hasegawa, S. Mouri, Y. Yamada and K. Tanaka, *J. Phys. Soc. Jpn.* **72** 41 (2003).
36. N. Tsuda, *Electronic Conduction in Oxides*, ed. N. Tsuda (Springer, Berlin, 2000), p. 47.
37. L. Mattheiss, *Phys. Rev.* **B6** 4718 (1972).
38. T. Hasegawa, M. Shirai and K. Tanaka, *Journal of Lumine.* **87–89** 1217 (2000).
39. H. Yasunaga, *J. Phys. Soc. Jpn.* **24** 1035 (1968).
40. R. Cohen, *Fundamental Physics of Ferroelectrics 2000*, AIP CP535 (AIP, New York, 2000).
41. K. Muller and H. Burkard, *Phys. Rev.* **B19** 3593 (1979).
42. R. Migoni, H. Bilz and D. Bauerle, *Phys. Rev. Lett.* **37** 1155 (1976).

43. V. Fridkin, *Ferroelectric Semiconductors* (Consultants Bureau, Plenum, New York, 1980).
44. G. Shirane and Y. Yamada, *Phys. Rev.* **177** 858 (1969).
45. H. Vogt, *Phys. Rev.* **B51** 8046 (1995).
46. T. Sekine, K. Uchinokura and E. Matsuura, *Solid State Comm.* **18** 569 (1976).
47. F. Luty, *Physics of Color Centers*, ed. W. Fowler (Academic Press, New York, 1986), p. 181.
48. I. Bersuker, *Phys. Lett.* **20** 589 (1966).
49. Y. Toyozawa, *Optical Processes in Solids* (Cambrigde University, Cambrigde, 2003), p. 326.
50. Y. Yacoby and Y. Girshberg, *Fundamental Physics of Ferroelectrics 2000*, AIP CP535 ed. R. Cohen (AIP, New York, 2000), p. 56.
51. R. Eglitis, E. Kotomin and G. Borstel, *J. Phys. CM* **14** (2002) 3735.
52. K. Nasu and Y. Toyozawa, *J. Phys. Soc. Jpn.* **50** 235 (1981).
53. K. Koshino and T. Ogawa, *J. Phys. Soc. Jpn.* **68** 2164 (1999).
54. O. Sakai, T. Ogawa and K. Koshino, *J. Phys. Soc. Jpn.* **71** 978 (2002); O. Sakai, M. Ishii, T. Ogawa and K. Koshino, *J. Phys. Soc. Jpn.* **71** 2052 (2002).
55. K. Matsuda, A. Machida, Y. Moritomo and A. Nakamura, *Phys. Rev.* **B58** R4203 (1998).
56. E. Fortin, S. Fafard and A. Mysyrowics, *Rhys. Rev. Lett.* **70** 3951 (1993).
57. T. Karasawa, H. Mino and M. Yamamoto, *J. Lumine.* **87–89** 174 (2000).
58. T. Inagaki, T. Iida and M. Aihara, *Phys. Rev.* **B62** 10852 (2000).

CHAPTER 2

TIME-RESOLVED SPECTROSCOPY OF THE DYNAMICS OF PHOTOINDUCED IONIC-TO-NEUTRAL PHASE TRANSITION IN TETRATHIAFULVALEN-P-CHLORANIL CRYSTALS

Katsumi Tanimura

The Institute of Scientific and Industrial Research
Osaka University, 8-1 Mihogaoka, Ibaraki
Osaka 567-0047, Japan
E-mail: tanimura@sanken.osaka-u.ac.jp

We discuss the primary mechanism of several steps involved in the dynamics of photoinduced ionic-to-neutral (I-to-N) phase transition in tetrathiafulvalen-p-chloranil (TTF-CA) crystals, based on recent experimental results of time-resolved and state-resolved spectroscopy. By analyzing the temporal evolution of reflection changes induced by 80-fs laser pulses that in turn induce charge-transfer (CT) transition in this crystal, three distinct primary steps have been extracted and resolved; these are precursor-forming step, the local proliferation of neutral molecular clusters and the final step of forming N-phase orders. The CT-exciton strings have been proposed as the precursor formed in the first step. The properties of the CT strings depend strongly on the photon energy of excitation and the dependence governs the state-sensitive features of phase transition. Upon generation of the strings above the critical density, they induce local proliferation to form neutral domains at the second step till 100 ps after excitation at 6 K. The local neutral domains can be described as a three-dimensional cluster consisting of one-dimensional neutral domains, which lacks a three-dimensional order of N phase. The third step follows the local proliferation to establish finally three-dimensional order of N phase and photoinduced N-phase domains are formed around 400 ps after excitation. Thus, the process of photoinduced I-to-N transition can be described as a cascade of generation, proliferation and crystallization of CT strings, analogous to the thermodynamics of I-to-N transition in TTF-CA, but each step of the cascade has been clearly resolved temporarily. Based on the microscopic picture of the process, we compare the processes induced by CT excited states and by molecular excited states reached by intra-molecular transition, and find a significant difference on the precursor-forming step. Also, impulsive acoustic phonon generation associated with the phase transition is discussed based on detailed analysis of the oscillatory modulation in reflection changes as a function of probe-photon energy and on the excitation intensity.

Contents

1. Introduction

1.1. *Photoinduced structural phase transitions*

Among several phenomena associated with the valence instability in solids, structural phase transition in low-dimensional systems is of great current interest.[1,2] Recently, it has been demonstrated that structural phase transitions in some low-dimensional crystals can be induced by optical excitation. Examples of such photoinduced structural phase transitions (PISPTs) are the ionic-to-neutral phase transition in a mixed-stack organic charge-transfer (CT) crystal,[3] color-phase transitions in polydiacetylene,[4] and in polythiophene.[5] These pioneering works have opened a new field of condensed-matter physics of photoinduced phase transitions.[2] The significance of this new phenomenon of PISPT lies in cooperative interactions during rclaxations in strongly coupled electron-lattice systems triggered by photoexcitation.

Structural phase transitions inherently include cooperative interactions in electronic and lattice systems. Under thermodynamical conditions, all degrees of freedom of a system concerned are equilibrated with temperature and pressure and the evolution is governed entirely by free energy to establish the lowest-energy state of the system under a given condition. In the case of PISPT, on the other hand, only the electronic system is excited instantaneously by light to trigger cooperative effects in other systems that were fixed under the original order. Therefore, it is essentially non-equilibrium and dynamical.[1] How such cooperative interactions can be induced is an important and interesting open question in condensed matter physics. Also, this subject has brought us a new point of view for studying the real-time dynamics of structural phase transitions that have been studied only under conditions of thermodynamical equilibrium. Optical detection of photo-induced phase transitions is a unique feature in the examples of PISPT mentioned above.[2–4] By using this technique, we can study the ultrafast dynamics of the system evolution after short-pulse excitation. Therefore, we can reveal several intermediate and short-lived states or "phases" during the transition from one phase to another. By extending time-resolved studies of PISPT, it may be possible to find a few hidden multi-stabilities of a given crystal.[1,6]

The excitation-induced structural changes in solids have been studied extensively for many different types of solids for a long time. A typical example is the lattice defect formation by valence excitation in alkali halides, where a halogen ion is displaced electronically by leaving an electron at its original site to an interstitial site to form a neutral Frenkel pair. In these phenomena, structural changes induced were essentially local; only a few atomic sites are subjected to change. This local nature is related directly to the properties of three-dimensional solids with small correlations. In such a crystal, the excitation-induced structural change needs localization of valence excitation, since the energy necessary to displace one atom from one site to the other, or to change the charge states of the basis is typically a few eV. The intrinsic localization by self-trapping can be induced when the electron-phonon interaction S exceeds the energy B gained by delocalization of charge carriers or excitons in a periodic lattice.[7] Therefore, the two states (localized and delocalized) can be classified critically as small and large polarons in a deformable lattice. In a system with S larger than B, the electronic excitation energy can be localized as a small polaron and hence can induce only local structural changes. On the other hand, the localization of the electronic excitation in low dimensional systems shows essentially different features from that in three-dimensional crystals as demonstrated theoretically; a localized state with much wider spatial distribution becomes possible.[8] Also, in low dimensional systems, several mechanisms (Peierls instability, spin-Peierls instability, etc.) can induce instability to generate distinct structural phases with small energy separations because of strong correlation and electron-phonon interaction, and the two charge states of the basis can become closer in energy, as in the case of mixed stack charge-transfer organic crystals.[9,10] For such low-dimensional systems, we may expect an essentially different structural response under electronic excitation. The PISPT may be one of typical examples.[2]

1.2. *The transition between neutral and ionic phases in TTF-CA*

The structural phase transition that will be discussed here is neutral to Ionic (NI) transition in a mixed-stack organic CT crystal of tetrathiafulvalene-p-chloranil (TTF-CA).[9,10] Such a class of organic crystals has electron donor (D) and acceptor (A) molecules, alternating along the same stack axis and are classified into two categories according to the degree of CT (ρ): quasi neutral (N) for $\rho < 0.5$ and quasi ionic (I) for $\rho > 0.5$. Since the electronic interaction among π electrons of D and A molecules is predominantly large along the stack axis, the crystals are viewed as quasi-one dimensional systems. In a simple picture, the relative stability of N and I states of a one-dimensional chain, with neglecting inter-chain interaction, can be discussed in terms of competition between the energy cost of forming ionic D^+A^- pairs, and the gain in electrostatic energy. The interface between I and N phases is defined at $I - E_A = a_M V_0$, where I is the ionization energy of the D molecule, E_A the electron affinity of the A molecule, a_M the Mardelung constant for a chain, and V_0 is the nearest-neighbor Coulomb interaction. Many crystals of this class have been classified into I- and N-phase crystals in terms of the criterion.[9] In more realistic descriptions, we have to take into account other factors such as a finite CT integral that leads to a partial degree of CT in each state and several other interactions among molecules and structural relaxations, together with inter-chain interaction.[11] In particular, dimerization of an ionic D^+A^- pair, which induces a symmetry breaking distortion, is one of the most important structural relaxations, and it governs magnetic (spin) and electric (dipolar) interactions of a crystal. The two phases in CT organic crystals undergo the phase transition (NI transition) thermodynamically by applying hydrostatic pressure[9] and by changing temperature.[10] The crystal of TTF-CA is an example showing temperature-induced NI transition at ambient pressure; the ionic phase with ρ of 0.7 below 81 K is transformed into the neutral phase with ρ of 0.3 above 84 K.[10,12]

The thermodynamics of the NI transition in TTF-CA has been studied extensively by neutron diffraction and nuclear quadrupole resonance.[13,14] Experimentally, three different phases have been identified in pressure-temperature phase diagram; these are ionic ferroelectric phase ($I_{ferro} = I$), ionic paraelectric phase (I_{para}), and neutral paraelectric phase ($N_{para} = N$). The I_{para} phase, which is detected at high-temperature and high-pressure regime, is characterized by the ionic-stack chains randomly arranged to give a paraelectric property. Therefore, it can be regarded as an intermediate between I and N phases. In the case of temperature-induced NI transition in TTF-CA at low-pressure regime, the transition displays a strong first order character, requiring sufficient inter-chain interactions. In the high-temperature N phase, the monoclinic unit cell contains two symmetry related neutral D^0A^0 pairs, which are located on inversion symmetry sites and form alternating regular stacks along the *a* (stack) axis. On the other hand, in the low-temperature I phase, an ionic D^+A^- pair is dimerized upon an enhanced molecular ionicity, and the dimerized stack chains form a ferroelectric arrangement between chains. Consequently, a loss of the screw-axis symmetry is resulted in I phase. Not only the symmetry-breaking distortion, but also, totally symmetric distortions are induced upon the transition; which result in abrupt changes in unit-cell parameters for the *b* and *c* axes.[13] A recent study of the NI transition has proven the importance of interstack elastic coupling that results in a multistep-transition character including mesoscopic phase coexistence around T_c.[14]

It has been pointed out that the transitions among these three phases are analogous to the transitions of solid-liquid-gas situation; solid, liquid and gas phases correspond to I, I_{para}, and N phases, respectively.[13,15] However, one of the significant aspects in the description of thermodynamics of NI transition in TTF-CA is cooperativity. As described above, TTF-CA is viewed as quasi-one-dimensional systems with strong electron correlation and electron-phonon coupling. Therefore, they can be approximated as a set of weakly interacting one-dimensional chains. Based on the electronic and structural aspects, it has been argued that the NI transition proceeds via a cascade of cooperative phenomena: the formation of CT-exciton strings, their one-dimensional lattice relaxation within a chain due to strong intra-chain interactions, and their three-dimensional condensation and ordering, both originating from weaker inter-chain interactions.[15] The CT-exciton string, which is a non-linear relaxed excited state, is the key species playing crucial role in the description of NI transition mentioned above. The CT string has first been identified by Kuwata–Gonokami *et al.* under optical excitation in a CT organic crystal (anthracene-PMDA).[16] Later, Collet *et al.* provided evidence of the thermal formation of the strings by x-ray diffuse scattering study in 2,6-dimethyltetrathiafulvalene-p-chloranil.[17] Although the exciton strings have not been proved yet in TTF-CA, several theoretical predictions[11,18–20] and successful description of thermodynamics of NI transition in terms of cooperativity[13,15] have provided sound basis of the crucial role of the strings.

1.3. *Photoinduced NI transition in TTF-CA*

The crystal of TTF-CA displays the unique feature of optical properties, which gives us a strong benefit to study the dynamics of NI transition.[3] The electronic properties of TTF-CA have been studied experimentally by measuring reflection spectra and the changes in the spectra have been correlated beautifully with the NI transition in this crystal.[12,21,22] In Fig. 1, I show the polarized reflection spectra of TTF-CA at several temperatures obtained by Tokura *et al.*[12] The reflection peaks due to CT transitions, shown in Fig. 1(a), are observed below 1.3 eV for the polarization with electric vector E parallel to the a axis ($E//a$). The energy and the intensity of the main peak (A_1) at 0.75 eV in I phase are roughly the same as those in the N phase, but the high-energy peak (A_2) around 1.15 eV is characteristic of only the I phase; the intensity is reduced significantly in the N phase.[12,22] On the other hand, the reflection spectrum measured for polarization with E perpendicular to the a axis ($E \perp a$) in the visible region shown in Fig. 1(b) is characterized by intra-molecular transitions of TTF molecules. In the spectrum, TTF^+ ions show two distinct peaks at 2.2 and 3.1 eV in the I phase; these are called, respectively, the B and C bands.[12,21] The peak energies of these two bands change abruptly upon NI transition to 2.4 and 3.3 eV in the N phase. These changes have been ascribed to the sensitive dependence of the transition energies on the degree of ionicity of TTF molecules. However, there is a significant difference between B

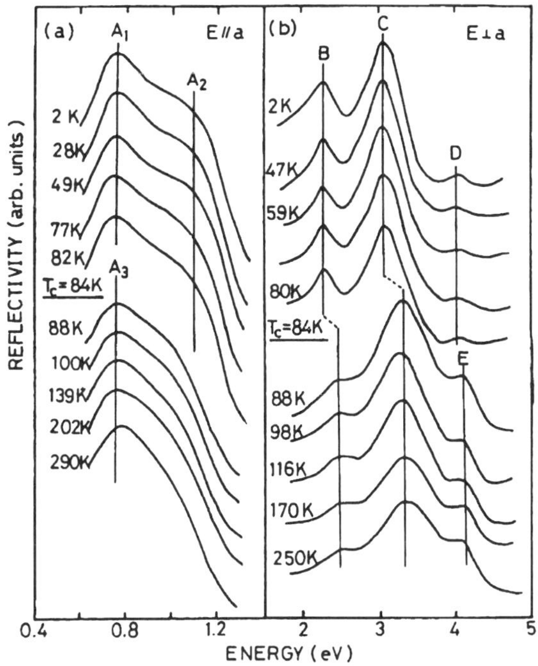

Fig. 1. Temperature changes of the charge-transfer excitation bands (a), and of the intramolecular excitation bands (b), across the neutral-to-ionic phase-transition temperature (after Ref. 12).

and C bands due to TTF^+; the B band is more localized in nature than the C band.[21] Therefore, the energy of the B band reflects sensitively the local ionicity of the donor molecules. Based on the correlation between the properties in reflection changes and the NI phase transition in TTF-CA, one can make use of the changes in optical reflection as a sensitive and fast probe of the transition in TTF-CA.

In fact, Koshihara *et al.* have first demonstrated the photoinduced transition from the I to N (I-to-N) phases at temperature lower than the transition temperature in TTF-CA by probing the reflection changes.[3] They excited I-phase crystals by visible (2.0–2.5 eV) light pulses and measured the induced changes in visible reflection. The observed change in the reflection spectrum was similar to that evaluated from reflection spectra at I and N phases by assuming that N-phasc domains are generated by light in I-phase crystal. Based on careful examinations of characteristics of reflection changes, they have concluded that a part of the I phase is converted into

N-phase domains by electronic excitation. The N-phase domain thus formed is metastable below transition temperature and it is transferred back into the original I phase within a few ms after excitation. Thus the formation and its decay of a transient phase can be probed by the optical method of detection.

Not only the transition from low-temperature I phase to high-temperature N phase, but also the transition from N to I phase can be induced by photoexcitation. Collet *et al.* have demonstrated by means of time-resolved X-ray diffraction technique that the I phase characterized by a macroscopic ferroelectric order is reorganized within 500 ps after exciting specimens of high-temperature N phase by 300-fs laser pulses[23]. The photoinduced transition from high-temperature to low-temperature phases, which cannot be due to laser-induced heating, has proven definitely that the photoinduced NI transition in TTF-CA is triggered by the excitation in electronic system of the crystal.

In the study of photoinduced I-to-N transition, Koshihara *et al.* have evaluated that as many as 160 D^+A^- pairs are converted to neutral pairs per absorbed photon. We show in Fig. 2 schematically the possible non-linear changes in electronic and structural systems in TTF-CA. Since the two phases can be classified by ρ, electronic excitation may induce the valence change from ionic to neutral configurations. Based on the concept of thermodynamics of NI transition, CT-exciton strings can be assumed as the initial product of photoexcitation and short-time relaxation. It is speculated that the region with photogenerated strings are then transformed into N-phase domains by inducing proliferation associated with the formation of three-dimensional order characteristic of the N phase at a later stage. The possible model of the photoinduced I-to-N transition in Fig. 2 has involved several fundamental questions about the process. First, what is the initial product of optical excitation and associated electronic and structural relaxation in a short temporal period? Since TTF-CA is a quasi-one dimensional system with strong correlation and electron-phonon interaction, the product may no longer be described as a simple picture of CT excitons as in the case of usual organic crystals. We need to identify the structural and electronic characteristics of the product, with reference to the key concept of CT-exciton strings established in a strongly correlated one-dimensional system. Then, the next question is how the formation of N-phase domain evolves after the fast relaxation leading to the initial product. In the thermodynamics of NI transition, generation, condensation and crystallization of CT strings have been introduced to describe the cascade

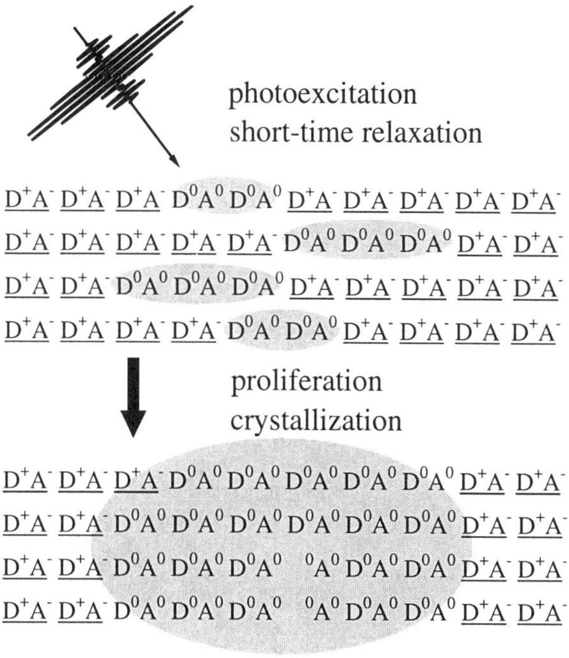

Fig. 2. A schematic presentation of a possible model of photoinduced ionic-to-neutral phase transition in TTF-CA induced by charge-transfer excitation. D^+ and A^- stand for the ionic donor and acceptor molecules, and D^0 and A^0 for the neutral donor and acceptor molecules, respectively.

of the process. In the case of photoinduced transition, we need to reveal what the primary steps involved in the evolution are and how they are distinguished experimentally and conceptually, by keeping one primary question in our mind i.e. whether the photogenerated phase is the same as the ordinary phase.

Previous studies of PISPT have shown that the phase transitions can be induced only under specific conditions of excitation wavelengths.[4,5] This feature indicates clearly that only those excited states that satisfy certain requirements can trigger the cooperative interaction and hence the phase transition. Therefore, the state-dependent property is one of the important and unique issues in the study of PISPT. This state-dependent feature has already been revealed in the photoinduced I-to-N transition in TTF-CA,[24] and will be described below as one of the main issues. Previous studies have also shown that the transition can be induced only above a certain threshold in excitation density,[4,5] revealing a nonlinear feature of PIPST. Such a threshold has also been detected in photoinduced I-to-N transition in TTF-CA.[24,25] Thus, the process of PISPT is highly nonlinear and state-dependent. In order to understand the microscopic origin of these characteristics, it is highly desirable to resolve primary steps involved in the process. Recently, the dynamics of the NI transition has been studied experimentally in the temporal domain of sub-ps,[25–28] and the three primary steps of the process have been resolved.[28] Below we describe recent experimental results showing state-sensitive, nonlinear and dynamical features of the photoinduced transition in TTF-CA, and we discuss possible microscopic processes of the transition in this crystal.

2. State-sensitive and Non-linear Features of Photoinduced NI Transition

In organic crystals, two different types of electronic excited states can be classified in general; one is essentially the excited state of constituent molecules and the other is the CT excited state. These two types of excited states have been studied so far in terms of Frenkel excitons and CT excitons in organic crystals, respectively. However, in quasi-one dimensional solids where the electron-electron correlation plays important roles, the concept of describing excited states may be revised significantly.[11] Here we cautiously use the term of molecular excited states and CT excited states to describe electronic excited states in TTF-CA, instead of Frenkel excitons and CT excitons.

2.1. *Threshold excitation density for metastable N-phase formation*

In the pioneering work by Koshihara *et al.*, the photon energies of laser light to induce the transition from I to N phases correspond to the intramolecular transitions of TTF^+ molecule, leading to molecular excited states.[3] The fast relaxation from the molecular excited stets to CT excited states was assumed to describe the transition in their study. Later, Suzuki *et al.* confirmed that the I-to-N transition could be induced upon generating CT excited states.[24] They excited a low-temperature I phase crystal with 1064-nm laser light, the photon energy of which falls on high-energy side of the peak energy of CT transition. The CT-transition nature of their excitation was confirmed by measuring the amount of N-phase domains generated as a function of polarization of the exciting light; N phases were

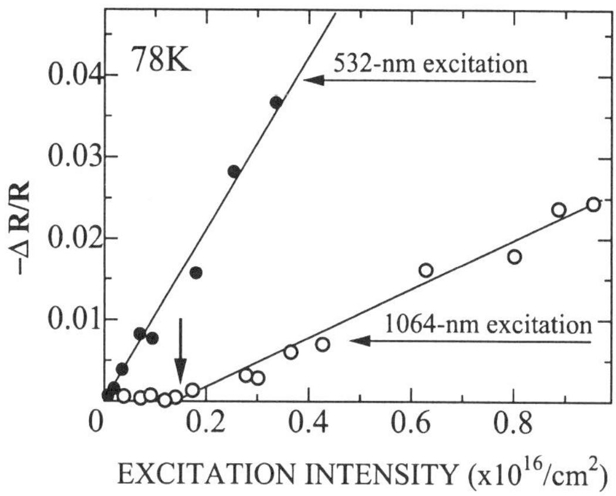

Fig. 3. Magnitudes of the reflectance changes at 3.0 eV induced by 1064-nm laser pulse, open circles, and by 532-nm laser pulse, solid circles, as a function of excitation intensity scaled in terms of fluence (after Ref. 24).

generated with the highest yield for $E//a$ polarization of 1064-nm light. Therefore, it was evident that the CT-excited states certainly induced the I-to-N transition, similarly to the case of intra-molecular excitation.

They studied the excitation-intensity dependence of the yield of photogenerated N phases for both molecular-excited states and CT-excited states, and revealed an interesting feature of state-dependent characteristics in inducing the transition.[24] Their result is shown in Fig. 3, where the yield of N-phase domains determined by probing reflection changes in $E \perp a$ polarization as a function of the excitation intensity. As clearly seen in the figure, the N-phase domain formation takes place almost in proportion to the excitation intensity in the case of molecular-excited states generated at visible photon energy. However, for the CT-excited states reached by 1064-nm light, the I-to-N transition can be induced only above a threshold-excitation intensity. The threshold in excitation intensity (I_{th}) indicates clearly that a single CT excitation cannot generate a macroscopic N-phase domain; the N-phase domain can be formed only via nonlinear processes of CT excited states above a certain excitation density. Therefore, the photoinduced I-to-N transition shows a strong nonlinear feature.

Tanimura and Akimoto have extended the study for the excitation wavelengths from 800 to 2000 nm to demonstrate the state-dependent feature more clearly.[25] They have shown that, similar to the case of 1064-nm, the I-to-N transition is induced only above the threshold excitation intensities in the whole range of CT transition and that the magnitude of I_{th} depends

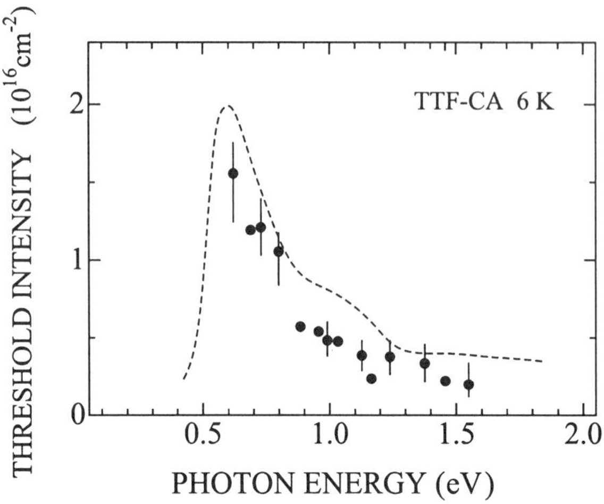

Fig. 4. The threshold excitation intensity for generating metastable N-phase domains as a function of the photon energy of exciting laser pulses. The broken curve shows the optical conductivity spectrum of charge-transfer transition at 45 K (after Ref. 22).

strongly on the photon energy of excitation. In Fig. 4, we show the result of the threshold in excitation intensity.[22] The intensity increases rapidly with decreasing photon energy of exciting laser light; the intensity at 0.62 eV is about 10 times larger than that at 1.55 eV. The threshold density, evaluated from I_{th} and reflectivity and absorption coefficient at a given energy shows similar dependence. The wavelength-dependent threshold in excitation density implies that CT excited states reached by optical excitation with a given photon energy are different with respect to the ability to induce NI transition. Thus, characteristics of photoinduced I-to-N transition are state-sensitive.

The threshold in excitation intensity described above is related to the free energy difference between two phases. Tanimura and Itoh have studied temperature dependence of I_{th} by using ns-pulse excitation.[29] At any temperatures, similar to the case in Fig. 3, the yield of N-phase domains were analyzed well in terms of a threshold following linear growth characterized by a slope. In the analysis, they found two features; one is that I_{th} decreases with increasing temperature and the other is that the slope increases with increasing temperature. In Fig. 5 are shown the results of the temperature dependence of I_{th} from 6 to 79 K. The threshold intensity decreases with temperature and reduces down to zero just at the transition temperature, where I phase is transferred into N phase without any optical excitation.

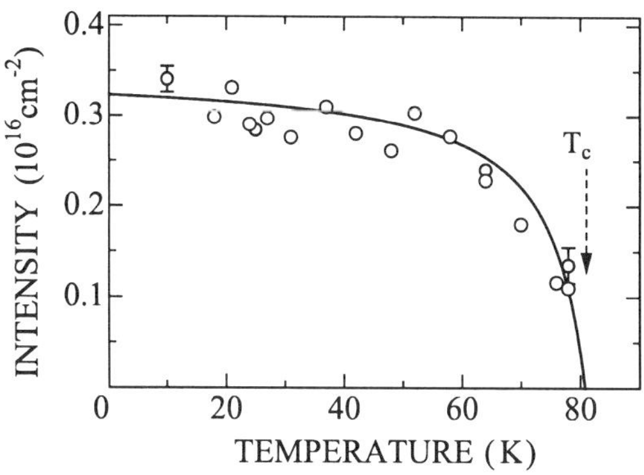

Fig. 5. The threshold excitation intensity for metastable N-phase domain formation of 1064-nm ns-laser pulses as a function of temperature (after Ref. 29).

Therefore, the result in Fig. 5 clearly shows that the threshold is the quantity that is related to the free energy difference between the two phases.

2.2. *The formation yield of N-phase domains vs. excitation intensity*

As described above, the characteristics of the photoinduced I-to-N transition in TTF-CA are dependent on the excitation wavelength, the intensity, and temperature. Hereafter we pay our main attention to the non-linear feature seen in excitation-intensity dependence, since difference in wavelengths and temperatures appears to change simply their magnitudes by keeping the same nonlinear characteristic essentially unchanged. In this respect, the molecular excited states can be regarded as an extreme case where I_{th} is small enough. Therefore, let us have an overview of the intensity dependence of photoinduced I-to-N transition for a wider intensity regime at a given excitation wavelength and fixed temperature. We plot in Fig. 6 the yield of photoinduced N-phase domains in the I-phase crystal as a function of the excitation intensity of 80-fs laser pulse of 810-nm light with $E//a$ polarization.[28]

For quantitative analysis of reflection changes, it is advantageous to probe a uniformly excited region in a specimen. In order to make this condition available, we need to excite a specimen at a wavelength where the absorption coefficient is smaller than those at other wavelengths where reflection changes are probed. Based on existing spectroscopic data,[9,12,22] we find that the wavelengths around 800 nm correspond to the region where

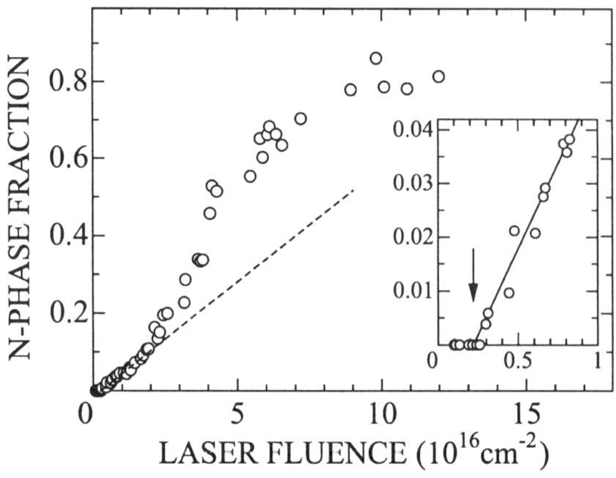

Fig. 6. The fraction of photoinduced metastable N-phase domains as a function of excitation intensity of 810-nm 80-fs laser pulses at 6 K. The broken curve shows the extrapolated linear growth above the threshold intensity. The inset shows the fraction at weak-intensity regime showing the threshold intensity of 2.5×10^{15} photon/cm^2 (after Ref. 28).

the absorption coefficients are smallest in the I phase crystal. The wavelength of 810 nm corresponds to the highest energy tail of the CT transition. At this wavelength, measurements of polarization-dependent yield of the photoinduced N-phase domains showed that the yield was the highest for the exciting light with $E//a$ polarization, confirming the generation of CT-excited states as a primary excitation. Under the selection of the excitation wavelength of 810 nm, I-phase crystals were excited with 80-fs laser pulses and the yield of photogenerated N-phase domains were measured at 500 ns after excitation. Since the pulse width of 80 fs is short enough, as will be shown later, we can exclude any possible effects of cascade excitation of excited states photogenerated that may cause other types of nonlinear effects. Also, the time delay of 500 ns is long enough to lead to the formation of the stationary N-phase domains via several dynamical processes, but is short enough to detect N-phase domains against their transient decays at ms regime after excitation. Therefore, the result in Fig. 5 displays purely the yield of the final product of this phase-transition process as a function of excitation intensity of the light pulse that induces CT transition.

As clearly seen in the figure, the formation yield of N-phase domains increases almost linearly above a threshold intensity at weak-excitation regime below 1×10^{16} cm^{-2} (see the inset), while the yield becomes super linear, followed by saturation, at intense excitation regime. The result reveals two different nonlinear processes involved; one is the process that results in the threshold excitation intensity at weak excitation regime and the other is the super linear increase at intense excitation regime. In order to understand the microscopic origin of two nonlinear processes revealed above, it is first important to clarify the primary steps involved in this PIPST and then to study the characteristics of each step. For clarifying and resolving the primary steps, studies of ultrafast dynamics of the transition are one of the most powerful methods of experimental studies.

3. Primary Steps in the Dynamics of Photoinduced NI Transition

As described in Sec. 1, the optical spectra of reflection, showing abrupt changes upon the NI transition in TTF-CA,[12] provide a sensitive probe to study ultrafast dynamics of the transition. In the previous studies of this phase transition, the changes in reflection at visible region measured by the probe light with $E \perp a$ polarization, which is featured by the intra-molecular transition of TTF^+ ions, have mainly used. Since the B-band transition is very local in nature,[21] it is certainly powerful for monitoring the change in ρ of TTF molecules locally. However, because of the strong local nature, it may be less sensitive to other long-range orders that are also important in the phase transition. As summarized in Sec. 1, not only the degree of charge transfer but also other three-dimensional crystalline orders change significantly upon NI transition.[14] Therefore, using other methods of optically probing the changes associated with the NI transition are desirable to detect temporal evolution of the whole process of transition; limited knowledge of the process may prevent us from comprehensive understanding of the dynamics. Not only the visible reflection with $E \perp a$ polarization, but also the visible and near infrared (NIR) reflection with $E//a$ polarization show abrupt changes in NI transition. Therefore, they can be used as probes of several changes associated with the transition. In particular, the visible reflection with $E//a$ polarization can probe the temporal change the concentration of CA^- ions that cannot be probed by other reflections.[12] Also, NIR reflection may probe differently the changes upon NI transition, since the transition nature is essentially different from those of intra-molecular transitions. Below, we first describe overviews of experimental results of fs time-resolved reflection study in (3.1). Some selected results that give us deeper insight into a few characteristics revealed in (3.1) are presented in (3.2). A phenomenological analysis of temporal evolution of reflection changes is given in (3.3) to describe the whole process in terms of three consecutive steps.

3.1. *Overview of the results of fs time-resolved spectroscopy*

First, let us confirm whether reflection changes in $E//a$ polarization can be used as a probe of photoinduced I-to-N transition as well. In Fig. 7(a)

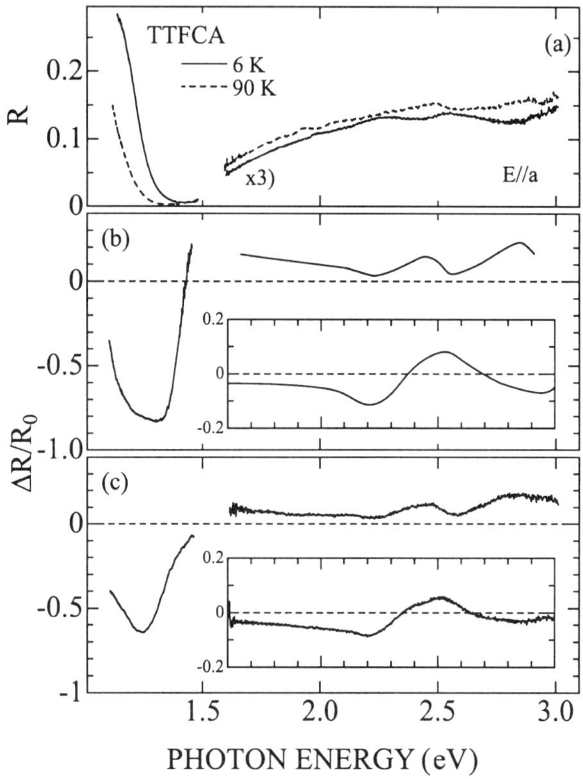

Fig. 7. (a) Reflection spectra at 6 K (solid curves) and 90 K (broken curves) measured for probe-light polarization with electric vector parallel to the stack axis ($E//a$). (b) Evaluated relative changes in reflectance upon temperature-induced transition from I to N phases of TTF-CA. Difference between the spectra at 90 K and 6 K was divided by that at 6 K. The inset shows the relative change in reflectance measured for probe-light polarization with electric vector perpendicular to the stack axis ($E \perp a$). (c) Relative changes $\Delta R/R_0$ in reflection spectra for $E//a$ polarization measured at 500 ps of excitation with laser pulses of 1.55 eV at 6 K. The inset shows the change in reflection spectra for $E \perp a$ polarization at 500 ps of excitation (after Ref. 28).

are shown the reflection spectra of a specimen measured with $E//a$ polarization at 6 and 90 K. The spectra at 6 K are representative of the spectra at I phase, while those at 90 K are characteristic of N phase. In the visible reflectance for $E//a$ polarization, the peak at 2.6 eV in 6-K spectrum has been assigned to the CA^- ions in I phase, which is absent in 90-K spectrum. On the other hand, the reflection below 1.5 eV due to the charge-transfer transition shows significant change in reflection intensity between two phases. Therefore, distinct changes in $E//a$ reflection are evident. The curves in Fig. 7(b) are the relative change $\Delta R/R_0$ in reflectance evaluated from the set of reflection spectra shown in Fig. 7(a), where R_0 is the spectrum of the I-phase crystal at 6 K and $\Delta R = R - R_0$. The curve in the inset is the similar change in the spectrum for $E \perp a$ polarization, which has been used in previous studies. In Fig. 7(c), we show the experimentally detected spectra of $\Delta R/R_0$ induced by exciting the I-phase crystal at 6 K with 1.55-eV 80 fs-laser pulses, and measured at 500 ps after excitation. It is clear that the measured spectra are similar to, but not exactly the same as, those evaluated from thermally induced changes in reflection spectra. In particular, spectral features of photoinduced change in NIR reflection show substantial difference, although those at visible reflectance are similar to those in Fig. 7(b). The similarity in the spectral features of $\Delta R/R_0$ indicates that the phase similar to the N phase is generated by photoexcitation. At the same time, the difference in the spectral shapes may suggest possible differences in some orders of photogenerated phase from those of ordinary N phase. Although this possible difference is an open question to be solved, we can confirm here that probes at wider spectral ranges are valuable for studying the dynamics of the transition.

How do these changes in reflection spectra evolve after fs-pulse excitation? In order to show the evolution from one side, we show in Fig. 8 the spectra measured at typical delay times for the three types of spectra described in Fig. 7. In visible reflection measured with $E \perp a$ polarization in Fig. 8(c), the spectrum measured at probe delay time τ of 1 ps is characterized be the negative change at 2.25 eV and positive changes at high and low energy sides. With increasing delay time, the spectrum changes progressively and the spectrum characteristic of N-phase formation is observed at $\tau = 300$ ps. In the visible reflection measured with $E//a$ polarization shown in Fig. 8(b), the initial change is characterized by two negative dips at 2.26 and 2.61 cV at 0.7 ps of excitation. With increasing delay time, a whole changes grow toward the positive side and the spectroscopic feature characteristic of N-phase formation is detected at 300 ps. The NIR reflection

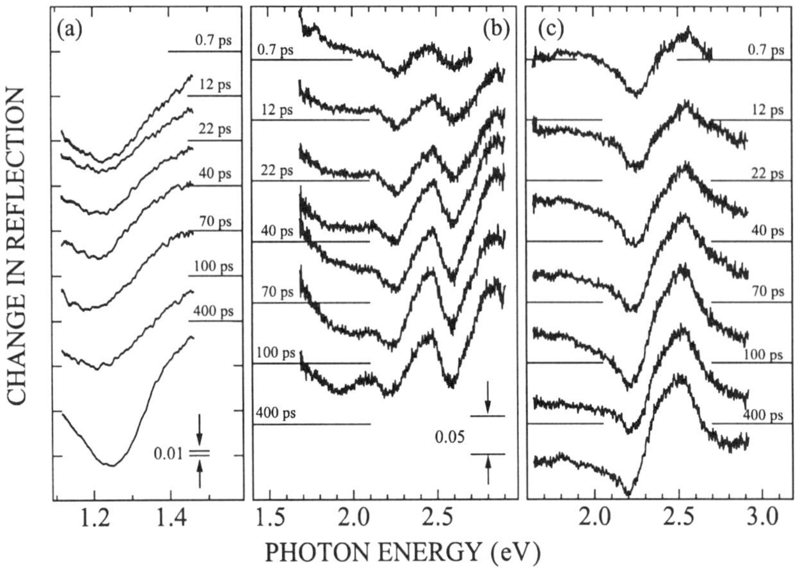

Fig. 8. Time series of the relative changes in reflecton spectra measured with probe-light polarization $E//a$ in (a) and (b), and that with probe-light polarization $E \perp a$ in (c) induced by 80-fs laser pulses of 804 nm. The excitation intensity is the same for the spectra shown in (a), (b), and (c). The number on the base line of each spectrum shows delay time after excitation (after Ref. 28).

measured with $E//a$ polarization in Fig. 8(c) shows the initial negative change peaked at 1.22 eV at $\tau = 0.7$ ps and then the amount of the change is recovered partially around $\tau = 10$ ps. With increasing delay time after 30 ps, the negative change grows again to display a peak at 1.25 eV around $\tau = 300$ ps. In these figures, it is clear that not only the magnitude of the changes but also the spectral shapes change progressively with delay times till about 400 ps after excitation.

In order to show the temporal evolution of reflection changes from the other side, we plot the relative change at typical photon energies as a function of delay time in Fig. 9. The changes at 2.25 and 2.50 eV in $E \perp a$ polarization, at 2.25 eV and 1.22 eV in $E//a$ polarization are shown in Figs. 9(a), (b), and (c), respectively. In these figures, $\Delta R/R_0$ measured for two different excitation intensities are shown for comparison. It is evident from the figure that the change in visible reflection is associated with a strong oscillatory modulation, although the change at 1.20 eV for $E//a$ polarization is almost free from this effect. It is also clear that that the mode

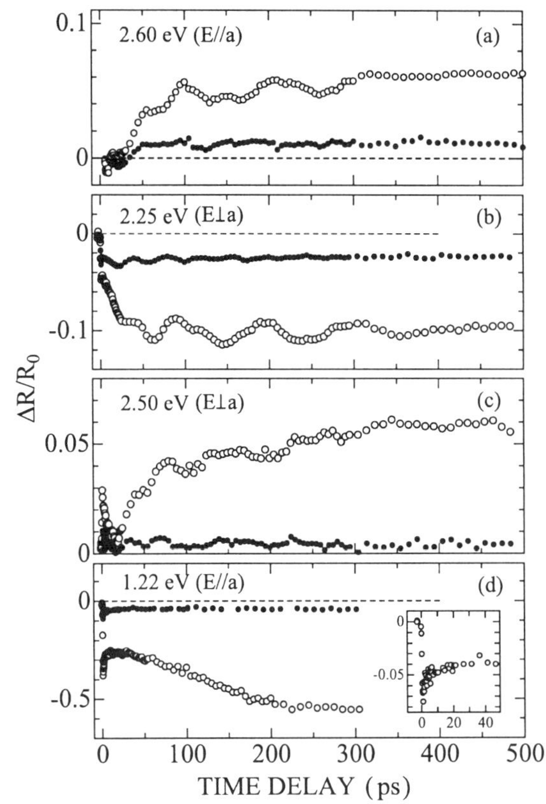

Fig. 9. Temporal evolution of relative changes in reflectance induced by 80-fs laser pulses of 804 nm for two different excitation intensities. In (a), (b), (c), and (d), are shown the changes measured at 2.25 eV with $E//a$ polarization, at 2.25 eV with $E \perp a$ polarization, at 2.50 eV with $E \perp a$ polarization, and 1.22 eV with $E//a$ polarization, respectively. In all figures, dots and open circles show the changes under the excitation of 2.4×10^{16} photon/cm^2 and of 5.2×10^{16} phton/cm^2 (after Ref. 28).

of the modulation is dependent on the excitation intensities, too. Therefore, the mode of oscillation is apparently dependent on probe-photon energy and excitation intensity.

Apart from the oscillatory modulations, it is evident that temporal evolutions of reflection change show two significant features. The first feature is that the mode of temporal evolution depends strongly on I_{ex}. In the case of weak excitation that induces less than 10% reduction in $\Delta R/R_0$ at 1.22 eV, for cxample, the amount of $\Delta R/R_0$ induced by photoexcitation shows a rapid partial recovery within a few tens ps of excitation as shown in the inset of Fig. 9(d), and the rapid recovery is followed by an almost constant value at longer delay times. Consequently, the magnitude of $\Delta R/R_0$ at 300 ps of excitation is less than that at the incidence of light pulse. On the other hand, in the case of strong excitation that induces more than 30% reduction in $\Delta R/R_0$, it is recovered partially within 10 ps of excitation, but is followed by the gradual increase in the amount of $\Delta R/R_0$. Consequently, the magnitude of $\Delta R/R_0$ at 300 ps becomes larger than that at 1 ps of excitation, showing a strong enhancement of induced changes. This characteristic of excitation-intensity dependent change is also clear in the changes probed at visible region, although the delay times that give the maximum magnitude of $\Delta R/R_0$ are dependent on the probe photon energies and polarizations.

The second feature is that the growth mode is dependent on the photon energy where the change is probed. The growth at 2.25 eV in $E \perp a$ polarization for the highest excitation intensity appears to be completed within 100 ps after excitation, whereas those induced by the same I_{ex} but monitored at different photon energies and polarization in Figs. 9(a) and (c) show persistent growth lasting over 200 ps. In order to examine further this characteristic of the growth mode dependent on probe-photon energy, we plotted $\Delta R/R_0$ monitored at several photon energies in the spectra measured with $E \perp a$ polarization in Fig. 10. At 1.70 eV, $\Delta R/R_0$ shows the growth resulting in the maximum within 100 ps, and then it keeps a constant value except a persistent oscillatory modulation. However, $\Delta R/R_0$ measured at other photon energies show rapid changes within 100 ps and persistent changes lasting over 300 ps after excitation. Since changes in reflection result from photoinduced changes in electronic and/or structural orders, it is clear that a stationary order can be reached only around 400 ps of excitation at 6 K. The sensitive dependence of the growth mode on probe-photon energy shown in Figs. 9 and 10 indicates clearly that several consecutive steps are involved in the photoinduced I-to-N transition and that these steps contribute differently to reflection change probed at a given photon energy.

From an overview of the results of time-resolved spectroscopy shown in Figs. 8–10, we can extract following three important features of the process. The first is the progressive evolution in the spectral changes in reflection lasting over 300 ps. This feature, which corresponds to a sensitive dependence of the growth mode on probe-photon energy, reflects consecutive changes in electronic and structural configurations triggered by photoexcitation. Therefore, the photoinduced transition is a multi-step process. The second is a nonlinear dependence of the final amount of reflection change on

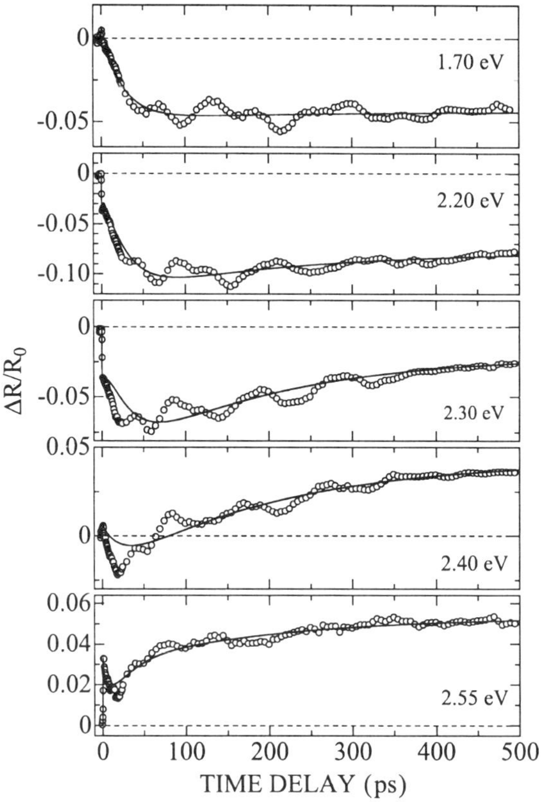

Fig. 10. Temporal evolution of relative changes in reflectance measured with $E \perp a$ polarization at different probe-photon energies. The excitation intensity is 5.2×10^{16} phton/cm^2. The solid curves in the figures are the result of simulation by using a rate-equation model (see the text).

the excitation intensity, as clearly seen in Fig. 9. The third is the oscillatory modulation of the reflection changes, the mode of which is dependent both on probe photon energy and on the excitation intensity. Below, we present more detailed results and the analysis on these three important features to have deeper insight.

3.2. *Quantitative results of three important characteristics in the dynamics*

(a) *Spectral analysis of the induced changes*

As seen in the spectra in Fig. 8, changes in reflection spectra are persistently evolved over 300 ps to lead to those similar to N-phase formation. In

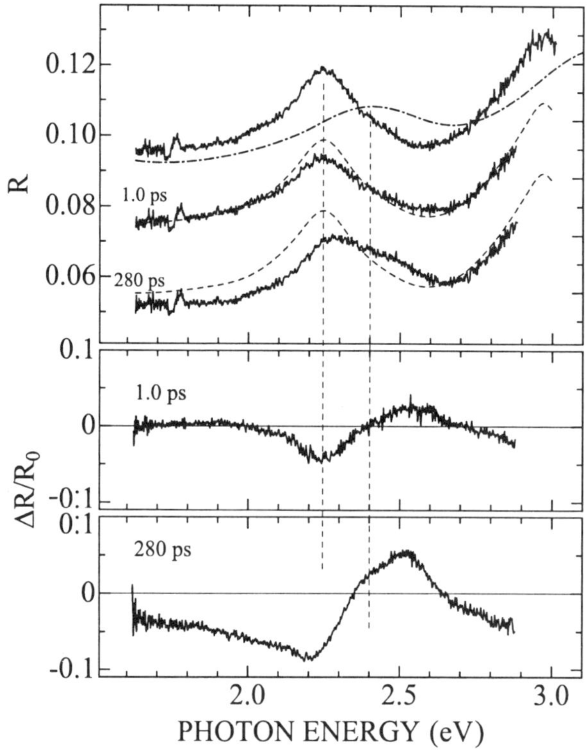

Fig. 11. The reflection spectra at 6 K measured with $E \perp a$ polarization before excitation and at 1 and 280 ps of excitation in (a), and the relative changes in reflectance measured at 1 ps (b), and at 280 ps (c) of excitation with 80-fs laser pulses of 804 nm. The chain-dot curve in (a) shows the reflection spectrum measured at 90 K. The spectrum before excitation at 6 K is shown by broken curve with shifted base line for comparison (after Ref. 28).

order to make the spectral feature of the changes at respective steps more transparent, we examine here the reflection spectrum itself. We show the reflection spectra with $E \perp a$ polarization measured before excitation and at two delay times in Fig. 11(a) and the corresponding relative changes of the latter two in (b) and (c). In Fig. 11(a), the chain curve shows the spectrum at 90 K, which is representative of the N-phase spectrum. In the figure, it is clear that the reflection peak at 2.25 eV due to the B band in I phase is bleached at 1 ps of excitation. This bleaching, by keeping the peak energy unchanged, is associated with broadening of the peak and this broadening is the origin of the positive change in $\Delta R/R_0$ in Fig. 11(b). This bleaching is not associated with the increment of the 2.4-eV peak in the reflection, which is characteristic of the B band in N phase. Therefore, it is clear that the laser excitation induces an abrupt reduction of the density of TTF^+ configurations, but does not convert it instantaneously into the TTF^0 configurations characteristic of N phase. The increase in reflection intensity at 2.40 eV, characteristic of the formation of N phase takes place at much longer delay times of about a few hundred ps after excitation as shown in Fig. 10. In fact, as seen in Fig. 11(c), the spectrum of $\Delta R/R_0$ at 280 ps shows clearly the B-band feature in N phase. The reflection spectrum at 280 ps in Fig. 11(a) certainly shows the peak characteristic of the B band in N phase. Therefore, we can conclude that the N phase similar to the ordinary N phase is formed only around 400 ps after the initial reduction of TTF^+ concentrations. Then, the progressive changes in the spectral shapes can be ascribed to the consecutive steps of the transition process, starting from an abrupt reduction in TTF^+ concentration and leading to the formation of N-phase domains.

The reduction of TTF^+ concentration means their conversion to other charge states, presumably to neutral TTF^0 state. Since this conversion is induced by CT excitation at 1.55 eV, similar conversion of CA^- may result. Here we examine the temporal change in CA^- concentrations to correlate it with the change in TTF^+ ions. The 2.61-eV peak in $E//a$ reflection has been assigned to CA^- ions in I phase,[12] and the evolution of $\Delta R/R_0$ at this peak energy has been shown in Fig. 9(a). However, as mentioned above, reflection changes at a given probe energy include several contributions of consecutive steps. In particular, a persistent changes are evident at the whole range in $E//a$ spectra shown in Fig. 8(b). Therefore, elimination of underlying changes may make the evaluation of the change in CA^- concentrations clearer. For this purpose, we take the difference of $\Delta R/R_0$ between 2.61 eV and 2.45 eV to represent the reduction of the height of 2.61-eV peak. Similarly, the difference of $\Delta R/R_0$ between 2.25 and 2.55 eV in $E \perp a$ spectrum was adopted as a measure of reduction of TTF^+ concentration for comparison. Both differences are plotted as a function of time delay and compared in Fig. 12. When we neglect oscillatory modulations in the two curves, it is clear that both quantities evolve in parallel with time: a rapid change within a few ps of excitation, a strong enhancement of the change till 100 ps, and a almost constant value after 100 ps. Therefore, the reduction of TTF^+-ion concentration is associated with the similar loss of CA^- ions after excitation. This result substantiates that D^+A^- pairs are bleached upon photoexcitation to convert them into other charge state and finally transferred to D^0A^0 pairs N-phase domains around 300 ps after excitation.

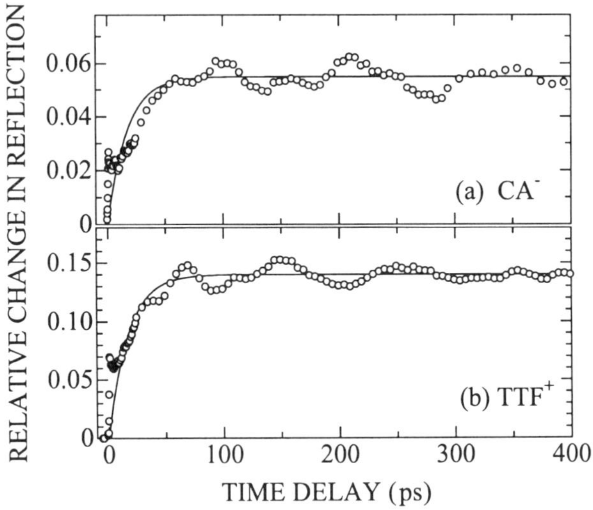

Fig. 12. The evaluated reduction in peak heights at 2.61 eV in $E//a$ spectrum due to CA^- ions (a), and at 2.25 eV in $E \perp a$ spectrum due to TTF^+ ions (b), as a function of delay time after excitation. The solid curves in the figure show the exponential growth with the rise time of 20 ps (after Ref. 28).

(b) *Excitation-intensity dependence of changes in* $\Delta R/R_0$

Here we examine quantitatively the excitation-intensity dependent properties of reflection changes. As seen in Fig. 9, one of the most important features is the enhancement of the change at longer delay times under intense excitation. We analyze mainly the changes in visible reflection with $E \perp a$ polarization and those in NIR with $E//a$ polarization, since these show typically the intensity-dependent features. We plot the magnitude of $\Delta R/R_0$ at 2.25 eV in $E \perp a$ spectrum measured at 1 and 60 ps of excitation as a function of excitation intensity in Fig. 13(a). Also the magnitudes of $\Delta R/R_0$ at 1.20 eV in $E//a$ spectrum measured at 1 and 300 ps of excitation are shown in Fig. 13(b) for comparison. The change at 1 ps of excitation is representative of initial photoinduced change, while that at 60 ps or at 300 ps is representative of the amount resulted by nonlinear process. It is clear from the figure that the induced change of $\Delta R/R_0$ measured at 1 ps is proportional to the excitation intensity, with a tendency of saturation at higher intensities. Therefore, the initial reduction is caused by the one-photon process. Contrary to this linear reduction, the $\Delta R/R_0$ measured at later time delays show a strong nonlinear characteristic. At intensities below 1×10^{16} cm^{-2}, the ratio of $\Delta R/R_0$ at 60 ps relative to that at 1 ps at

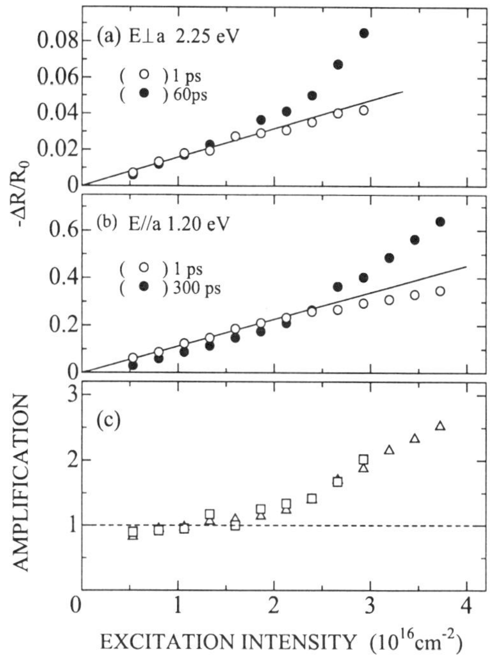

Fig. 13. The magnitudes of the relative change $\Delta R/R_0$ at 2.25 eV in $E \perp a$ reflection spectra measured at 1 and 60 ps of excitation (a), and those at 1.22 eV in $E//a$ spectra at 1 and 300 ps of excitation (b), as a function of excitation intensity. In (c), the ratio of the magnitude at 60 ps relative to that at 1 ps 2.25 eV in $E \perp a$ reflection spectra is plotted by open squares as a function of the intensity, and is compared with the similar ratio (open triangle) at 300 ps to that at 1 ps at 1.22 eV in $E//a$ spectra. The ratio at 1.22 eV is multiplied by a factor of 1.4 for comparison (after Ref. 28).

2.25 eV, which is defined as the multiplication factor, is around unity, but it increases superlinearly above the critical value of the intensity. Similarly, the ratio of $\Delta R/R_0$ at 300 ps relative to that at 1 ps in the change at 1.20 eV show essentially the same dependence on the intensity. The results indicate that the enhancement of the magnitude of $\Delta R/R_0$ probed at two different photon energies and polarizations are correlated with each other, although the nonlinear enhancement is detected at a different delay time.

(c) *Oscillatory modulation in reflection changes*

As shown in Figs. 9 and 10, a strong oscillatory modulation is associated with reflection changes induced by photoexcitation. Such a modulation has been observed in previous studies.[24,26,30] The new feature to be stressed here is that the mode of oscillation is dependent on the excitation intensity.

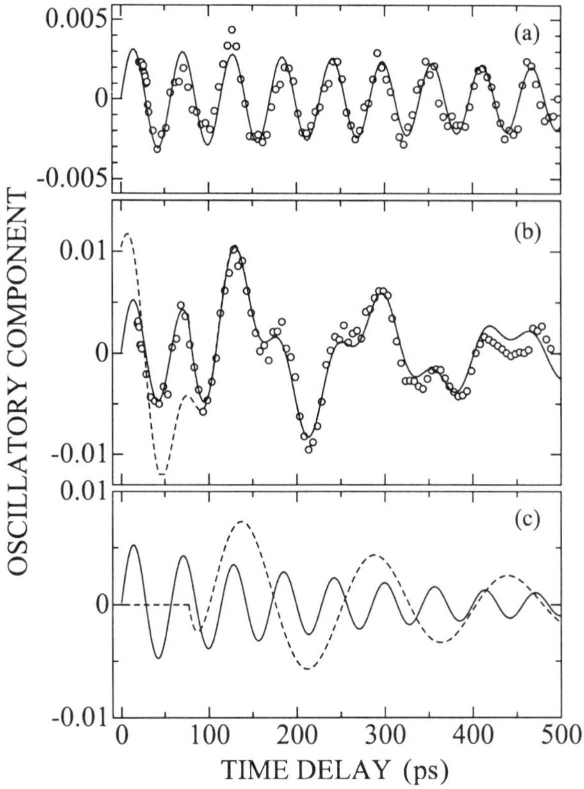

Fig. 14. The oscillatory component monitored at 1.70 eV with $E \perp a$ polarization for the excitation intensity of 2.8×10^{16} photon/cm^2,(a), and for the intensity of 5.2×10^{16} photon/cm^2, (b). The solid curve in (a) shows the fitted dumped oscillation of a single component with frequency of 112 GHz, and the slid curve in (b) shows the fitted dumped oscillations of two components with frequencies of 112 and 54 GHz shown by solid and broken curves in (c) (after Ref. 28).

To demonstrate this feature more clearly, we show in Fig. 14 the oscillatory component $r(t)$ in $\Delta R/R_0$ measured at 1.70 eV with $E \perp a$ polarizations for two different excitation intensities. The $r(t)$ was evaluated by subtracting a smooth growth with a form of $A\{1-\exp(-Bt)\}$ from the observed result, by using appropriate constants of A and B. In the case of weak excitation, the oscillation is essentially characterized by a single mode with the frequency of 112 GHz as seen in Fig. 14(a). On the other hand, the oscillation induced under strong excitation shows clearly significant interference effects with other frequency components.

When we introduce a second mode with a low frequency, which is almost half of the primary frequency, then the mode of oscillation can be described satisfactorily, as shown by solid curves. In this analysis, $r(t)$ was fitted by a function;

$$r(t) = \sum_{i=1}^{2} A_i \exp(-t/\tau_i) \sin(\omega_i t + \phi_i),$$

where A_i, τ_i, ω_i, and ϕ_i are the amplitude, angular frequency, the phase of a oscillatory component and τ_i is the time constant which characterizes the damping of amplitude. In this analysis we found that the delay of about 70 ps in the evolution of the low-frequency component is essential for the reasonable fitting in the whole temporal region, although the high-frequency component starts instantaneously upon excitation. The broken curve in Fig. 14(b) shows the result of simulation with no delay in the generation of the low-frequency component; it fails to reproduce the observed oscillation in temporal domain from 1 to 100 ps. At any excitation intensities and at different probe-photon energies for two polarizations in the visible reflectance, the oscillatory components could well be analyzed as a superposition of the two damped oscillators. In Fig. 15, I plot the frequencies obtained by the analysis as a function of photon energy of probe beams. It is clear that the newly introduced frequency of the low-frequency mode is also proportional to the photon energy, or the inverse of wavelength, similar to the case of the high-frequency mode. This linear dispersion is one of the

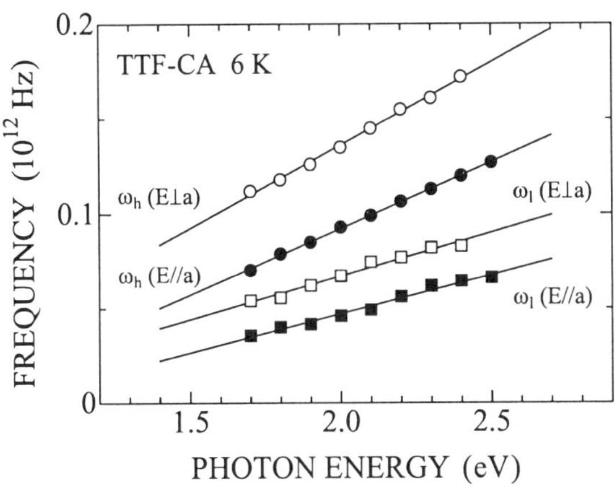

Fig. 15. Frequencies of oscillatory modulations in visible reflectance changes as a function of inverse of wavelength. Solid and open simbols show the frequencies measured for $E \perp a$ and $E//a$ polarizations (after Ref. 28).

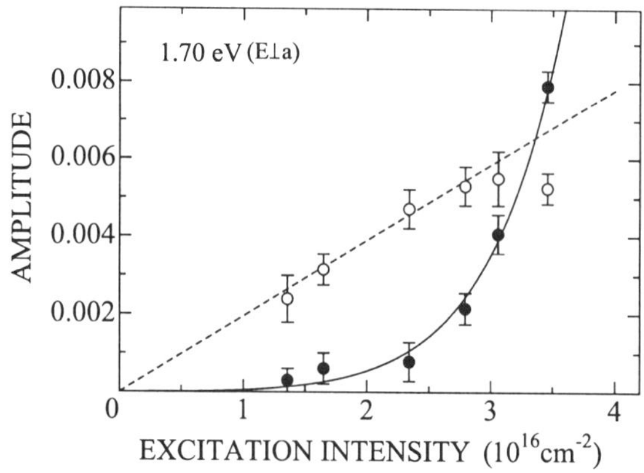

Fig. 16. Amplitudes of oscillatory modulations of high- and low-frequency components at 1.70 eV in reflectance changes measured with $E \perp a$ polarization as a function of excitation intensity (after Ref. 27).

characteristics of acoustic mode of phonons.[30] Therefore, both oscillations are related to the phonons. The oscillation frequencies of the two modes for $E//a$ polarization are lower than those at the same photon energy for $E \perp a$ polarization. This difference may be ascribed to the smaller magnitude of the reflective index for $E//a$ polarization.

The best-fit values, with estimated maximum error, of the amplitude of the two modes at 1.70 eV are shown as a function of excitation intensity in Fig. 16. The amplitude of the high-frequency mode is roughly in proportion to the excitation intensity in the weak-intensity region, but it saturated or even reduced in the intense regime. On the other hand, the amplitude of the low-frequency mode increases superlinearly. The analysis of the results at other probe-photon energies has shown essentially the same result that the low-frequency component is activated superlinearly under intense excitation. The superlinear dependence of the amplitude of the low-frequency component shows essentially the same dependence as the amplification factor described in Fig. 13. Therefore, generation of the low-frequency phonons is related to the process of amplification of $\Delta R/R_0$ characteristic of intense excitation.

3.3. *Phenomenological analysis of the temporal evolution of NI transition*

Based on the results presented in (3.1) and (3.2), it is apparent that the process of I-to-N transformation after fs-pulse excitation include at least three distinct steps. The first is from the onset of excitation till a few tens of ps. In this first step, excitation induces an abrupt change in reflection followed by a partial rccovcry, as most clearly detected in NIR reflection. Visible reflections for both polarizations also show this feature, as seen in Figs. 9, 10, and 12. The spectroscopic features of the changes are not the same as those characteristic of N-phase formation, as shown in Fig. 11. Also, the change in reflection is almost linear as a function of the excitation intensity. This first step is followed by a second step of the process till ~100 ps, which is characterized by the nonlinear enhancement of $\Delta R/R_0$ effectively monitored around 2 eV in $E \perp a$ spectra. Temporal evolution probed at several photon energies in $E//a$ spectra have revealed the third step, which typically last over 300 ps. As shown in Fig. 11, N-phase domains are finally generated at the end of the third step.

Before discussing the microscopic mechanism of the photoinduced transformation consisting of several steps, we first analyze temporal evolution of the whole process phenomenologically with neglecting oscillatory modulations, in order to confirm the consistency of describing the process in terms of the three distinct steps mentioned above. For this purpose, we introduce a set of rate equations;

$$\begin{aligned} dn_1/dt &= g(t) - k_1 n_1, \\ dn_i/dt &= k_{i-1} n_{i-1} - k_i n_i \quad (i > 1), \end{aligned}$$

where $g(t)$ is a generation function of the initial change governed by the shape of fs-laser pulses, n_1, and n_i represent the molar fractions of consecutive changes in the transformation process. Since the process of phase transformation is clearly nonlinear, the rates that characterize the transition from one state to the other may not be well defined. So, the use of the rate-equation model may not be vilified. However, for phenomenological description, the model may be useful. In fact, we have found empirically that solutions of five sequential rate equations could describe consistently overall changes in $\Delta R/R$ at any probe-photon energies and for any excitation intensities. The solutions of the five rate equations are grouped into three as

$$\begin{aligned} n_p &= n_1(t) + \alpha n_2(t), \\ n_l &= n_3(t) + \beta n_4(t), \\ n_N &= n_5(t), \end{aligned}$$

where α and β are appropriate constants. By using the three functions with fixed appropriate rate constants of k_i (i = 1–5), overall changes in

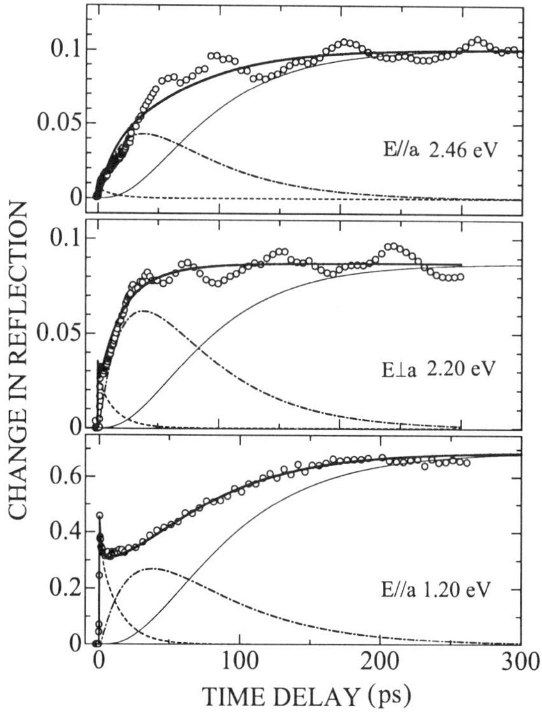

Fig. 17. Simulation of temporal evolution of reflectance changes in terms of rate-equation model consisting of three components shown by broken, chain-dot and thin solid curves at different probe-photon energy (after Ref. 28).

reflectance can be described reasonably as

$$\Delta R/R_0 = c_P n_P + c_L n_L + c_N n_N,$$

with the same set of rate constants. Only different contributions of c_P, c_L and c_N at different probe-photon energies and polarizations are assumed. Typical results of the description are shown by solid curves in Fig. 17, where temporal changes at 2.46 eV ($E//a$), 2.20 eV ($E \perp a$), and 1.20 eV ($E//a$) for the same excitation intensity are compared with the simulations. It is clear that the temporal evolution of $\Delta \mathrm{R}/\mathrm{R}_0$ measured at these typical photon energies, which show superficially different growth modes, are well simulated by the three-step model. The solid curves in Fig. 9 also show the results of a similar simulation. The reasonable fit may substantiate that the whole process can be analyzed as three consecutive steps. In this analysis, we find that the time constants of k_2 and k_4 that govern the decay of n_P

and n_L, respectively, become larger for higher excitation intensities. This may reflect the nonlinear feature of the transformation in these steps of the process.

In the three steps that are resolved above, the first step, which is characterized by fast change in $\Delta R/R_0$ at the incidence of laser pulse followed by rapid partial recovery, can be ascribed to the formation of the precursor of N-phase domain. As shown in Figs. 11 and 12, the fast change is accompanied by the decrease in ionic donor concentration in the phase-I crystal. However, the state formed at this step is not the N phase as discussed above. Since the time constant for the partial recovery in $\Delta R/R_0$ in the first step is rather long, lasting over ~10 ps, it is not simply the population bleaching due to intense fs-pulse excitation. Instead, it is reasonable to consider that a certain metastable state is generated. Since the final amount of the N-phase domains depends sensitively on the yield of this metastable state, the term "precursor" is appropriate to describe the metastable state. The second step of the process is characterized by nonlinear enhancement of $\Delta R/R_0$ effectively monitored below 2.3 eV with $E \perp a$ polarization. However, this step is not related directly to the N-phase forming process, since the peak intensity at the B band (2.4 eV) of the N phase increases only at the third step. Therefore, we may assign the third step to the process of forming the N-phase order. Since it is evident that the loss of D^+A^- pairs, or the generation of D^0A^0 pairs, is enhanced at the second step, we tentatively ascribe the second step to the local proliferation of neutral domains with lacking three-dimensional N-phase order.

4. Microscopic Mechanisms of the Primary Processes in the Dynamics

When discussing the microscopic processes of photoinduced I-to-N transition, we should take into account the requirement that the processes are consistent with the present understanding of the temperature- and pressure-induced IN transition in this crystal.[26] As summarized in Sec. 1, three different phases have been identified in the pressure-temperature phase diagram; these are ionic ferroelectric phase (I_{ferro}), ionic paraelectric phase (I_{para}), and neutral paraelectric phase (N_{para}). The transition among the three phases has been described in terms of a cascade of cooperative phenomena, in which CT strings play crucial roles.[13,15] This description of NI transition may certainly provide a basis on which we discuss the photoinduced phase transformation from I to N phases. However, one new aspect we keep in our mind is that the photoinduced I-to-N transition is induced under non-equilibrium conditions in a system where the I phase is more stable than the N phase.

In the case of temperature-induced I-to-N transition near the interface, neutral D^0A^0 pairs on a stack chain can be formed via thermal energy to form strings with desired lengths, and neutral domains consisting of one-dimensional strings can be progressively grown to transform them into the N phase. In this case, the transition process can be categorized as a cascade of formation, condensation and crystallization of CT strings. However, in the photoinduced I-to-N transition, the I phase is still more stable than the N phase at temperatures well below T_c. Therefore, CT strings (if we assume them as a species playing the role) cannot be generated spontaneously, but their formation, properties, and densities are strictly restricted by the excitation condition and a finite free-energy difference of the system. Then, the mechanisms of proliferation of photogenerated species and of forming the N-phase order under the restriction are key issues to be discussed. Thus, we can set the problems as a cascade of generation, proliferation, and crystallization of CT strings in the case of photoinduced I-to-N transition. Below we discuss possible mechanisms, based on experimental results summarized in Secs. 2 and 3 and on the present understanding of thermodynamics of NI transitions mentioned above.

4.1. *Precursor-forming process*

As shown in Fig. 13, the amount of $\Delta R/R_0$ measured at 1 ps after excitation is proportional to the excitation intensity, or the number of absorbed photons. However, the process of forming the precursor is essentially nonlinear in the sense that one absorbed photon leads to the conversion of a few tens of ionic D^+A^- pairs in the case of 800-nm excitation, as shown below. From the results shown in Fig. 13, we can evaluate quantitatively the magnitude of $\Delta R/R_0$ at 2.25 eV per incident photon. Therefore, it is possible to estimate the concentration of D^+A^- pairs converted into non-ionic charge state per absorbed photon, which is defined as the proliferation factor η_p, by correlating the magnitude of $\Delta R/R_0$ with molar fraction of D^+A^- pairs photobleached. Since the precursor is not N phase as shown in Fig. 11, we cannot use the spectrum of $\Delta R/R_0$ calculated from temperature-induced changes in reflection; we need a different measure.

Iwai *et al.* have studied the change in $\Delta R/R_0$ at 2.25 eV as a function of excitation intensity of 0.65-eV photons and evaluated that about $8D^+A^-$

pairs are converted to non-ionic state per absorbed photon by analyzing the space-filling effect.[27] In their result, $\Delta R/R_0$ at low-intensity region is in proportion to the excitation intensity, giving the slope of 2.2×10^{-18} per incident photon. The corresponding slope in the result in Fig. 13 is 1.7×10^{-18} per photon for the excitation at 1.55 eV. By using the spectroscopic data of wavelength dependence of the reflectivity and absorption coefficient,[22] we find that η_p under 1.55-eV excitation is about 3 times larger than that under 0.65-eV excitation; about $24D^+A^-$ pairs are converted into non-ionic charge states by absorbing one 1.55-eV photon.

Mizouchi *et al.* have studied theoretically the proliferation of excitons in a strongly correlated system.[18,19] They have shown that the strong Coulomb interaction, with the aid of excess energy upon exciton creation, can act as a nonlinear mechanism to proliferate one photoexcited exciton into clusters of excitons during relaxation. Also, Iwano has shown theoretically that a small cluster of a one-dimensional excited-state domain can be formed in a coupled-chain system under intense excitation.[20] Since TTF-CA crystal has a one-dimensional feature characterized by strong correlation and strong electron-phonon interaction, some one-dimensional nonlinear excitations within the stack chain may be formed upon photo-excitation, similar to the case of Anthracene-PMDA.[16] Therefore, the CT-exciton strings may be one of the most plausible candidates of the non-linear relaxed excited state. Then, we may conclude that the 24 D^+A^- pairs, converted to neutral pairs per one photon, take the form of one-dimensional CT strings. As mentioned above, η_p at 0.65 eV, which is about 8 at 6 K, is smaller than at 1.55 eV. Since the degree of proliferation during lattice relaxation may be a function of excess energy upon CT exciton generation,[18,19] it may be reasonable to presume that 1.55-eV excitation, which injects a large amount of excess energy, can generate longer exciton strings than the case of 0.65-eV excitation. Therefore, η_p and hence the length of the CT exciton strings can be a function of photon energy of excitation. This conclusion, implying a sensitive dependence of the string properties on excitation-photon energy, may be consistent with existing results showing state-dependent characteristics of photoinduced I-to-N transition, as discussed below.

As seen in Fig. 9, the amount of the precursor governs the mode and magnitude of the latter processes and how much precursor is produced at the incidence of light pulse is a crucial factor to measure the amount of finally generated N-phase domains. This result suggests the presence of some critical densities of neutral D^0A^0 pairs that governs the later processes. By using this concept of the critical density, we first discuss the state-dependent threshold excitation intensity for inducing the metastable N phases that can be detected around 1 μs after excitation. The threshold intensity in Fig. 4, which is essentially proportional to the number ($n_{\rm th}$) of absorbed photons during 3-ns temporal width, increases rapidly with decreasing photon energy of exciting laser light; it is almost inversely proportional to the photon energy. This dependence can be understood at least qualitatively, when we introduce the critical density of D^0A^0 pairs that forms the metastable N-phase domains detectable around 1 μs after excitation and the photon-energy dependent η_p mentioned above, since $n_C \approx \eta_p n_{\rm th}$. However, quantitatively, the dependence of $I_{\rm th}$ in Fig. 4 is too strong; the length of the strings at 0.65 eV is only about one-third of that at 1.55 eV, whereas n_{th} at 0.65 eV is about 10 times more than that at 1.55 eV. This quantitative discrepancy suggests other state-dependent properties. The important one may be the stability of the strings generated at different excitation energies.

As seen in Fig. 9, the change in $\Delta R/R_0$ once induced at the lowest excitation intensity does not show significant proliferation not only in visible reflectance but also in NIR reflectance. Therefore, the temporal evolution might be representative of the stability of the strings formed at 1.55-eV excitation. The decay time of the change estimated is rather long; it is about a few ns. Since the excitation intensity is already higher than the threshold intensity in Fig. 6, the observed long lifetime in Fig. 9 may not be the case when the intensity is below the threshold. However, Akimoto *et al.* have shown that the magnitude of $I_{\rm th}$ at 1.55 eV is essentially the same for both 80-fs pulses and 3-ns pulses.[31] This is only the case where the photogenerated states have lifetimes of about a few ns, supporting the long lifetime of the precursor generated under 1.55-eV excitation. On the other hand, Iwai *et al.* have shown that the initial product of the string generated by 0.65-eV photon decays with the time constant of 300 ps,[26] which is much shorter than the case of 1.55-eV excitation. Therefore, it appears that the length of the strings affects the stability of the state. In fact, Iwano has supported this conclusion theoretically. He studied the stability of one-dimensional excited-state domain, and provided the result that longer domains have higher stabilities against radiative decay.[20] Although the length dependence of the stability against non-radiative decay has not been studied, one can expect higher stability for longer strings based on his theoretical results.

When we take into account this state-dependent stability, then we can solve the quantitative discrepancy mentioned above for the threshold intensity shown in Fig. 4. Since the threshold intensity was measured for the laser pulses with temporal width (w) of 3-ns, the density during pulse excitation is affected strongly when the photogenerated states have a lifetime (τ) shorter than the pulse width. In such a case, the number of incident photons Φ to reach a certain critical density n_C at an excitation photon energy with the absorption coefficient k is given by $\Phi = (n_C/k)(w/\tau)$. The state generated under 0.65-eV excitation is less stable than that under 1.55-eV excitation. Therefore, to reach the critical density, we need fluence higher than that expected from the simple difference in length of strings. We can give more quantitative argument on this point, based on existing data. The lifetime of the strings generated by 0.65-eV photon is 0.3 ns. Therefore, the threshold excitation intensity shown in Fig. 3 may give the magnitude 10 times larger than that of short-pulse excitation. Iwai *et al.* have found the excitation intensity of 80-fs pulses to generate metastable N phase under 0.65-eV excitation is about $0.2 \times 10^{16}\,\mathrm{cm}^{-2}$, which is just smaller than the value in Fig. 4 by a factor of 10, as we expect. This quantitative argument may substantiate the proposal that the CT-exciton strings generated under different photon energy of excitation have different stability, which may be related to the different length governed by the excess energy upon CT transition.

Although the precise knowledge of the state-dependent stability of the photogenerated species under CT excitation has not been fully obtained experimentally and theoretically, existing data suggest strongly the important correlations among excess energy of exciton photogeneration, lengths of strings formed, and their stabilities. These correlations, which are essentially non-linear, may provide the basis on which we can understand the state-dependent characteristics of photoinduced I-to-N transition from microscopic point of view.

The density of precursors governs not only the threshold of metastable N-phase formation, but also the super-linear enhancement of the N-phase domains as evidenced in Figs. 9 and 13. In the former case, the fractions of neutral D^0A^0 pairs surviving as metastable N phase appears to be smaller than that generated at the incidence of light pulse in the case of 1.55-eV excitation as seen in Figs. 9 and 13. However, in the latter case of nonlinear behavior, the fraction increases super-linearly at longer delay-time region around 100 ps beyond the initial photoinduced amount, showing a different type of proliferation from that during excitation and short-time relaxation to form exciton strings. Evidently, the super-linear growth of the yield of N phase shown in Fig. 6 results from this proliferation. We discuss possible mechanisms of this nonlinear feature in (4.2).

4.2. *Local proliferation of neutral domains*

In the second step of the photoinduced I-to-N transformation, which we ascribed to the local proliferation, the magnitude of $\Delta R/R_0$ monitored around 2 eV in $E \perp a$ polarization show nonlinear enhancement. Based on the fact that both TTF^+ and CA^- ions show essentially the same non-linear reduction at this step shown in Fig. 11, we can conclude that the concentration of D^0A^0 pairs is non-linearly enhanced at this temporal domain. Spectroscopic feature around 40 ps shown in Fig. 8(c) is similar to, but not the same as, that of the formation of N-phase domains. Therefore, the domains consisting of D^0A^0 pairs, which have similar electronic properties as N phase, is formed at this step.

Cailleau *et al.* have proposed a scenario for the photoinduced I-to-N transformation, in which they assumed a step of forming soliton-antisoliton pairs as a result of relaxation of CT-exciton strings in temporal domain from 1 to 100 ps.[15] This formation of the pair results in the I_{para} phase as an intermediate from I to N phases. In the process of the conversion from a CT exciton string to a pair of soliton and antisoliton, the number of D^0A^0 molecules on a chain once formed by photoexcitation may be reduced substantially. However, as seen in the results shown in Fig. 12, the number of D^+A^- pairs is reduced, or the D^0A^0 concentration is enhanced at this temporal region. Therefore, the second step of the process is not related to the proposal that the soliton-antisoliton pairs are formed to result in I_{para} phase.

We propose here that the second step corresponds to the process of forming three-dimensional domains consisting of one-dimensional neutral domains (ODNDs) with still lacking three-dimensional N-phase order, based on the following arguments. As discussed in (4.1), an isolated small cluster of DA molecules, or the CT-exciton strings which we assumed as the structure of the precursor, may be unstable in I phase at low temperature, because of electronic energy loss of forming N-I domain walls (NIDWs) on a chain at the boundary and of electronic and elastic energy loss via interchain interaction around the boundary. However, when they are densely populated above a certain critical density mentioned in (4.1), they can be collapsed into larger neutral domains by converting intervening D^+A^- ionic configurations into D^0A^0 configurations. This conversion could be induced when the energy, which is gained by reducing the boundary sites and by enhanced electrostatic energy in a larger cluster, exceeds the energy cost to convert intervening D^+A^- pair to D^0A^0 pairs, although no theoretical estimates are available on this energetic.

This proposal is based on the fact that the proliferation at this step is related to the free-energy difference between I and N phases. The energy stored at the boundary and the number to be converted from ionic to neutral pairs may be governed by the energy difference of the system. Therefore, the degree of such a proliferation (if it is responsible for the step) is expected to be a strong function of temperature. This is certainly the case as shown below.[29]

As shown in Fig. 13, the enhancements of $\Delta R/R_0$ at the second step probed by visible reflectance and those at the third step probed by NIR reflectance are strongly correlated; both give essentially the same quantitative presentation of amplification factor. Therefore, we can study the proliferation by monitoring $\Delta R/R_0$ at 1.20 eV with $E//a$ polarization, which is almost modulation free as shown in Fig. 8; it is easier to analyze the effect of proliferation quantitatively. In Fig. 18, the temporal evolution of

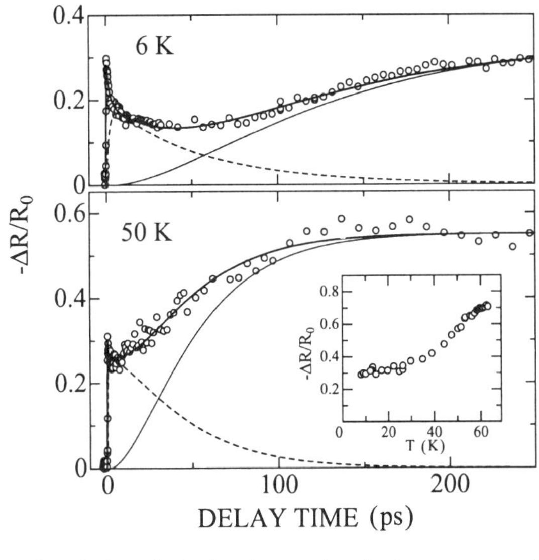

Fig. 18. Temporal evolution of relative changes in reflectance measured at 1.22 eV with $E//a$ polarization at 6 and 50 K. The excitation intensity is the same for both temperatures. The inset in (b) shows the amount of relative change measured at 200 ps of excitation under the same excitation intensity as a function of temperature (after Ref. 29).

$\Delta R/R_0$ is compared for two different temperatures but under the same excitation intensities. As clearly seen in the figure, the change at the incidence of laser pulses is the same at both temperatures, but the mode of evolution at later time delays are different significantly. At 50 K, the time constant that govern the enhancement at the third step becomes faster, and the amount of $\Delta R/R_0$ at 200 ps of excitation, which correspond to N-phase domains formed, becomes larger than those at 6 K. Therefore, the proliferation depend sensitively on temperature for the same amount of precursors formed. The inset shows the magnitude of $\Delta R/R_0$ measured at 200 ps after excitation at temperatures below 77 K under the same excitation conditions; temperature-dependent enhancement is evident. The temperature-dependent increase is characterized by an activation energy of about 10 meV, which is almost the same as free-energy difference between N and I phases. Therefore, we can conclude that the proliferation is related to the free energy difference between I and N phases. This result of temperature-dependent enhancement of proliferation is consistent with the result shown in Fig. 5, where the threshold excitation intensity to induce I-to-N transition under CT excitation has become down to zero at the transition temperature.

4.3. *N-phase order formation*

This second step is followed by the third step where the three-dimensional order for N phase is finally established. The three-dimensional order in this I-to-N phase transformation may be governed by the inter-chain interaction that has both electronic and elastic origin. In the case of I phase, electronic interaction becomes important, since the two possible modes (A_u and B_u) of the dimerization patterns for two interacting chains are possible. However, in the case of one-dimensional N domains if fully relaxed, the molecules are located on inversion symmetry sites by releasing dimerization distortions and form alternating regular stacks along the a axis. Therefore, the important factor to establish ordinary N-phase order for three-dimensional clusters of ODNDs is to solve the mismatch in the inter-chain configurations. In particular, expansion of the lattice constants along the *b* and *c* axes may be the most essential change in this step. Therefore, we assume that the final step of photoinduced I-to-N transition, which starts around 50 ps of excitation and end up around 300 ps, is the macroscopic structural relaxation associated with lattice expansion to establish three-dimensional order of N phase.

Thus, local proliferation of neutral clusters and N-phase order formation takes place at distinct temporal domains. A recent precise study on the temperature-induced NI transition process in TTF-CA has demonstrated a multistep-transition character; N phase is formed from I phase via an intermediate state characterized by the coexistence of N and I phases at mesoscopic scale over more than 130 unit cells.[14] This coexistence of the two phases (or the mesoscopic phase separation) has been ascribed to a large elastic strain between the chains, mainly along the b-axis direction. Although evolution of the photo-induced structural transformation may not be exactly the same as that of temperature-induced transformation, it may be reasonable to assume that the mesoscopic phase separation is induced after the local proliferation at the second step, before establishing the three-dimensional N-phase order in the process of photo-induced NI transition. Because of the local nature of the B band of TTF, reflection changes measured with $E \perp a$ polarization at visible region may not be sensitive to distinguish the phase-separated order and ordinary N-phase order. However, reflection changes measured with $E//a$ polarization at both NIR and visible regions may detect the difference sensitively, showing persistent changes over 300 ps at the third step.

4.4. *Oscillatory modulations in reflectivity changes*

In the temporal evolution of $\Delta R/R_0$ monitored at visible region, the change is always associated with oscillatory modulations. As revealed in (3.2), at least two modes with different frequencies modulate the reflectance changes: the high-frequency mode the amplitude of which is proportional to the excitation intensity at weak-intensity regime, and the low-frequency mode, which is super-linear in amplitude with respect to the excitation intensity. Iwai *et al.* have found a similar low-frequency mode in their reflection studies at 77 K.[27] They ascribed the low-frequency mode to the coherent motion of the NI domain boundary over a macroscopic scale, based on the less pronounced dispersion of the mode. However, as clearly seen in Fig. 15, both modes show dispersion that is characteristic of the acoustic phonon modes; the frequency is a linear function of the inverse of wavelength, or the wavevector. In the previous report, the origin of the modulation has been ascribed to the impulsively generated acoustic phonons due to the sudden appearance of the stress associated with structural phase transition.[24] Here we discuss possible origins of the two modes, based on results and discussions described above.

In the I phase, the three-dimensional ferroelectric order is established among one-dimensional stacks; there is a significant interstack interaction of attractive Coulomb type along the c axis. On the other hand, the interaction in N phase is mostly a van der Waals type.[13] This difference may be the origin of an abrupt jump in the lattice constant of the c axis (about 0.1%), which has been detected upon the NI transition.[13] When photoexcitation generate neutral CT strings on a stack chain in the first step of the process, abrupt change in the force balance may result at the boundary along the *b* and *c* axis. This sudden appearance of the stress at boundaries may generate a packet of coherent acoustic-phonon waves along the direction, which may induce the oscillatory modulation in reflectivity changes.

So far as the fraction of neutral-molecular clusters is small, the properties of the acoustic wave thus generated may be characterized by the dispersion determined by the force constant among the ionized stack chains in I phase. We presume that the high-frequency component of the oscillation reflect this wave characteristic of I phase, since the oscillation with this frequency can be induced even at low-excitation intensities and the delay of generating this component is almost zero, as seen in Fig. 14. On the other hand, when neutral clusters are generated densely as a result of the local proliferation at the second step, we may expect that the force constant among stacks may be significantly different from that in the I phase and consequently the frequency of acoustic waves that propagate the region may change: The fraction of the ODND clusters increases non-linearly with increasing excitation intensity as shown in Fig. 12. Also, the amplitude of the low-frequency oscillation shows essentially the same dependence on the intensity as the fraction. Therefore, we can ascribe the low-frequency modulation to the acoustic waves closely related to the formation of mesoscopic phase separation.

5. CT Excited State vs. Molecular Excited State

The photoinduced I-to-N transition can be induced by both intra-molecular excitation and CT excitation. The critical difference observed previously is the threshold excitation intensity to induce the transition; it is induced without detectable threshold intensity in the case of intra-molecular excitation.[24] On the basis of the results and discussions for the transition induced by CT excitation we described so far, the difference can be discussed now from microscopic point of view. Before going into some details of the discussion, let us compare a few experimental results obtained by fs time-resolved studies.

As shown in Fig. 11, the initial spectrum of $\Delta R/R_0$ with $E \perp a$ polarization measured at 1 ps of excitation is characterized by a negative dip at 2.25 eV and positive changes at high and low energy sides in the case of CT excitation at 1.55-eV. In the case of intra-molecular excitation at 3.06 eV, essentially the same spectrum has been detected at 1 ps after excitation.[25] Also, the initial spectra of $\Delta R/R_0$ at NIR with $E//a$ polarization measured at 1 ps after excitation were essentially the same for the two photon energies of excitation. These results confirms that intra-molecular excitation to generate molecular excited states in TTF-CA also leads to the formation of the precursor which is similar to that formed under CT excitation. These may be the CT-exciton strings, but presumably with different lengths depending on the excess energy of generating a CT exciton. This result provides a support to the right assumption made in the pioneering work by Koshihara *et al.* that the molecular-excited states are relaxed into CT excitations for inducing I-to-N transition.[3]

Next, we compare the temporal evolution of the changes induced by CT excitation and intra-molecular excitation. In Fig. 19, $\Delta R/R_0$ measured at 2.25 eV with $E \perp a$ polarization is compared for the two types of excitation. In the measurement, the intensity of the 402-nm fs-laser pulses with $E \perp a$ polarization was chosen so that the observed initial change of $\Delta R/R_0$

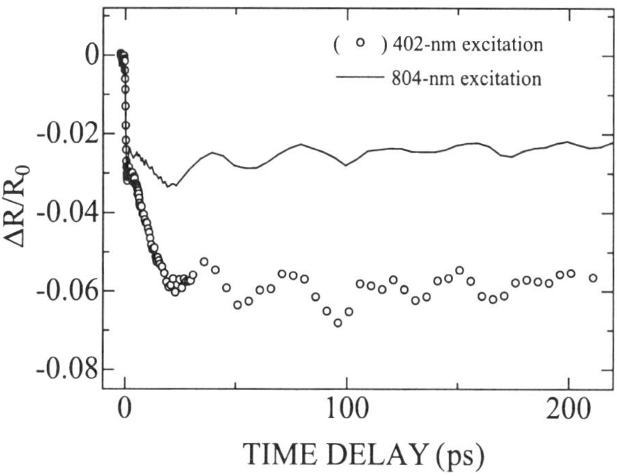

Fig. 19. Temporal evolution of relative changes in reflectance measured at 2.25 eV with $E \perp a$ polarization induced by 804-nm excitation, solid curve, and by 402-nm excitation, open circles. The magnitude of $\Delta R/R_0$ at 1 ps of excitation is the same for both cases.

at 1 ps of excitation was the same for both excitations. For the same amount of the initial change, effects of proliferation at later delay times are different. In the case of 402-nm excitation the change is enhanced significantly around 60 ps, whereas it is reduced monotonically in the case of 804-nm CT excitation. This result shows superficially the effective generation of N-phase domains in the case of intra-molecular excitation. However, before obtaining any definite quantitative conclusion for the effectiveness, careful examinations are needed to evaluate the density of D^0A^0 pairs really formed at the first step for the two cases by taking into account the difference in absorption coefficients at wavelengths of excitation and at that of probing the reflection change. Available spectroscopic data have shown that the coefficient at 2.25 eV is smaller than at 3.05 eV, but that it is larger than that at 1.55 eV. Therefore, the probe at 2.25 eV for the case of 3.05-eV excitation monitors the excited region with a strong concentration gradient, although it probes the uniformly excited region in the case of 1.55-eV excitation. Therefore, the initial fraction of $\Delta R/R_0$ in the case of 402-nm may be underestimated; the real fraction may be larger than that in the case of 804-nm excitation.

On the other hand, it is clear that the mode of temporal evolution of $\Delta R/R_0$ induced by intra-molecular transition at 3.08 eV is essentially the same as those observed under dense CT excitation shown in Fig. 8. Therefore, we may conclude that the process of forming N-phase domains by intra-molecular excitation is the same as that by CT excitation after an efficient relaxation into CT excited states of the form of CT exciton strings. The difference may lie in the effectiveness of inducing proliferation via the length of strings generated by the relaxation. Based on the result shown in Figs. 17 and 18, together with available spectroscopic data at 405 nm,[9,32] we estimated that about 500 D^+A^- pairs per absorbed photon are instantaneously (within 1 ps of excitation) converted to neutral states. This value is much larger than under CT excitations. Such an efficient conversion, which makes it easier to reach the critical density for the I-to-N transition for less excitation intensity, together with larger absorption coefficient, may reduce the threshold intensity of I-to-N transition to the amount smaller than those under CT excitations.

6. Photoinduced vs. Temperature-induced NI Transitions

We have discussed so far the process of photoinduced I-to-N transition in TTF-CA in terms of the consecutive three-step process consisting of the precursor-forming step, the step of local proliferation, and three-dimensional-order forming step. When we compare the features revealed above with those in the thermally induced NI transitions, there are some differences and similarities. In order to make the comparison easier, we replace the naming of the three steps of the transition by generation, proliferation, and crystallization of CT-exciton strings, similar to the case of thermodynamics of NI transition.

As in the case of temperature-induced NI transition, the generation of CT strings may be the starting point of the transition. In the case of temperature-induced transition, CT strings can be formed at any site (or density) with any lengths near the transition temperature governed by the free energy of the system. However, in the photoinduced transition, formation of the CT-exciton strings has two strict limitations on the site (or density) where they are formed and on the length they can be stable. These limitations are certainly related to the condition that I phase is still stable than N phase in the present case. Under this energetic restriction, the generation of CT strings shows an important feature that the properties (length and stability) of CT strings are governed by the photon energy of excitation via excess energy. As we discuss in detail in (4.1), the photon-energy dependence of the properties of the strings plays crucial roles on the state-dependent features of this phase transition. This is one of the unique features of photoinduced structural phase transition.

The density of CT strings is governed by the excitation intensity in the case of photoinduced transition, and the density affects critically the formation of metastable N phase and the proliferation (or condensation) of the strings at later stage. As discussed in detail, the proliferation is induced above the critical density, showing a strong nonlinear feature. This nonlinear feature may correspond to the spontaneous proliferation at the transition temperature in the case of temperature-induced transition. In the temperature-induced NI transition at low-pressure regime, condensation and crystallization of CT exciton strings are correlated, showing a strong first-order character for the transition. However, in the case of photoinduced transition, the two steps of proliferation and crystallization can be resolved temporally. Consequently, we can detect the cluster of ODNDs before establishing the three-dimensional order (crystallization). This cluster, which may comprise neutral regular stacks but with three-dimensional order of phase I, may be viewed as a new "phase" of TTF-CA resolved only temporally in the study of photoinduced I-to-N transition of this crystal.

Acknowledgments

The author would like to thank Dr. I. Akimoto, Prof. C. Itoh, Prof. K. Nasu, Prof. S. Koshihara, and Prof. Y. Tokura for valuable discussions. This work is supported by the Grant-in-aid of the Japanese ministry of Education, Science and Culture, the priority area (B) "Photoinduced Phase Transition and Their Dynamics".

References

1. *Relaxation of Excited States and Photo-induced Structural Phase Transitions*, ed. K. Nasu, Vol. 124 of Springer Series in Solid State Sciences (Springer, Berlin, 1997).
2. Phase Transitions Vol. 75, Section B: *Proc. of the 1st Int. Conf. on Photoinduced Phase Transitions, Their Dynamics and Precursor Phenomena* (Taylor & Francis, Hampshire, 2002).
3. S. Koshihara, Y. Tokura, T. Mitani, G. Saito, and T. Koda, *Phys. Rev.* **B42**, 6853 (1990).
4. S. Koshihara, Y. Tokura, K. Takeda, and T. Koda, *Phys. Rev. Lett.* **68**, 1148 (1992).
5. N. Hosaka, H. Tachibana, N. Shiga, M. Matsumoto, and Y. Tokura, *Phys. Rev. Lett.* **82**, 1672 (1999).
6. T. Tayagaki and K. Tanaka, *Phys. Rev. Lett.* **86**, 2886 (2001).
7. M. Ueta, H. Kanzaki, K. Kobayashi, Y. Toyozawa and E. Hanamura, *Excitonic Processes in Solids* (Springer, Berlin,1986).
8. H. Sumi and A. Sumi, *J. Phys. Soc. Jpn.* **63**, 637 (1994).
9. J. B. Torrance, J. E. Vazquez, J. J. Mayerls, and V. Y. Lee, *Phys. Rev. Lett.* **46**, 253 (1981).
10. J. B. Torrance, A. Girlando, J. J. Mayerle, J. I. Crowley, V. Y. Lee, P. Batail, and S. J. LaPlace, *Phys. Rev. Lett.* **47**, 1747 (1981).
11. N. Nagaosa, *Solid State Commun.* **57**, 179 (1986), and *J. Phys. Soc. Jpn.* **55**, 3488 (1986).
12. Y. Tokura, T. Koda, T. Mitani, and G. Saito, *Solid State Commun.* **43**, 757 (1982).
13. M. H. Lemee-Cailleau, M. Le Cointe, H. Cailleau, T. Luty, F. Moussa, J. Roos, D. Brinkmann, B. Toudic, C. Ayache and N. Karl, *Phys. Rev. Lett.* **79**, 1690 (1997), and M. Le Cointe, M. H. Lemee-Cailleau, H. Cailleau, B. Toudic, L. Toupet, G. Heger, F. Moussa, P. Schweiss, K. H. Kraft and N. Karl, *Phys. Rev.* **B51**, 3374 (1995).
14. M. Buron-Le Cointe, M. H. Lemee-Cailleau, H. Cailleau, B. Toudic, A. Moreac, F. Moussa, C. Ayache and N. Karl, *Phys. Rev.* **B68**, 064103 (2003).
15. H. Cailleau, E. Collet, T. Luty, M. H. Lemee-Cailleau, M. Buron-Le Cointe, S. Koshihara, *Multiphoton and Light Driven Processes in Organics: New Phenomena, Materials and Applications*, ed. F. Kajzar and M. V. Agranovich (Kluwer Academic, Netherlands, 2000), p. 451.

16. M. Kuwata-Gonokami, N. Peyghambarian, K. Meissner, B. Flugel, Y. Sato, K. Ema, R. Shimano, S. Mazumdar, F. Guo, T. Tokihiro, H. Ezaki, E. Hanamura, *Nature*, **367**, 47 (1994).
17. E. Collet, M. H. Lemee-Cailleau, M. Buron-Le Cointe, H. Cailleau, S. Ravy, T. Luty, J. F. Berar, P. Czarnecki, and N. Karl, *Europhys. Lett.* to be published.
18. H. Mizouchi, and K. Nasu, *J. Phys. Soc. Jpn.* **69**, 1543 (2000).
19. H. Mizouchi, P. Huai, and K. Nasu, *Phase Transitions* **74**, 1 (2001).
20. K. Iwano, *Phys. Rev.* **B65**, 024302 (2002).
21. Y. Tokura, T. Koda, G. Saito, and T. Mitani, *J. Phys. Soc. Jpn.* **53**, 4445 (1984).
22. C. S. Jacobsen and J. B. Torrance, *J. Chem. Phys.* **78**, 112 (1983).
23. E. Collet, M. H. Lemee-Cailleau, M. Buron-Le Cointe, H. Cailleau, M. Wulff, T. Luty, S. Koshihara, M. Meyer, L. Toupet, P. Rabiller and S. Techert, *Science*, **300**, 612 (2003).
24. T. Suzuki, T. Sakamaki, K. Tanimura, S. Koshihara, and Y. Tokura, *Phys. Rev.* **B60**, 6191 (1999).
25. K. Tanimura and I. Akimoto, J. Lumin, **94–95**, 483 (2001).
26. S. Koshihara, Y. Takahashi, H. Sakai, Y. Tokura, and T. Luty, *J. Phys. Chem.* **B103**, 2529 (1999).
27. S. Iwai, S. Tanaka, K. Fukiwara, H. Kishida, H. Okamoto, and Y. Tokura, *Phys. Rev. Lett.* **88**, 057402 (2002).
28. K. Tanimura, to be published.
29. K. Tanimura and C. Itoh, to be published.
30. M. Fiebig, K. Miyano, Y. Tomioka, and Y. Tokura, *Appl. Phys.* **B71**, 211 (2000).
31. I. Akimoto, D. Baba, and K. Tanimura, un-published (2002).
32. S. Iwai, private communications.

CHAPTER 3

STUDY ON THE COOPERATIVE PHOTOINDUCED LOW-SPIN TO HIGH-SPIN STATE CONVERSION PROCESSES

Osamu Sakai

Department of Physics, Tokyo Metropolitan University
Hachioji 192-0397, Japan
sakai@phys.metro-u.ac.jp

Tetsuo Ogawa

Department of Physics, Osaka University
Toyonaka, Osaka 560-0043, Japan

The photo induced cooperative low-spin (LS) to high-spin (HS) conversion phenomena in spin-crossover complexes are studied theoretically. The volume of molecule usually increases when its transition element ion turns into HS states. Therefore there arises interaction between complexes (ion-ion interaction) through the coupling with lattice distortions. The effective energy of HS state is lowered by this ion-ion interaction when the fraction of HS increases. Since the quantum mechanical potential barrier becomes high, the relaxation rate of HS to LS states decreases rapidly when the energy of HS decreases. A model which shows cooperative conversion is presented by taking into account the above facts. A relatively short-range ion-ion interaction is caused by optical modes, while very long-range interaction is caused by acoustic modes. A Monte Carlo (MC) method is developed to examine the LS-HS transition in such a model by including the spatial and temporal fluctuation effects. The existence of the threshold light intensity and the incubation period before the steep conversion, as observed in experiments, is demonstrated. The incubation time corresponds to the nucleation time of the HS domain. It is pointed out that the nucleation barrier is surmounted by the fluctuation caused by the repetition of excitation and de-excitation, instead of the energy fluctuation in the usual temperature-induced transition. The time scale of incubation period can be estimated following the usual domain formation theory, and it accounts grossly for the MC results in three-dimensional cases. The macroscopic phase separation of HS and LS domains during the conversion process is caused by the volume striction effect through coupling with acoustic modes. A quantity which describes the coexistence condition of two phases in such photoinduced conversion process is presented. The abrupt step-like conversion observed in experiments is ascribed to the de-pinning effect of the volume striction.

Contents

1. Introduction

The photoinduced low-spin (LS) state to high-spin (HS) state conversion in transition metal complexes has been extensively studied.[1,2] Recently, the photoinduced cooperative LS to HS conversion has been demonstrated experimentally for $[Fe(2pic)_3]Cl_3 \cdot EtOH$.[3–8] The LS state is considered as the $Fe^+(t^6)^1A_1$ and the HS state as $Fe^+(t^4e^2)^5T_2$. Hereafter we call these HS and LS states of an ion as the spin state, not the spin multiplet states within the HS state.

This compound shows also cooperative spin state transition at $T_C \sim 120\,K$ from the LS phase to the HS phase as the temperature is increased.[1,9] When the cw light is irradiated in the LS phase at low temperature, for example at $T \sim 2\,K$, the spin state converts cooperatively to the HS one. Notable features are summarized;[3,4] (i) the existence of the threshold light intensity (I_{th}) for the cooperative LS $\rightarrow$ HS phase conversion, (ii) the existence of the incubation period (τ_{inc}) before the cooperative LS $\rightarrow$ HS conversion, (iii) the steep decrease of the LS fraction after the incubation period, (iv) phase separation during the conversion process and (v) the rapid step-like decrease of the LS fraction under the intense light irradiation. It was noted that the photoinduced HS phase can be re-converted to the LS phase by further irradiation of light at another absorption band, about 1.5 eV.[3] Therefore the observed LS $\rightarrow$ HS conversion cannot be attributed to the heating effect of crystal beyond T_C by light irradiation.[10]

It has been pointed out that the relaxation rate from the HS to LS states is strongly dependent on their energy difference.[11,12] When the HS state changes to the LS state, the excess energy which is illustrated as the energy difference between a and c, or b and c in Fig. 1 spontaneously transfers to the phonon energy which is illustrated as the energy of a′ or b′ relative to the c point in the LS state. Denoting the number of excited phonons by n, the overlapping factor of phonon state is given as $e^{-S}S^n/n!$, where S is the Huang–Rhys factor[1] due to the displacement of the equilibrium point of lattice for the two spin states. Therefore the relaxation rate decreases drastically as the energy difference between LS and HS states is small. This fact is illustrated in Fig. 1 as the a $\rightarrow$ a′ thick dashed arrow (not slow relaxation) and the b $\rightarrow$ b′ thin dashed arrow (slow relaxation).

The effective energy of HS state itself decreases when the HS fraction increases through the ion-ion interaction mediated by the lattice distortion.[1,13] (The interaction energy is given to be dependent on the mutual spin states. We call this type of interaction the ion-ion interaction hereafter. The exchange interaction energy between the HS states is expected to be small.[1]) In Fig. 1, the HS state energy is decreased from a to b since the HS energy decreases from ε_{HS}^{allLS} to ε_{HS}^{allHS} when the LS fraction decreases from 1 to 0. The relaxation from HS to LS is rapidly suppressed due to this energy change.

Taking into account these two facts, Koshino and Ogawa presented a model for the dynamical process of the photoinduced LS $\rightarrow$ HS conversion.[14] They derived a simple rate equation by neglecting the spatial

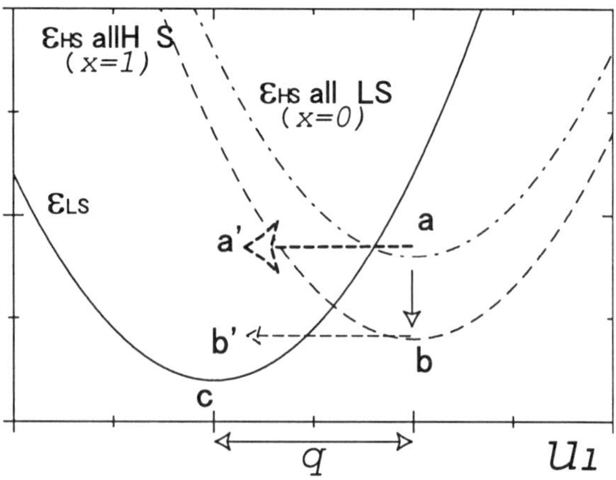

Fig. 1. Energy diagram of LS state ($\varepsilon_{\mathrm{LS}}$, solid line), and HS state ($\varepsilon_{\mathrm{HS}}$, dashed and dot-dashed lines) with lattice distortion energy. The horizontal shift q illustrates the shift of stable position. The vertical shift represents the energy difference between LS and HS states. The energy decrease from a $\left(\text{of } \varepsilon_{\mathrm{HS}}^{\mathrm{allLS}}\right)$ to b $\left(\text{of } \varepsilon_{\mathrm{HS}}^{\mathrm{allHS}}\right)$ illustrates the energy decrease due to the ion-ion interaction in the HS phase. The arrows a → a′ and b → b′ indicate the HS to LS transition. The b → b′ transition rate is small compared with a → a′ transition. Note that the barrier height is higher for the b → b′ transition as the energy difference b′ − c is smaller than that of a′ − c.

and temporal fluctuations in the conversion process. The equation qualitatively reproduces the facts (i)–(iii). The model is presented in Sec. 2 and the scenario based on their equation is explained in Sec. 3. However as noted by the fact (iv), the conversion process is considered as the HS domain formation in the sea of the LS state, thus it is necessarily accompanied by the spatial fluctuation of LS fraction. Fact (v) will also relate to the spatial inhomogeneity of HS distribution. There arise problems: (a) how the spatial and temporal fluctuation effects affect the conversion process, (b) whether Koshino–Ogawa's model can explain the facts (iv) and (v), or not, when we consider the fluctuation effects.

A possible way of taking into account such fluctuation effects is the Monte–Carlo simulation (MC) method.[15] Koshino–Ogawa's model was recently examined by the MC method to check points (a) and (b).[16] It was shown that the time evolution of LS fraction is qualitatively similar to that of the rate equation though the relevant parameter range is shifted. Therefore facts (i) to (iii) are reproduced also by the MC simulation, in gross meaning. The coexistence of the HS and LS domains is observed in the MC simulation during the conversion process, but macroscopic phase separation of LS and HS states is not obtained. It was concluded that other extra effects are necessary to reproduce facts (iv) and (v). The calculation is explained in Sec. 4. We should note that the similarity between the calculated and the experimental results is used in a gross meaning: when the experimental result with step-like jumps is smoothened, both curves show a similar sigmoid nature. In the calculation, the LS fraction decreases steeply but continuously to almost zero after the incubation period.

In the experiment, the conversion to HS state is suppressed at some stage (suppressed stage), and macroscopic phase separation of HS and LS regions takes place. After some time, the LS fraction abruptly decreases to some extent. The incubation time seems to correspond, in some meaning, to the time to break the suppressed stage. Therefore we think that it is necessary to lead the suppressed stage to reproduce the experiment results.

The suppressed stage is similar to the phase separation state of the first order transition of the conserving system in the usual thermal phase transition. On the other hand, the transition of Koshino–Ogawa's model is similar to that of the non-conserving system. The ion-ion interaction is attractive and monotonic decreasing function of distance of the pair in the model, because only the optical modes are considered. In recent papers,[17,18] we have proposed a model in which the coupling with the acoustic mode is also included. This coupling gives the attractive interaction for near neighbor pairs, but gives the repulsive interaction for pairs with large distance through the volume striction effect when the volume of the system is fixed.

On the basis of MC in Sec. 8, we demonstrate that the volume striction effect prevents the full conversion of crystal to the HS state and leads the macroscopic phase separation (iv) when the volume cannot change freely. On the other hand, the step-like jump (v) cannot be reproduced as the intrinsic behavior of the present model. The averaged LS fraction of large size sample gradually approaches to the value of the phase separated state after showing a relatively steep decrease, and does not show any abrupt jump, though the growth of the individual HS domain seems to be not continuous. We ascribe fact (v) to the de-pinning effect of the volume striction in Sec. 9. The importance of the coupling with the volume striction has been claimed in several studies for the spin state transition by thermal effect.[1,19] It is also stressed in experiments that the crystal surface swells in the HS region.[6]

In the usual theory of domain formation,[15,20] the energy barrier which is classified as the free energy of the droplet with critical size plays a crucial role in the rate of domain formation. This energy barrier is surmounted by the energy fluctuation characterized by the temperature. However, the present photoinduced HS conversion takes place at very low temperature. There arises the question as to what fluctuation drives it to surmount the nucleation barrier. We note in Sec. 5 that the repetition of the excitation and the de-excitation acts as the driving power on the basis of the Fokker–Planck equation for the master equation. Defining the quantity corresponding to the effective temperature, time scales of the incubation period are estimated following the usual domain formation theory, and compared with the result of MC simulation. Koshino–Ogawa's rate equation is re-derived and the inter-relation of it and Fokker–Planck approach is discussed.

To describe the suppressed stage, we present a quantity, which gives an account of the condition of the phase coexistence in the photoinduced transition as Gibbs's free energy does in the usual thermodynamics. The estimated phase separation condition based on the quantity is compared with ones obtained by MC simulation in Sec. 8. In Sec. 10, summary and discussions are given. In the summary, a general view of the photoinduced phase transition (PIPT) phenomena is presented. The photoinduced cooperative LS to HS transition is classified as the non-equilibrium phenomena which is induced by light as the transition from the stable state to the metastable one. It is noted that we have another type of PIPT which is classified as the transition from the metastable to the stable one. The later one is not studied here, but also has much variety accompanied by the phenomena classified as the photoinduced domino processes.

Studies based on the MC method on the similar problem with the photoinduced LS to HS transition have already been carried out by Romstedt *et al.*[21] They used a short range ion-ion interaction model and a simplified transition probability and studied mainly the LS $\rightarrow$ HS relaxation process after the interruption of the light irradiation. We note that the ion-ion interaction mediated by lattice distortion acts between pairs even when their distance is large. Several interesting problems have been studied recently based on the MC method.[22–24] The relaxation process from HS to LS states has been studied by several methods of taking into account the ion-ion correlation effect.[25,26,28] The switching effect between bistable photoinduced transition has been also studied.[29,30]

Recently, Tayagaki *et al.* have studied the photoinduced LS $\rightarrow$ HS transition by various observation methods.[7,8] They concluded that the photoinduced HS phase is accompanied by the symmetry transformation contrasted to the temperature induced HS phase and suggested the Jahn–Teller (JT) deformation in the photoinduced phase. Such a detailed form of the coupling with lattice distortion is not taken into account in this paper, which concentrates on the dynamics of the domain formation process. The quantum dynamical relaxation proccss[31,32] is also not considered here because we expect the very strong electron-phonon interaction to be necessary in this cooperative conversion process.

2. Formulation

2.1. *Master equation and the ion-ion interaction*

Following Ref. 14, we adopt a master equation,

$$\frac{\partial P(\{n_1,\ldots,n_N\})}{\partial t} = \sum_{\{n\}}\sum_{\ell}\left(\prod_{\ell'(\neq\ell)}\delta_{n_{\ell'}n'_{\ell'}}\right)(\delta_{n_{\ell,0}}-\delta_{n_{\ell,1}}) \times [n'_\ell\hat{\omega}_\ell(\{n'\})-(1-n'_\ell)I]P(\{n'\}). \quad (1)$$

Here $P(\{n_1,\ldots,n_\ell,\ldots,n_N\})$ is the multi-point probability function of the ion configuration $\{n_1,\ldots,n_\ell,\ldots,n_N\}$, and $n_\ell = 1$ (0) indicates the HS (LS) state at the ℓth point. N is the number of ion sites, and I is the excitation intensity. $\hat{\omega}_\ell(\{n\})$ is the relaxation probability of the HS state to the LS state and it depends on the configuration of neighbor ions through the interaction energy mediated by lattice distortion.

The energy of ions and the lattice distortion is given by the following expression,

$$E = \sum_\ell \varepsilon^0 n_\ell + \frac{\kappa}{2}\sum_\ell (u_\ell - qn_\ell)^2 + \frac{\kappa}{4}\sum_{\ell\neq\ell'} K_{\boldsymbol{R}_\ell-\boldsymbol{R}_\ell}(u_\ell-u_{\ell'})^2 - \frac{\kappa}{2}\sum_\ell q^2 n_\ell. \quad (2)$$

The quantity ε^0 is the bare excitation energy from LS to HS state and u_ℓ is the lattice distortion. In Fig. 1, the energy difference between c and a corresponds to ε^0. The second term is the local distortion energy of lattice, whose equilibrium point shifts q at the HS site from that of the LS site. The parabolic lines in Fig. 1 symbolize distortion energy. The third term gives the coupling between the local lattice distortions, and $\boldsymbol{R}_\ell$ is the lattice vector of the ℓ site. To simplify the discussion, in this section, we restrict ourselves to the case where we have only one representative distortion mode for each magnetic ion. Some more realistic cases including the coupling with the acoustic mode will be studied in later sections. By introducing the

Fourier transformation, $n_k = \frac{1}{\sqrt{N}}\sum_\ell n_\ell e^{-\mathrm{i}\boldsymbol{k}\boldsymbol{R}_\ell}$ and $u_{\mathrm{k}} = \frac{1}{\sqrt{N}}\sum_\ell u_\ell e^{-\mathrm{i}\boldsymbol{k}\boldsymbol{R}_\ell}$, the energy is represented as follows:

$$E = \frac{\kappa}{2}\sum_{\boldsymbol{k}}(1+K(\boldsymbol{k}))\left|u_{\boldsymbol{k}} - \frac{qn_{\boldsymbol{k}}}{1+K(\boldsymbol{k})}\right|^2 - \frac{\kappa q^2}{2}\sum_{\boldsymbol{k}}\frac{1}{1+K(\boldsymbol{k})}n^*_{\boldsymbol{k}}n_{\boldsymbol{k}} + \sum_\ell \varepsilon^0 n_\ell, \tag{3}$$

with $K(\boldsymbol{k}) = \sum_h K_{\boldsymbol{R}_h}(1-e^{\mathrm{i}\boldsymbol{k}\boldsymbol{R}_h})$. The first term of the right hand side of Eq. (3) indicates that the lattice vibrations are considered as the displaced phonon modes, if $n_{\boldsymbol{k}}$ is regarded as the classical quantity, or its time evolution is very slow. The excess energy caused by the HS → LS transition transfers to these vibrations, and we assume that the relaxation of the vibrations to their ground state (with given $n_{\boldsymbol{k}}$) is fast.[31] The ion-ion interaction energy will be given by the term grouped as $n^*_{\boldsymbol{k}}\, n_{\boldsymbol{k}}$ in the second part of (3). By inverting Fourier transform, we get the interaction in the site representation,

$$\Delta E = -\frac{1}{2}\sum_{\ell(\neq)\ell'} J_{\boldsymbol{R}_\ell - \boldsymbol{R}_{\ell'}} n_\ell n_{\ell'} + \sum_\ell \varepsilon_0 n_\ell, \tag{4}$$

where the effective single site energy $\varepsilon_0 = \varepsilon^0 - \frac{\kappa q^2 G(0)}{2}$ and the interaction constant $J_{\boldsymbol{R}_\ell - \boldsymbol{R}_{\ell'}} = \kappa q^2 G(\boldsymbol{R}_\ell - \boldsymbol{R}_{\ell'})$ are given by using Green's function:

$$G(\boldsymbol{R}_h) = \frac{1}{N}\sum_k \frac{\mathrm{e}^{\mathrm{i}\boldsymbol{k}\boldsymbol{R}_h}}{1+K(\boldsymbol{k})}. \tag{5}$$

The effective energy level of HS state at site ℓ is given as

$$\varepsilon_\ell = \varepsilon_0 - \sum_\ell J_{\boldsymbol{R}_\ell - \boldsymbol{R}_{\ell'}} n_{\ell'}, \tag{6}$$

which depends on the configuration around the ion. We note that the ion-ion interaction $J_{\boldsymbol{R}_h}$ is not zero even for pairs with larger distance for which the quantity $K_{\boldsymbol{R}_h}$ is zero. The decrease of energy from a to b in Fig. 1 originates from this interaction term.

We note the relation $\sum_h J_{\boldsymbol{R}_h} = \kappa q^2(1-G(0))$. When all sites are in the HS state, the energy of HS state is given as $\varepsilon^{\mathrm{allHS}} = -\kappa \mathrm{q}^2(1-\mathrm{G}(0)) + \varepsilon_0$ shifted from $\varepsilon^{\mathrm{allLS}} = \varepsilon_0$ of the case that all sites are in the LS state.

In (2) we have neglected the kinetic energy term of the lattice distortion. This simplification will be not so bad in the strong ion-lattice interaction case, i.e. large q^2 case.[31,32] In weak ion-lattice interaction case, Green's function is replaced by a form mediated by phonons.

2.2. *Transition probability*

The next step is to give the relaxation probability of HS state to the LS state. The magnitude of the matrix element of electronic part, t_e, which causes the HS to LS transition is assumed to be very small compared with the characteristic frequency of phonons. Let us denote the energy of the HS state relative to the LS state by ε_ℓ at site ℓ. When the ion state changes to the LS state, this energy transfers to phonons under the ion configuration in which the site is changed to the LS state. The transition probability from HS to LS state includes the overlapping integral of phonons,

$$\hat{\omega}(\varepsilon_\ell) = \frac{2\pi}{\hbar} t_e^2 \sum_{\{n_{\boldsymbol{k}}^{(p)}\}} \prod_{\boldsymbol{k}} \frac{e^{-S_{\boldsymbol{k}}} S_{\boldsymbol{k}}^{n_{\boldsymbol{k}}^{(p)}}}{n_{\boldsymbol{k}}^{(p)}!} \delta\left(\varepsilon_\ell - \sum_{\boldsymbol{k}} \hbar\omega_{\boldsymbol{k}} n_{\boldsymbol{k}}^{(p)}\right), \tag{7}$$

where $n_{\boldsymbol{k}}^{(p)}$ is the excited phonon number in the transition. This simplified treatment will be allowed in large q^2 case.[31,32] The phonon frequency with $\boldsymbol{k}$ is given as $\omega_{\boldsymbol{k}} = \sqrt{1+K(\boldsymbol{k})}\omega_o$, where ω_o is the frequency of local mode. The quantity $S_{\boldsymbol{k}}$ is given by $S_{\boldsymbol{k}} = \frac{1}{N\hbar\omega_{\boldsymbol{k}}} \frac{\kappa q^2}{1+K(k)}$. By introducing a quantity, $S(x) = \sum_{\boldsymbol{k}} S_{\boldsymbol{k}} \delta(x - \hbar\omega_{\boldsymbol{k}})$, $\hat{\omega}(\varepsilon_\ell)$ is given as

$$\hat{\omega}(\varepsilon_\ell) = \frac{t_e^2}{\hbar} \int_{-\infty}^{\infty} ds \exp\left(is\varepsilon_\ell - \int_0^\infty dx\, S(x)(1 - e^{-isx})\right). \tag{8}$$

The expression indicates that the lattice relaxation is mainly shared by the interaction mode. When the Huang–Rhys factor for the interacting mode, $S = \int_0^\infty dx\, S(x)$, is large, we may use an approximate form for $\hat{\omega}(\varepsilon_\ell)$,

$$\omega(\varepsilon_\ell) = \frac{2\pi t_e^2}{\hbar} \frac{1}{\sqrt{2\pi}\Gamma} e^{-\frac{1}{2}\left(\frac{\varepsilon_\ell - \Delta}{\Gamma}\right)^2}, \tag{9}$$

where $\Delta = \int_0^\infty dx\, S(x)x$, and $\Gamma^2 = \int_0^\infty dx\, S(x)x^2$. We note that Δ is given as $\Delta = \kappa q^2 G(0)$.

The transition probability used in Ref. 14 has been obtained by freezing the displacement of neighboring sites. This method of estimation for the transition probability gives appreciably different magnitude from that given by (9) in quantitative sense, but gives very similar results when t_e is readjusted.[16]

3. Temporal Evolution of LS Fraction based on Koshino–Ogawa's Approach

Let us denote the average density of LS fraction, x. The relaxation probability will be proportional to the factor, $x_{HS} = 1 - x$, to lead $(1-x)\gamma_{nrt}(x)$, where $\gamma_{\rm nrt}(x) = \hat{\omega}(\varepsilon_0 - \kappa q^2(1-x))$.[35] On the other hand, the excitation probability by light will be proportional to the factor Ix. Taking the above two processes into account, the temporal evolution of the density will be given by the following rate equation,

$$\frac{\rm d}{{\rm d}t}x = \gamma_{\rm nrt}(x)(1-x) - Ix. \tag{10}$$

In this derivation, the spatial and temporal fluctuation have been neglected. Koshino and Ogawa noted that this simplification is justified in very long range interaction case. In a later section, we will re-drive this equation from the Fokker–Planck equation.

The solution of Eq. (10) with the initial condition $x = 1$ (LS phase) is shown in Fig. 2 for three typical values of I. The figure indicates the existence of the threshold excitation rate $I_{\rm th}$. If the excitation rate is lower than the threshold value ($I < I_{\rm th}$), the LS fraction never reaches a HS-dominant value but only relaxes to the LS-dominant stationary fraction $x = \bar{x}^{\rm LS} \sim 1$, even if the complexes are irradiated for an infinitely long time. If the excitation rate is very much higher than the threshold value ($I_{\rm th} < I$), x decreases rapidly to $x = \bar{x}^{\rm HS} \sim 0$. On the other hand, when I is larger than $I_{\rm th}$, but very near it, x decreases at first, and next, it decreases very slowly to around a value $x_{\rm S}$. After the incubation period $\tau_{\rm incu}$, the LS fraction suddenly decreases to reach the HS-dominant fraction $\bar{x}^{\rm HS} \sim 0$,

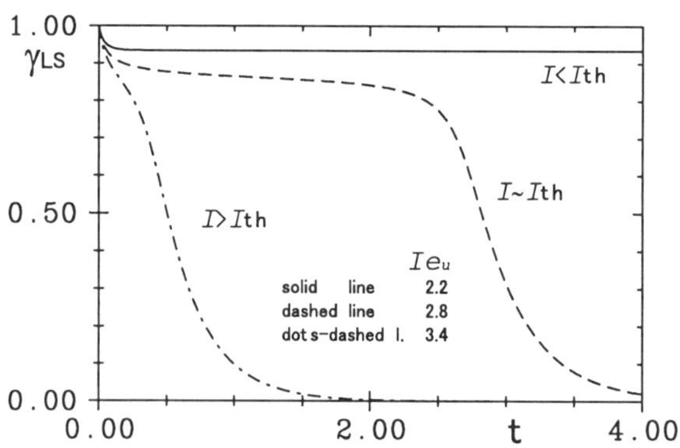

Fig. 2. Temporal evolution of the LS fraction x under three different excitation rates: $I \ll I_{\rm th}$ (solid line), $I \gtrsim I_{\rm th}$ (dashed line), and $I \gg I_{\rm th}$ (dot-dashed line). The figure indicates the existence of the threshold excitation rate $I_{\rm th}$ to reach a HS-dominant value. Under a large enough excitation rate to satisfy $I > I_{\rm th}$, the LS fraction makes a sudden decrease after an incubation period $\tau_{\rm incu}$, which depends on the excitation rate I. The quantity e_u gives the scale of the intensity, see the text for definition.

where the HS $\to$ LS decay is slow. Similar behaviors have also been studied also in other situations.[36,37]

We can understand this behavior of the solution of Eq. (10) with the help of Fig. 3. The quantity $\gamma_{\rm nrt}(x)(1-x)$ is plotted by the thick three-dots-dashed curve (strong interaction case) and the thin four-dots-dashed curve (weak interaction case). The stationary solutions of Eq. (10) are given by the intersections of the straight lines of Ix with these curves. We note here that $\gamma_{\rm nrt}(x)(1-x)$ has a positive value at $x=0$, and zero at $x=1$. In the weak interaction case, we have only one solution for any values of I, and it continuously changes from large x to small x as I increases.

Even in the strong interaction case, we have also only one solution with $x \sim 1$ when I is very small. This solution is denoted as LS-dominant solution $\bar{x}^{\rm LS}$. When I is very large, we have also only one solution with very small x, i.e. HS-dominant solution, $\bar{x}^{\rm HS}$. It becomes bistable under the intermediate excitation rate $I'_{\rm th} < I < I_{\rm th}$ as shown by the solid straight line in Fig. 3. (The $\bar{x}^{\rm HS}$ solution cannot be easily identified in the figure, but it certainly

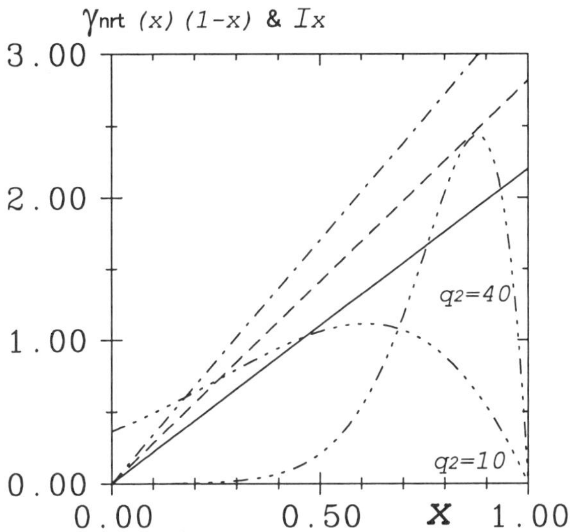

Fig. 3. Schematic diagram which explains the solutions of Eq. (10). The stationary solutions are given by the intersections of $\gamma_{\rm nrt}(x)(1-x)$ and Ix. The three-dots-dashed curve (four-dots-dashed curve) is the curve of $\gamma_{\rm nrt}(x)(1-x)$ in the strong (weak) interaction case. The solid (dot-dashed) line is classified to the high (low) intensity case of $I \gg I_{\rm th}$ ($I \ll I_{\rm th}$) for $q^2 = 40$. The dashed line is the case of high, but very near the threshold intensity, $I_{\rm th}$.

exists because $\gamma_{\text{nrt}}(x)(1-x) > 0$ at $x = 0$.) The threshold intensity of $\bar{x}^{HS}$ is denoted as I_{th}, and also the threshold intensity for $\bar{x}^{\text{LS}}$ is by I'_{th}.

Taking into account that the initial phase is the LS phase ($x = 1$), the slowing down of the LS $\rightarrow$ HS conversion takes place for $I \gtrsim I_{\text{th}}$ (dashed line) in Fig. 3, because the right hand of Eq. (10) has a form, $-A^S(x - x_{\text{S}})^2 - \Delta^{\text{S}}$ with positive A^{S}, and small $\Delta^{\text{S}} \sim (I - I_{\text{th}})$. x_{S} is the LS fraction of the tangential point. This is the origin of the incubation period. The dependence of the incubation period τ_{incu} on the excitation rate I is given by

$$\tau_{\text{incu}} \sim \Delta_{\text{S}}^{-1} \sim (I - I_{\text{th}})^{-1}. \tag{11}$$

In the weak-interaction case, the dependence of the HS decay rate γ_{nrt} on x is weak. Then the stationary solution changes continuously as the excitation rate is increased. The critical interaction strength k_{c} which separates the weak and strong interaction cases is estimated as the tangential condition of Ix to $(1-x)\gamma_{\text{nrt}}(x)$, i.e. $\frac{1}{x(1-x)} - \frac{\partial}{\partial x} \ln \gamma_{\text{nrt}}(x) = 0$ and also the first derivative of this quantity must be zero. This gives roughly

$$k_{\text{c}} \sim \frac{4}{\left(\kappa q^2 - \varepsilon^{\text{allHS}}\right)}. \tag{12}$$

Peculiarities such as the threshold behavior of the excitation rate and the existence of the incubation period can be explained grossly by the present nonlinear rate equation. The nonlinearity originates from the LS fraction-dependent alteration of the diabatic potential, which results in the longer lifetime of the HS state for smaller LS fraction.

4. Monte Carlo Simulation

4.1. *Method of simulation*

The theory described in the previous section neglects the spatial and temporal fluctuation of LS fraction. Therefore it cannot describe the phase separation and the step-like decrease of the LS fraction. In this section, we adopt the Monte Carlo (MC) simulation method, and check whether Koshino–Ogawa's model can explain such problems when the fluctuation effects are considered.

The MC simulation is carried out following a procedure corresponding to the master equation (1). First we give an initial configuration, then the following processes are made;[15]

(1) choose a site by using a random number,
(2) if the ion at the site is LS, determine whether it is excited or not by comparing $I\Delta t_{\mathrm{MC}}$ with a random number,
(3) choose a site (we call it as ℓ) by using a random number,
(4) if the ion at the site is HS, determine whether or not it is relaxed by comparing $\omega(\varepsilon_\ell)\Delta t_{\mathrm{MC}}$ with a random number,
(5) return to 1.

When this loop is done N times (where N is the number of sites), it is ascribed to a unit time of the Monte Carlo simulation (1 MC step). The quantity Δt_{MC} is a time corresponding to the 1 MC step. We choose it sufficiently small to fulfill the condition $\max(\omega(\varepsilon_\ell)\Delta t_{\mathrm{MC}}, I\Delta t_{\mathrm{MC}}) \ll 1$ to ensure that the change of the state is small in the 1 MC step.

4.2. *Results in two-dimensional case*

In this subsection, we will show the result of MC simulation in two-dimensional lattice. The interaction constant $K_{\boldsymbol{R}_h}$ is chosen to be constant for pairs within the 5th neighbor and zero for the pairs outside of it in the two-dimensional case. ($K_{\boldsymbol{R}_h}$ has finite range, but $J_{\boldsymbol{R}_h}$ is not zero even for pairs with larger distance.) This model is called the long range one, and used when it is not exceptionally noted.

The typical time evolution of LS fraction γ_{LS} in the two-dimensional case with 100×100 sites is shown in Fig. 4. The local phonon energy $\hbar\omega_o$ is used as the energy unit hereafter, and the unit of time τ_u is parametrized as $\frac{2\pi t_e^2}{\hbar^2\omega_o}\tau_u \equiv e_u$. We choose $Ie_u \sim 2$ in Fig. 4. The time interval Δt_{MC} is chosen to be 0.25×10^{-3}.

The thick solid line is the case which shows steep LS to HS conversion after some incubation time. In the initial stage, the LS fraction decrease very rapidly with time scale, $\tau_{\mathrm{rap}} \sim (Ie_u)^{-1}$, then very slowly in the intermediate region, and after incubation period it decreases steeply to almost zero value in the final stage. The thick one-dot dashed line is the case that the excitation intensity is weak. In this case the steep conversion to HS state in the final stage has not taken place within the calculation time. The thick dashed line is the case that the excitation intensity is strong. The LS fraction decreases steeply without appreciable incubation time. We plot the thick two-dots-dashed line in a different time scale as a function of $t/2$, i.e. 4.0 of the horizontal axis correspond to $t = 8$. This line has incubation time about twice that of the solid line, but it does not largely

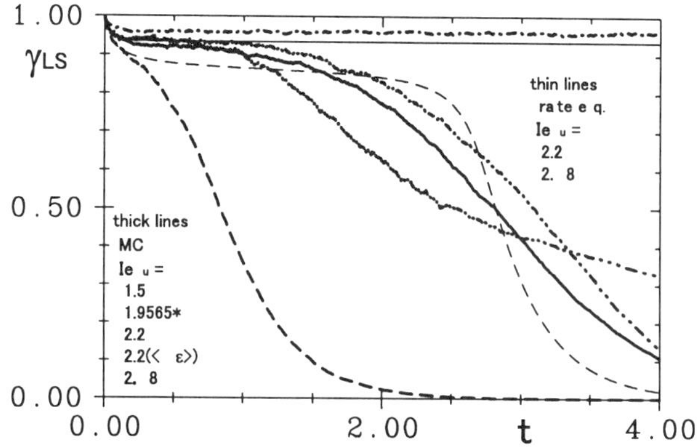

Fig. 4. Time evolution of the low-spin (LS) fraction γ_{LS} for typical cases in the two-dimensional lattice with 100×100 sites. The thick solid line is calculated for the excitation intensity $Ie_u = 2.2$, the thick dot-dashed line is for $Ie_u = 1.5$, the thick dashed line is for $Ie_u = 2.8$, and the thick two-dots-dashed line is for $Ie_u = 1.9565$ by the Monte Carlo (MC) method with $e_u = 10^4$. The $Ie_u = 1.9565$ line is plotted as a function of $t/2$, i.e. 4.0 of the horizontal axis corresponds to $t = 8.0$. The three-dots-dashed line gives the average of the effective energy of the HS site, $\langle\varepsilon_\ell\rangle_{HS}/\varepsilon_0$ for $Ie_u = 2.2$. The parameters are $k = 0.5$, $q^2 = 40$, $\varepsilon^{allHS} = 10.00$ ($\varepsilon_0 = 36.37$), and $\kappa = 2.0$. For the definition of these parameters, see the text. The thin solid line is calculated by the rate equation (10) for $Ie_u = 2.2$, and the thin dashed line for $Ie_u = 2.8$.

deviate from the latter in the late stage when the time scale is changed. The thin lines are results given in Fig. 2. The rapid decrease in the initial stage overlaps very well on that of the thick lines with same excitation intensity, but they deviate largely in the later stages. Qualitative behavior is similar if the parameters are replaced, for example, the thick solid line and the thin dashed line.

The snapshot of the spatial distribution of HS sites is shown in Fig. 5. At time 0.1, which is just after the steep decrease of the initial stage, HS sites are almost randomly distributed. Small loose clusters of HS sites are also observed, but almost of them are repeating dissociation and aggregation processes. At about the mid time of the incubation period, at time 0.5, some of the HS clusters become tight, and change to nucleation centers of HS domain. In Fig. 4, the average of the effective energy of the HS site, $\langle\varepsilon_\ell\rangle_{HS}/\varepsilon_0$, is plotted by the thick three-dots-dashed line. It starts to decrease at about $t = 1.0$ before the solid line shows the steep decrease. At this time, many nucleation centers of HS domain appear as seen from Fig. 5. At time 1.8 near the starting region of the steep decrease of x_{LS}, many HS domains start to grow. Domains become very large and amalgamate in the mid time of the steeply decreasing region, at time 2.6. In the

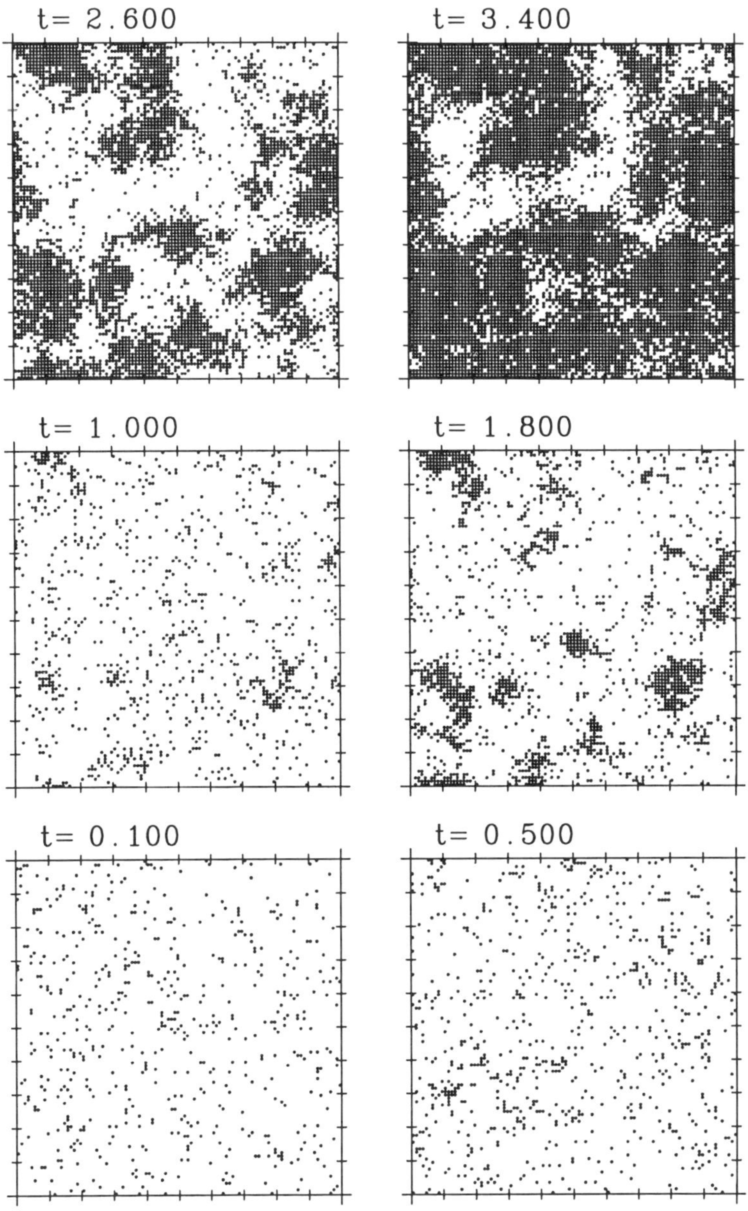

Fig. 5. The snapshot of the ion configuration at various time. The dots indicate the HS sites. The parameters are the same as those of the thick solid line in Fig. 1, $Ie_u = 2.2$, $k = 0.5$, $q^2 = 40$, $\varepsilon^{\rm allHS} = 10.00$, $e_u = 10^4$, and $\kappa = 2.0$. The numbers in the figure indicate the time whose unit is the same as that of Fig. 4.

final stage of the steep decrease, at time 3.4, the HS domains combine into one connected region, and the left LS regions shrink rather slowly.

From the observation of Fig. 5, we see that the time evolution of HS states can be considered as the growing process of the HS domain. In particular the incubation time corresponds to that of the HS nucleation. This means the transition to the HS state seems to be very similar to the transition induced by the temperature in the non-conserving system.[15] In the usual thermal transition, the nucleation rate has temperature factor in the exponential function because the barrier energy of the nucleation is surmounted by the thermal activation energy.[20] However the present master equation (1) does not include the temperature, because we have neglected it assuming that the temperature is low. The problem arises as to what fluctuation drives it to surmount the nucleation barrier. This will be discussed in Sec. 5.

As noted previously the transition probability is estimated in Ref. 14 by freezing the distortion of neighboring lattice. Such a case has also been studied in Ref. 16. The change from the incubation stage to the steep decrease becomes relatively gentle.

4.3. *Three-dimensional case*

In Fig. 6, we show the time evolution of LS fraction in three-dimensional case. Overall features are similar to those in the two-dimensional case, but the transition from the incubation stage to the final steep decrease stage is relatively sharp, even when the time scales of the initial rapid decrease, $\tau_{\rm rap}$, and the incubation time, $\tau_{\rm inc}$, are comparable to those in Fig. 4. This time evolution by MC simulation shows the tendency approaching to the behavior given by the dashed line in Fig. 2. However, we note the dashed line is obtained in the special case by choosing I to be very near the threshold value $I_{\rm th}$. The steep change is widely seen in MC simulation lines for various incubation time cases in Fig. 6.

The n.n. interaction model gives relatively slow change to the final decreasing regime, as seen from the thin four-dots-dashed line. We put $K_{\boldsymbol{R}_\ell}(=k/6)$ as nonzero only for the nearest neighbor site in the n.n. model, while we put $K_{\boldsymbol{R}_\ell}(=k/126)$ as nonzero within the 9th neighbor for the long range model in the three-dimensional case.

In later sections, we note that the surface energy of the nucleation center is proportional to the square root of the factor $J'' = -\frac{\partial^2 J(k)}{a^2\partial^2 k}$, where $J(k)$ is the Fourier transform of the ion-ion interaction. The constant J'' is given as $\kappa q^2\frac{k}{3}$ and $\kappa q^2\frac{125\,k}{63}$, respectively, for n.n. and the long-range model.

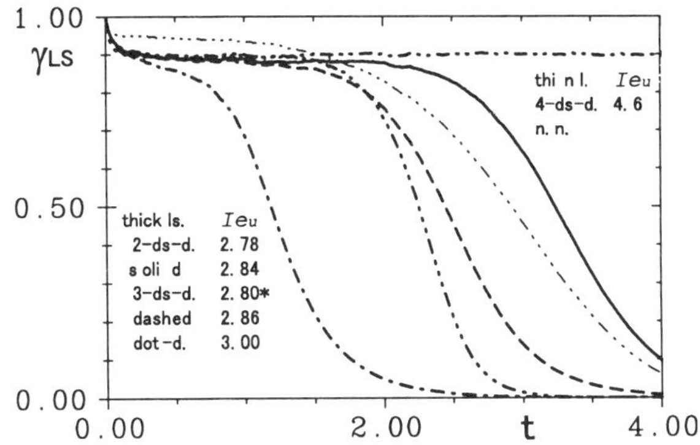

Fig. 6. Time evolution of the low-spin (LS) fraction $\gamma_{\rm LS}$ in the three dimensional lattice with $40 \times 40 \times 40$ sites. The two-dots-dashed line is calculated for the excitation intensity $Ie_u = 2.78$, the solid line is for $Ie_u = 2.84$, the dashed line is for $Ie_u = 2.86$, and the dot-dashed line is for $Ie_u = 3.0$, with $(k = 0.5, q^2 = 40, e_u = 1 \times 10^4, \varepsilon^{\rm allHS} = 9.78, \kappa = 2.0)$. The three-dots-dashed line is calculated for $Ie_u = 2.80$ with the same parameter values except I. The horizontal axis is $t/2$ for this line, i.e. 4.0 in the figure is $t = 8.0$. The four-dots-dashed line is calculated for the n.n. model with $Ie_u = 4.6$ with $(k = 0.5, q^2 = 40, e_u = 5 \times 10^4, \varepsilon^{\rm allHS} = 9.78, \kappa = 2.0)$.

The three-dots-dashed line has incubation time about twice that of the dashed line. We plot it with the time unit so that the steep decrease can be seen in the figure.

5. Difference and Similarity with the Temperature Induced Transition

5.1. *The Fokker–Planck equation*

To analyze the result of MC calculation, we derive Fokker–Planck equation from the master equation (1). After making several simplifying approximation steps, we get the following Fokker–Planck equation:

$$\frac{\partial P(\{x\})}{\partial t} = \sum_\ell \frac{\partial}{\partial x_\ell}\left[\left(x_\ell \omega\left(-Jx_\ell - \frac{a^2}{2}J''\nabla_\ell^2 x_\ell\right) - (1-x_\ell)I\right)\right.$$
$$\left. + \frac{I\omega(-Jx_\ell^*)}{I+\omega(-Jx_\ell^*)}\frac{\partial}{\partial x_\ell}\right] P(\{x\}). \tag{13}$$

Here $P(\{x\})$ is the probability function for the configuration that $\{n_\ell\}$ takes $\{x_\ell\} \equiv \{x\}$. The HS state energy at ℓ site, $\varepsilon_\ell = \varepsilon_0 - \sum_{\ell'} J_{\boldsymbol{R}_\ell - \boldsymbol{R}_{\ell'}} x_{\ell'}$, has been approximated as $\varepsilon_0 - Jx_\ell - \frac{a^2}{2}J''\nabla_\ell^2 x_\ell$ in Eq. (13). The quantity

x_ℓ^* indicates a representative average value of the HS fraction ratio, and we have re-expressed $\omega(\varepsilon_\ell)$ as $\omega(\varepsilon_\ell - \varepsilon_0)$.

We define a quantity corresponding to the effective potential density $v(x_\ell)$, and the "kinetic energy $k(x_\ell)$",

$$v(x_\ell) = -\int^{x_\ell} dx'_\ell (x'_\ell(\omega(-Jx'_\ell)) - (1 - x'_\ell)I), \tag{14}$$

$$k(x_\ell) = -Da^2\nabla_\ell^2 x_\ell. \tag{15}$$

Here $D = \frac{1}{2}(x_\ell^{**}\omega(-Jx_\ell^{**})')J''$, and we again define a representative average value of the HS fraction ratio in the $k(x_\ell)$ term.

The simplified Fokker–Planck equation is re-written as

$$\frac{\partial P(\{x\})}{\partial t} = \sum_\ell \frac{\partial}{\partial x_\ell}\left[-\frac{\partial v(x_\ell)}{\partial x_\ell} + k(x_\ell) + T_{\text{eff}}\frac{\partial}{\partial x_\ell}\right] P(\{x\}). \tag{16}$$

Here the effective temperature is defined:

$$T_{\text{eff}} = \frac{I\omega(-Jx_\ell^*)}{I + \omega(-Jx_\ell^*)}. \tag{17}$$

In the usual Fokker–Planck equation treating the thermal fluctuation effect,[20] there appears a coefficient which is proportional to the temperature in the second term of the right hand side of (16). The fluctuation effect similar to that of the temperature appears through the excitation and de-excitation processes by themselves in the present master equation. The effective force has the following relation to the effective potential term:

$$f(x_\ell) = -\frac{\partial v(x_\ell)}{\partial x_\ell}. \tag{18}$$

5.2. *Rate equation for the LS fraction*

Next, we derive the Langevin equation for the average $\langle x_{\text{HS}}\rangle = \langle x_\ell\rangle = x_{\text{HS}}$ neglecting the $k(x_\ell)$ term,[38]

$$\frac{\mathrm{d}x_{\text{HS}}}{\mathrm{d}t} = -x_{\text{HS}}\omega(-Jx_{\text{HS}}) + (1 - x_{\text{HS}})I + f_{\text{random}}, \tag{19}$$

where f_{random} is the random force, and we have approximated as $\langle -x_{\text{HS}}\omega(-Jx_{\text{HS}}) + (1 - x_{\text{HS}})I\rangle \to -x_{\text{HS}}\omega(-Jx_{\text{HS}}) + (1 - x_{\text{HS}})I$. This equation corresponds just to Eq. (10) in Sec. 3 when we neglect the random force. The right hand side term except the random force is given by the expression $f(x_{\text{HS}})$.

Following Ref. 14, we replace Eq. (19) to that of the LS fraction, $x_{LS} = 1 - x_{HS}$, the force term is replaced as $\bar{f}(x_{LS}) \equiv -f(1 - x_{LS}) = (1 - x_{LS})\omega(-J(1 - x_{LS})) - x_{LS}I$. As noted in Sec. 3, $\bar{f}(x_{LS} = 1) = -I < 0$ and $\bar{f}(x_{LS} = 0) = \omega(-J) > 0$. Therefore we always have at least one solution. In Fig. 7, $\bar{f}(x_{LS})$ for typical values of I is shown. The equation $\bar{f}(x_{LS}) = 0$ is classified as three cases depending on I. (1) $I < I'_{\rm th}$ where it has only one solution ($x_{\rm LS}^{(1,1)} \sim 1$), (2) $I'_{\rm th} < I < I_{\rm th}$ where it has three solutions ($x_{\rm LS}^{(2,1)} < x_{\rm LS}^{(2,2)} < x_{\rm LS}^{(2,3)}$), (3) $I_{\rm th} < I$ where it has only one solution ($x_{\rm LS}^{(3,1)} \sim 0$). The solid and the one-dot-dashed lines in Fig. 7 are classified to the case (2). The dashed line ($Ie_u = 2.8$) is classified to the case (3), but very near the boundary with (2). At this point the pair of solutions, $x_{\rm LS}^{(2,2)}$ and $x_{\rm LS}^{(2,3)}$ have disappeared.

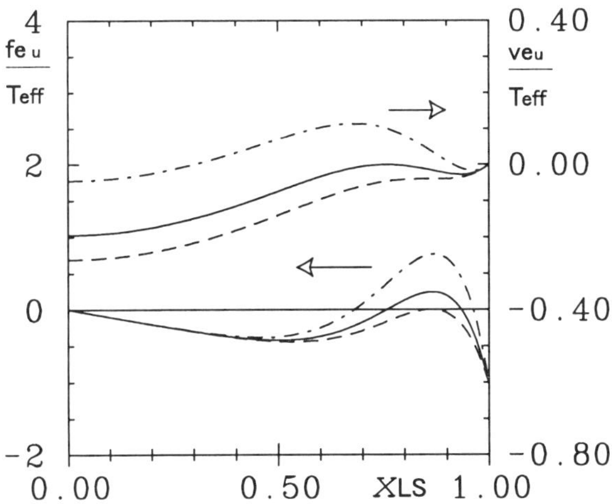

Fig. 7. The force $\bar{f}(x_{\rm LS})$ and the potential $v(x_{\rm LS})$ as a function of $x_{\rm LS}$ for several cases of I. The parameters are the same as those in Fig. 4. The quantity $\bar{f}(x_{\rm LS})$ is very small but positive at $x_{\rm LS} = 0$, since $\bar{f}(x_{\rm LS} = 0) = \omega(-J) > 0$. The solid line is for the excitation intensity $Ie_u = 2.2$, the dot-dashed line is for $Ie_u = 1.5$, and the dashed line is for $Ie_u = 2.8$. $k = 0.5$, $q^2 = 40$, $\varepsilon^{\rm allHS} = 10.00$, $e_u = 10^4$, and $\kappa = 2.0$. The solid and the dot-dashed lines are cases classified to (2) in the text, i.e. they have three solutions $x_{\rm LS}^{(2,1)} < x_{\rm LS}^{(2,2)} < x_{\rm LS}^{(2,3)}$. Among them, one of $x_{\rm LS}^{(2,1)}$ and $x_{\rm LS}^{(2,3)}$ is usually metastable point and another is the absolute stable point of the potential. The value $x_{\rm LS}^{(2,1)}$ is very small because $\omega(-J)$ is extremely small. The dashed line is classified to the case (3), but very near the boundary [2,3], i.e. $\Delta^{\rm S}$ is very small, 0.90×10^{-3} and $x_{\rm LS}^{\rm S} \sim 0.86$. The quantities $\bar{f}$ and v are normalized by $T_{\rm eff}$, where the representative value x_ℓ^* is chosen to be $x_{\rm HS}^{(2,3)} = 1 - x_{\rm LS}^{(2,3)}$.

The solutions $x_{\rm LS}^{(2,1)}$ and $x_{\rm LS}^{(2,3)}$ give the stable solution of Eq. (10), but the point $x_{\rm LS}^{(2,2)}$ is the unstable solution. Of course $x_{\rm LS}^{(1,1)}$ and $x_{\rm LS}^{(3,1)}$ are the stable points of Eq. (10). Calculated results on the basis of Eq. (10) have already been presented in Sec. 3.

5.3. *Potential barrier and nucleation of HS domain*

In Fig. 7 we also show the potential energy $v(x_{\rm LS})$ for several I cases. The potential has minimum or maximum at the points which satisfy the condition $\bar{f}(x_{\rm LS}) = 0$. Even when the potential is lower at $x_{\rm LS}^{(2,1)}$ than the potential at $x_{\rm LS}^{(2,3)}$, the solution is trapped at the $x_{\rm LS}^{(2,3)}$ point if we start from $x_{\rm LS} = 1$ in Eq. (10) because we have neglected the random force. However when returning to the Fokker–Planck equation, the random force is caused by the diffusion term including the $T_{\rm eff}$ factor and the potential barrier is surmounted by the fluctuation. The domain of $x_{\rm LS}^{(2,1)}$ phase will grow in the sea of $x_{\rm LS}^{(2,3)}$ phase.

Let us denote the potential differences, $\delta\mu = v(x_{\rm LS}^{(2,3)}) - v(x_{\rm LS}^{(2,1)}))$. Following the theory for the first order phase transition,[20] the nucleation rate in d-dimensions is given as follows in Ref. 16:

$$I^{(d)} = \frac{\Gamma_c^{(d)} e^{-\beta_{\rm eff} h_c^{(d)}}}{D^{(d)}}, \tag{20}$$

with $D^{(2)} = 2\pi\sqrt{\beta_{\rm eff}\delta\mu}$ and $D^{(3)} = 4\pi\sqrt{\beta_{\rm eff}\sigma}$. We have defined $\beta_{\rm eff} = 1/T_{\rm eff}$, $\sigma = y(v(x_{\rm LS}^{(2,2)}) - v(x_{\rm LS}^{(2,3)})\xi$ is the effective "surface energy", with $y = 16/3$ and ξ is the characteristic length of the boundary.[16] Here d denotes the dimension of space, $h_c^{(d)}$ is the energy of droplet with critical radius $h_c^{(2)} = \pi\frac{\sigma^2}{\delta\mu}$ and $h_c^{(2)} = \frac{16\pi\sigma^3}{3(\delta\mu)^2}$ with critical radius $R_c^{(d)} = \frac{(d-1)\delta\sigma}{\delta\mu}$. $\Gamma^{(d)}$ characterizes a rapid time scale.

Using these quantities, the characteristic time scale to show the sigmoidal shape has been estimated roughly as[20] $\tau_{\rm sig}^{(d)} = \left[\Gamma_c^{(d)}(R_c^{(d)}/a)^d(x_{\rm HS}^{(3,2)}I^{(d)}/((d+1)\Gamma_c^{(d)}))^{1/(d+1)}\right]^{-1}$. This quantity drastically changes when the intensity I changes because $\delta\mu$ in $h_c^{(d)}$ in the exponential function changes. We may identify this time as the time scale of the incubation period.

This quantity is estimated as 0.8 and 6.2, respectively, for the solid line ($Ie_u = 2.2$) and two-dots-dashed line ($Ie_u = 1.9565$) cases in Fig. 4. The ratio of these values is about 8, while the ratio of the incubation time for the two lines is expected to be about 2. The expression does not give an account of the MC result. However, the estimated order of magnitude is

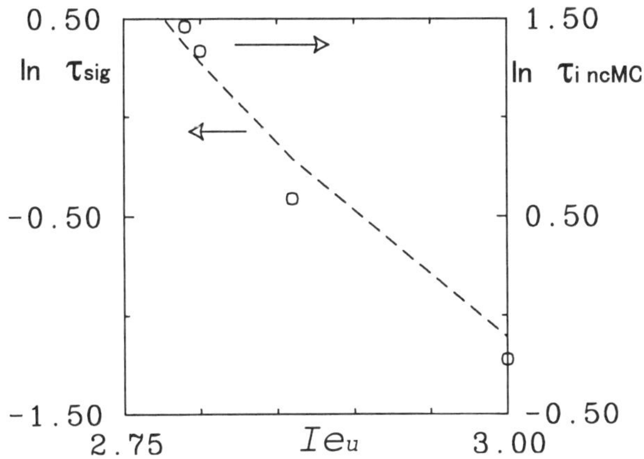

Fig. 8. Logarithms of $\tau_{\rm sig}$ and $\tau_{\rm inc}^{\rm MC}$ as a function of Ie_u for the three-dimensional case. The dashed line gives $\ln \tau_{\rm sig}$ and the circles give $\ln \tau_{\rm inc}^{\rm MC}$, where $\tau_{\rm inc}^{\rm MC}$ is defined as the time at which $\gamma_{\rm LS}$ takes the value 0.8. The parameters are ($k = 0.5, q^2 = 40$, $e_u = 1 \times 10^4, \varepsilon^{\rm allHS} = 9.78, \kappa = 2.0$).

not so far from the value expected from the simulation, even though the expression has a very rapidly varying exponential factor.

The time scale $\tau_{\rm sig}^{(3)}$ for the 3-dimensional case is estimated as 0.3, 0.8 and 1.3, respectively, for the dot-dashed line ($Ie_u = 3.00$), for the dashed line ($Ie_u = 2.86$) and for the three-dots-dashed line ($Ie_u = 2.80$) cases. In Fig. 8, we plot the logarithms of these values and also the logarithms of the incubation times estimated by the MC calculation as a function of Ie_u. The incubation time is defined as the time at which $\gamma_{\rm LS}$ takes the value 0.8. We note that the scale of the right vertical axis is shifted by 1 in Fig. 8. An approximate relation $\tau_{\rm inc}^{\rm MC} \sim \tau_{\rm sig} \times e \sim 3\tau_{\rm sig}$ is expected from the figure. The difference 3 in the numerical factor may be ascribed to the crudeness of the method to derive $\tau_{\rm sig}^{(d)}$ from $I^{(d)}$. In the three-dimensional case, the estimation of the incubation time based on the usual domain formation theory can give rather good account of the MC result.[39]

6. Miscellaneous Problems

6.1. *The Zeeman field*

The relaxation probability is sensitive to the energy of HS state, therefore the conversion process to the HS will be affected by the Zeeman field. In Fig. 9, we show the incubation time $\tau_{\rm inc}$ as a function of Zeeman field. The incubation time is tentatively defined as the time at which the LS

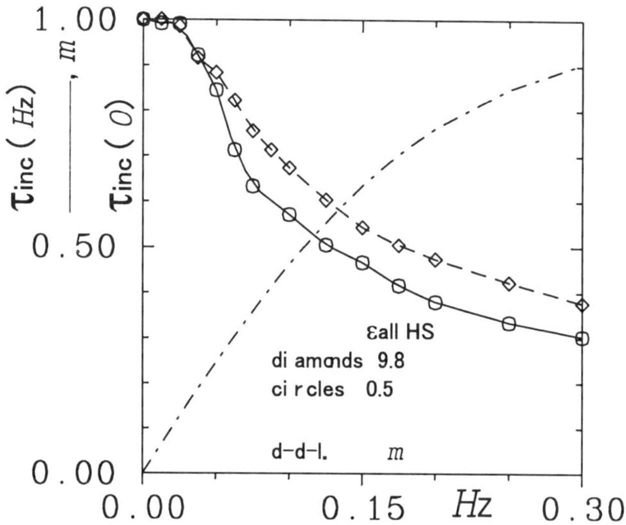

Fig. 9. Incubation time as a function of Zeeman field. The incubation time is normalized by the value for the $H_Z = 0$ case. The solid line with circles is calculated for $I = 8.40 \times 10^{-6}$ with $e_u = 2 \times 10^5$ and $\varepsilon^{\text{allHS}} = 0.5$. The dashed line with diamonds is calculated for $I = 2.90 \times 10^{-4}$ with $e_u = 1 \times 10^4$ and $\varepsilon^{\text{allHS}} = 9.78$. The other parameters are $k = 0.5$ and $q^2 = 40$. The spin polarization is calculated assuming the two-valued model with energy $H_Z = -H_Z\mu$ ($\mu = \pm 1$) and the temperature $T = 0.2$. The dot-dashed line gives the magnetization calculated for this model.

fraction decreases to 0.8. It starts to decrease with a slope of zero, then steeply in the region with the magnetization of $\frac{1}{4}$ of the saturation value. In the solid line cases, the energy gain from the Zeeman interaction is chosen as comparable to the energy $\varepsilon^{\text{allHS}} \sim 0.5$. But we found that the qualitative behavior is not so different from the dashed line with larger $\varepsilon^{\text{allHS}} \sim 10$.

In Fig. 9, the decrease of the incubation time shows the tendency of saturation when it decreases down to about 0.3 of the $H_Z = 0$ case. This is partly ascribed to the fact that the time scale of the initial rapid decrease τ_{rap} is about $\frac{1}{10}$ of the incubation time for the $H_Z = 0$ case. The incubation time will decrease down to a time scale of τ_{rap} when the Zeeman field increases extremely.

6.2. *Intermittent excitation*

The potential barrier of the critical droplet is surmounted by the fluctuation due to the excitation and de-excitation processes as noted

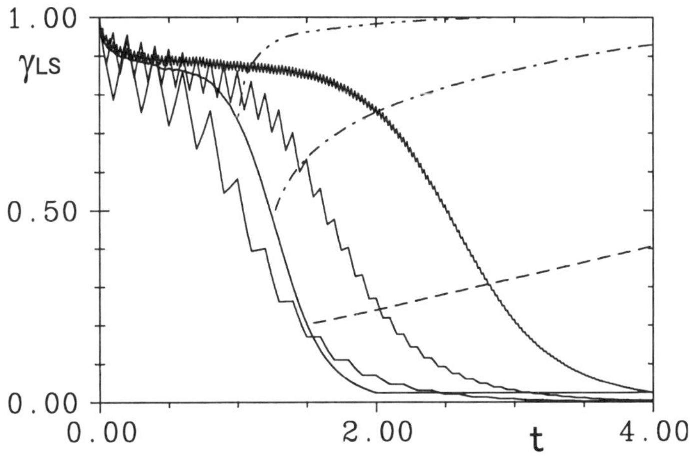

Fig. 10. Time evolution of the low-spin (LS) fraction $\gamma_{\rm LS}$ under intermitted excitation, and break of excitation for three-dimensional lattice by MC calculation. The zig-zag solid lines are calculated for the intermitted excitation with the time interval $t_{\rm int} = 0.0125$, 0.05 and 0.1. The excitation with intensity of twice of the original value is continued during a interval $t_{\rm int}$, then it is stopped in the next interval of $t_{\rm int}$. The parameters are $I = 2.90 \times 10^{-4}$, $e_u = 1 \times 10^4$, $\varepsilon^{\rm allHS} = 9.78$, $k = 0.5$ and $q^2 = 40$. The smooth solid line ($t_{\rm break} = 4$), dashed line ($t_{\rm break} = 3$), dot-dashed line ($t_{\rm break} = 2.5$) and two-dots-dashed line ($t_{\rm break} = 2$) are the results that the excitation light is interrupted. The scale of the horizontal axis for these lines is $t/2$, i.e. 2.0 of the horizontal scale in this case means $t/2 = 2.0$.

previously. The average effect is given by the effective temperature $T_{\rm eff} = Ix_{\rm peak} + \omega(1 - x_{\rm peak}) \sim \frac{I\omega}{I+\omega}$. We calculate by the MC method the time evolution of LS fraction under the intermittent excitation in Fig. 10. The excitation light with twice the intensity of the original value is applied during a time interval $t_{\rm int}$, then it is stopped in the next interval of $t_{\rm int}$. When $t_{\rm int}$ is shorter than the time scale of the initial rapid decrease $\tau_{\rm rap}$, the average value of $\gamma_{\rm LS}$ is not changed as much as we can see from the comparison with the $t_{\rm int} = 0.0125$ line in Fig. 10 and the dashed line in Fig. 6.[40] We should use the averaged intensity with the time scale $\tau_{\rm rap}$ for I in the expression of $T_{\rm eff}$. When $t_{\rm int}$ becomes comparable to $\tau_{\rm rap}$, the conversion to the HS state is accelerated.

6.3. *Interruption of excitation*

In Fig. 10, we also show the time evolution when the excitation is interrupted at some time. When excitation is stopped before the fraction $\gamma_{\rm LS}$ is not so decreased, initially it increases with time scale of $\tau_{\rm rap}$, then gradually to $\gamma_{\rm LS} = 1$. When the excitation is stopped after $\gamma_{\rm LS}$ is decreased near to zero, the restoring speed is very slow, as seen from smooth solid line. A similar fact has already been pointed out in Ref. 21. The slowdown of the relaxation rate will be drastic for systems showing the cooperative LS $\rightarrow$ HS conversion.

7. Coupling with Volume Striction Modes

The model proposed by Koshino and Ogawa in Ref. 14, Eq. (2), was examined in previous sections on the basis of MC simulation. The coexistence of the LS and HS regions was observed during the conversion process, but macroscopic phase separation (iv) was not reproduced. The step-like decrease of LS fraction (v), was also not reproduced.

As noted in the introduction, the coupling between the volume striction and the spin state is important to reproduce the macroscopic phase separation of HS and LS regions. The energy of ions and lattice distortion will be given by the following expression when we take into account the volume striction effect:

$$E = \sum_{\nu\nu'} \sum_{\boldsymbol{l}\boldsymbol{\delta}_\nu} \Big[\frac{\kappa_1}{2} \delta_{\nu\nu'} \big(v^{(\nu)}_{\boldsymbol{l}+\boldsymbol{\delta}_\nu} - v^{(\nu)}_{\boldsymbol{l}}\big)^2 + \frac{\kappa_2}{2}(1-\delta_{\nu\nu'})\big(v^{(\nu')}_{\boldsymbol{l}+\boldsymbol{\delta}_\nu} - v^{(\nu')}_{\boldsymbol{l}}\big)^2 - \kappa_1 q \delta_{\nu\nu'}\big(v^{(\nu)}_{\boldsymbol{l}+\boldsymbol{\delta}_\nu} - v^{(\nu)}_{\boldsymbol{l}-\boldsymbol{\delta}_\nu}\big) n_{\boldsymbol{l}} \Big] - \sum_{\boldsymbol{l}} f_{\ell_x} \kappa_1 q \big(v^{(x)}_{\boldsymbol{l}+\boldsymbol{\delta}_x} - v^{(x)}_{\boldsymbol{l}}\big) + \sum_{\boldsymbol{l}} \varepsilon^0 n_{\boldsymbol{l}}. \tag{21}$$

The quantity $v^{(\nu)}_{\boldsymbol{l}}$ denotes the ν component of the lattice displacement at site $\boldsymbol{l}$, and $\boldsymbol{\delta}_\nu$ is the translation vector to the nearest neighbor site along the ν direction.

Quantities κ_1 and κ_2 give the force constants for the longitudinal and the transverse displacements. The third term in the bracket denotes the coupling between the lattice stretching mode and the HS ion. The quantity q denotes the strength of the coupling. The fourth term is the coupling between the lattice stretching mode and the stress field f_{ℓ_x} along the x direction, where the quantity ℓ_x denotes the x-component of $\boldsymbol{l}$. By the field f_{ℓ_x}, the lattice spacing between the ℓ_x+1-th and the ℓ_x-th layers is stretched (if f_{ℓ_x} is positive). This term is introduced to control the strain of crystal. Hereafter we denote the site $\boldsymbol{l}$ by ℓ to abbreviate expressions when it does not cause confusion. By introducing the Fourier transformation,

$n_{\boldsymbol{k}} = \frac{1}{\sqrt{N}} \sum_\ell n_\ell \mathrm{e}^{-i\boldsymbol{k}\boldsymbol{R}_\ell}$ and $v_{\boldsymbol{k}}^{(\nu)} = \frac{1}{\sqrt{N}} \sum_\ell v_\ell^{(\nu)} \mathrm{e}^{-i\boldsymbol{k}\boldsymbol{R}_\ell}$, where $\boldsymbol{R}_\ell$ is the lattice vector at site ℓ, the energy is represented as follows:

$$\begin{aligned} E = \sum_{\nu,\boldsymbol{k}\neq 0} \Bigg[& \frac{\kappa_1}{2}(A^{(\nu)}(\boldsymbol{k}) + \kappa' C^{(\nu)}(\boldsymbol{k})) \\ & \times \left| v_{\boldsymbol{k}}^{(\nu)} + q \frac{iB^{(\nu)}(\boldsymbol{k}) n_{\boldsymbol{k}} + (D(\boldsymbol{k}) + iE(\boldsymbol{k}))\delta_{\nu,x} f(\boldsymbol{k})}{A^{(\nu)}(\boldsymbol{k}) + \kappa' C^{(\nu)}(\boldsymbol{k})} \right|^2 \\ & - \frac{\kappa_1 q^2}{2}(A^{(\nu)}(\boldsymbol{k}) + \kappa' C^{(\nu)}(\boldsymbol{k})) \\ & \times \left| \frac{iB^{(\nu)}(\boldsymbol{k}) n_{\boldsymbol{k}} + (D(\boldsymbol{k}) + iE(\boldsymbol{k}))\delta_{\nu,x} f(\boldsymbol{k})}{A^{(\nu)}(\boldsymbol{k}) + \kappa' C^{(\nu)}(\boldsymbol{k})} \right|^2 \Bigg] + \sum_\ell \varepsilon^0 n_\ell, \end{aligned} \tag{22}$$

with $A^{(\nu)}(\boldsymbol{k}) = 2(1 - \cos k_\nu), B^{(\nu)}(\boldsymbol{k}) = 2\sin k_\nu, C^{(\nu)}(\boldsymbol{k}) = 2\sum_{\nu'}(1 - \cos k_{\nu'})(1 - \delta_{\nu\nu'}), D(\boldsymbol{k}) = (1 - \cos k_x)\delta_{\boldsymbol{k}_\perp,0}, E(\boldsymbol{k}) = \sin k_x \delta_{\boldsymbol{k}_\perp,0}$ and $\kappa' = \frac{\kappa_2}{\kappa_1}$. The quantity $\boldsymbol{k}_\perp$ denotes the vector perpendicular to the x-axis. $N = L^d$ is the number of site, where d is the dimension of the space. We get the interaction in the site representation,

$$\Delta E_{i-i} = -\frac{1}{2} \sum_{\ell(\neq)\ell'} J^v_{\boldsymbol{R}_\ell - \boldsymbol{R}_\ell} n_\ell n_{\ell'}. \tag{23}$$

The interaction constant $J^v_{\boldsymbol{R}_\ell - \boldsymbol{R}_\ell} = \kappa_1 q^2 G^v(\boldsymbol{R}_\ell - \boldsymbol{R}_\ell)$ is given by using Green's function:[17]

$$G^v(\boldsymbol{R}_h) = \frac{1}{N} \sum_{\nu\boldsymbol{k}\neq 0} \frac{B^{(\nu)}(\boldsymbol{k})^2}{A^{(\nu)}(\boldsymbol{k}) + \kappa' C^{(\nu)}(\boldsymbol{k})} \mathrm{e}^{i\boldsymbol{k}\boldsymbol{R}_h}. \tag{24}$$

We note the relation $\sum_h J^v_{\boldsymbol{R}_h} = -\kappa_1 q^2 G^v(0)$. This means that the $\boldsymbol{k} = 0$ component of the interaction is repulsive. The $\ell = \ell'$ term of (23) gives the single site energy, thus modifies the HS energy as $\varepsilon_0 = \varepsilon^0 - \frac{1}{2}\kappa_1 q^2 G^v(0)$.

In Fig. 1 of Ref. 17, we have shown $G^v(\boldsymbol{R}_h)$ for typical cases. The overall features of $G^v(\boldsymbol{R}_h)$ are not so changed even when the parameter κ' has been changed. We put $\kappa' = 0.1$, hereafter. In the two-dimensional case, the interaction is attractive (i.e. ferro-type coupling) for the first site, but is repulsive for sites after the second one in the (1,0) direction. The interaction in the (1,1) direction is attractive, but is very weak. In the three-dimensional case, the coupling constant in the (1,0,0) direction shows similar behaviors to those of the (1,0) direction of the two-dimensional case.

It is attractive for the first site and is repulsive for sites after the second. The interaction in the (1,1,0) and (1,1,1) directions is attractive, but very weak, similarly to the (1,1) direction in the two-dimensional case.

The effective energy of ions is changed by stress field. It will be given by the terms including $n^*_{\boldsymbol{k}} f(\boldsymbol{k})$ or $f^*(\boldsymbol{k}) n_{\boldsymbol{k}}$ in the second term of Eq. (22). The ion-field interaction energy is given by

$$\Delta E_{i-f} = \sum_{(\ell_x, \boldsymbol{l}_\perp)} \kappa_1 q^2 \left\{ \frac{2}{L} \sum_{\ell'_x} f_{\ell'_x} - (f_{\ell_x - 1} + f_{\ell_x}) \right\} n_{\boldsymbol{l}}, \tag{25}$$

which we denote as $(\ell_x, \boldsymbol{l}_\perp) = \boldsymbol{l}$. We note that the term $\frac{2}{L}\sum_{\ell'_x} f_{\ell'_x}$ in (25) originates from the exclusion of $k_x = 0$ in the summation. The effective energy level of HS state at site ℓ is given as $\varepsilon_\ell = \varepsilon_0 + \frac{\partial}{\partial n_\ell}(\Delta E_{i-i} + \Delta E_{i-f})$.

In the above derivation, we have neglected the total volume change and used the periodic boundary condition. When the system is stretched along the x-direction, the lattice displacement is expressed as $v^{(x)}_\ell = \varepsilon R^{(x)}_\ell$. Substituting $v^{(x)}_\ell$ into Eq. (21), the energy change is calculated as

$$\begin{aligned} E^{(V:x)} &= \frac{\kappa_1 a^2 N}{2} \left[\varepsilon - \frac{q}{a} \left\{ \frac{2}{N} \sum_\ell n_\ell + \frac{1}{L} \sum_{\ell_x} f_{\ell_x} \right\} \right]^2 \\ &\quad - \frac{\kappa_1 q^2 N}{2} \left\{ \frac{2}{N} \sum_\ell n_\ell + \frac{1}{L} \sum_{\ell_x} f_{\ell_x} \right\}^2, \end{aligned} \tag{26}$$

where a is the lattice constant, and we put $R^x_\ell = a\ell_x$.

When the total volume is relaxed, the energy of ions decreases as

$$\Delta E^{(V)}_{i-i} = -2d\kappa_1 q^2 N \left(\frac{1}{N} \sum_\ell n_\ell \right)^2. \tag{27}$$

Here we assume the condition $\sum_{\ell_x} f_{\ell_x} = 0$, and sum over d-dimensional directions. The expression (27) just means that we add $\boldsymbol{k} = 0$ term in Eq. (24) as $\boldsymbol{k} \to 0$ limit with $\frac{C^{(\nu)}(\boldsymbol{k})}{A^{(\nu)}(\boldsymbol{k})} = 0$. If the volume of system is not fixed, the $\boldsymbol{k} = 0$ component of the ion-ion interaction is modified as $\sum_h J^v_{\boldsymbol{R}_h} \Rightarrow \kappa_1 q^2 (4d - G^v(0)) = J$, and changed into attractive interaction (i.e. $J > 0$).

In this section, we treat only the lattice distortion of the acoustic type explicitly. However, the interaction with optical modes, which is studied in the previous sections, will also be important. They give rise to the attractive ion-ion interaction of a rather short-range type. Practically, the HS domain hardly becomes stable in the fixed volume case when we consider coupling only with the acoustic modes. In actual calculation, a short range ferro-type coupling term is added to enhance the stability of HS domain by partly compensating the repulsive $\boldsymbol{k} = 0$ component in the full HS phase under the fixed volume condition. This interaction will be ascribed to that caused by the optical modes. The added attractive coupling is shared by pairs within the 5th shell in the two-dimensional case, and 2th shell in the three-dimensional case. The total magnitude of the added interaction is expressed as $\kappa_1 q^2 G^v(0) \times \eta_J$.

In the previous section, we have shown that the relaxation probability is approximated by the Gaussian function: $\omega(\varepsilon_\ell) = \frac{2\pi t_e^2}{\hbar}\frac{1}{\sqrt{2\pi}\Gamma}\mathrm{e}^{-\frac{1}{2}\left(\frac{\varepsilon_\ell - \Delta}{\Gamma}\right)^2}$. The parameters, Δ and Γ are given as $\Delta = \eta_G \Delta_0$ and $\Gamma^2 = \eta_G \Gamma_0^2$. Here, $\Delta_0 = N^{-1}\sum_{\boldsymbol{k}} \hbar\omega_{\boldsymbol{k}} S_{\boldsymbol{k}}$ and $\Gamma_0^2 = N^{-1}\sum_{\boldsymbol{k}}(\hbar\omega_{\boldsymbol{k}})^2 S_{\boldsymbol{k}}$. The effective phonon frequency is expressed as $\omega_{\boldsymbol{k}} = \omega_0\sqrt{A^{(\nu)}(\boldsymbol{k}) + \kappa' C^{(\nu)}(\boldsymbol{k})}$, and the effective Hung–Rhys factor, $S_{\boldsymbol{k}}$, is given as $S_{\boldsymbol{k}} = \frac{\kappa_1 q^2}{\hbar\omega_{\boldsymbol{k}}}\frac{B^{(\nu)}(\boldsymbol{k})^2}{A^{(\nu)}(\boldsymbol{k}) + \kappa' C^{(\nu)}(\boldsymbol{k})}$. The unit of time τ_u has been parametrized as $\frac{2\pi t^2}{\hbar^2\omega_0}\tau_u \equiv e_u$. The time unit of 1 MC is usually chosen to be $\Delta t_{\mathrm{MC}} = 0.25 \times 10^{-3}$. Hereafter we put $\kappa_1 = 1$. The quantity $\eta_G \geq 1$ is the modification factor of the transition probability due to the extra modes which has been already mentioned as the origin of the short range attractive coupling in the previous paragraph.

Let us denote by $\Delta\varepsilon$ the energy decrease of HS state when the phase changes from the LS to the HS domain. The quantity $\omega(\varepsilon_0)/\omega(\varepsilon_0 - \Delta\varepsilon)$ must have considerable magnitude to lead the cooperative conversion. In the present model, quantities Δ, Γ^2 and also the interaction constant which is the origin of $\Delta\varepsilon$ are proportional to q^2. The proportional constants are not so changed by other parameters. If we restrict $\eta_G = 1$, the cooperative conversion can take place almost only when the effective HS energy in the HS domain becomes negative, i.e. HS state in the HS domain is stable compared with the LS state ($\varepsilon_0 - \Delta\varepsilon < 0$). By the quantity η_G, we control parameters of the relaxation probability (Δ and Γ^2) and the interaction constant $\kappa_1 q^2$ separately. When we use $\eta_G > 2{\sim}3$, the photoinduced LS to HS conversion occurs in the parameter region where LS is stable in equilibrium. The coupling with optical modes seems to be necessary in realistic model. But in this review it is modeled by introducing the short range ferro-type interaction (η_J) and the modification factor η_G to simplify the calculation.

In Koshino–Ogawa's model, the ratio $\omega(\varepsilon_0)/\omega(\varepsilon_0 - \Delta\varepsilon)$ can be changed almost freely[41] because the model includes the ion-lattice coupling of the on site type, which contributes to Δ and Γ^2, but does not to $\Delta\varepsilon$.

8. Macroscopic Phase Separation

8.1. *MC results*

In Fig. 11, we show the time evolution of LS fraction for models with the fixed- and the free-volume cases in the two dimensional lattice. We use parameters $\eta_J = 1$ and $\eta_G = 3$. The dot-dashed line is the result for free-volume case. It deceases to almost zero after a incubation time. In this case the $k = 0$ component of the interaction [Eq. (27)] is added to the ion-ion interaction given in Eqs. (23) and (24). The $\boldsymbol{k} = 0$ component of the total interaction becomes attractive. The macroscopic phase separation does not take place, and the LS fraction decreases near to zero. The result is almost equivalent to the one given in the previous section for Koshino–Ogawa's model.

The solid line is the result of the fixed volume case. After the initial rapid decrease, it decreases to a value of intermediate LS fraction. The snapshot of the HS domain growing process will be shown later. HS domains aggregate gradually and grow to macroscopic HS regions coexisting with the LS phase. In Fig. 12 the snapshot of the HS and LS configuration after a

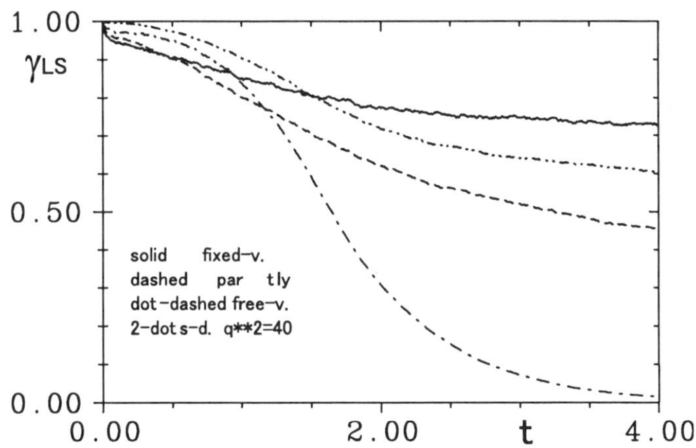

Fig. 11. Time evolution of the low-spin (LS) fraction $\gamma_{\rm LS}$ for the fixed and the free-volume cases in the two-dimensional lattice with 100×100 sites. The solid line is calculated for the fixed volume case with $Ie_u = 1.4$. The dot-dashed line is calculated for the free volume case with $Ie_u = 3$. The dashed line is calculated by adding a half of Eq. (27) in the ion-ion interaction for $Ie_u = 2.2$. Other parameters are $q^2 = 20$, $\kappa' = 0.1$, $\varepsilon_0 = 105$, $e_u = 2 \times 10^3$, $\eta_G = 3$, and $\eta_J = 1$ for these lines. The two-dots-dashed line is calculated for $q^2 = 40$, $\kappa' = 0.1$. $\varepsilon_0 = 100$, $Ie_u = 3$ with $e_u = 5.0 \times 10^4$, $\eta_G = 2$ and $\eta_J = 1.5$. The sigmoid nature of the initial stage is conspicuous because the ferro-type short range interaction is relatively large, and also the ion-lattice coupling factor is large.

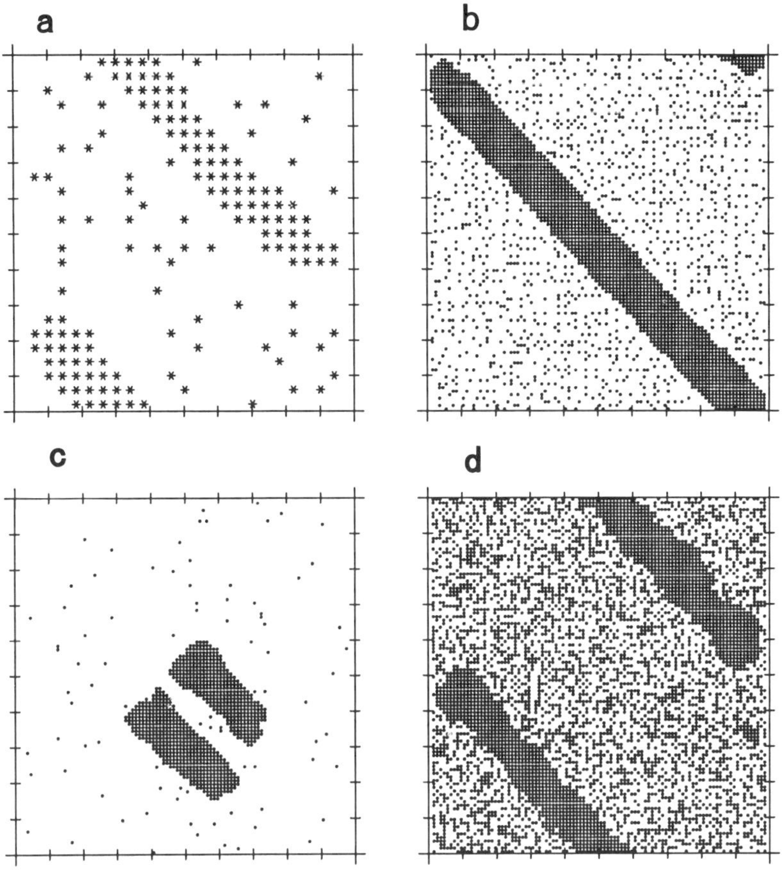

Fig. 12. The snapshot of the ion configuration after long period $t = 40$ ($e_u = 2 \times 10^4$). Dots indicate the HS sites. (a) is the result for lattice with 25×25 and (b) is one for the lattice with 100×100. Other parameters for (a) and (b) are the same to those of the solid line in Fig. 2, $I = 1.5 \times 10^{-3}$. (c) is the result for very weak light intensity ($I = 0.3 \times 10^{-3}$) and (d) is for very strong intensity ($I = 15 \times 10^{-3}$).

very long period is shown. We can observe HS and LS domains, which have, respectively, dense and dilute HS concentration. The HS domain has almost full HS configuration. Figure 12a is calculated for $L = 25$ and Fig. 12b for $L = 100$ under the condition in which other parameters are equal. The size of the phase separation pattern becomes almost same when it is normalized by the lattice size. This indicates that the pattern is not the microscopic long period one but the macroscopic one determined by the macroscopic boundary condition. As noted previously, the interaction has a rather large repulsion term in the second neighbor site for the (1,0) direction. Reflecting this fact, the HS domain has a stripe structure in the (1,1) direction to avoid the repulsive coupling.

In the solid line of Fig. 11, the incubation period is not distinguishable because the excitation intensity is selected to be strong to obtain the large domain in the restricted calculation time. The dashed line has an incubation period and also the macroscopic phase separation because the excitation intensity is not so strong. We calculate the dashed line by adding half of the volume term (i.e. half of Eq. (27)) to the interaction. By the two-dots-dashed line we show more strong coupling case $q^2 = 40$ and relatively strong short range attractive interaction case. The incubation period, which corresponds to the nucleation period of HS domain centers, becomes more conspicuous.

8.2. *Phase coexistence condition*

As discussed in the appendix, the equilibrium ratio of the coexistence phase will be determined by the stationary condition of the following function.

$$G(\varepsilon_1, \varepsilon_2, x_1, x_2, \gamma, c) = \sum_i^2 c_i \left[W(\varepsilon_i) - \varepsilon_i x_i + (\varepsilon_0 - J_v(\gamma - 1)) x_i + \frac{J}{2}(x_i - 1)^2 \right] + \frac{J_v}{2}\gamma^2. \tag{28}$$

The function $W(\varepsilon)$ is defined by Eq. (30) in the Appendix, and c_i is the fraction of the phase characterized by the local LS fraction x_i. The constant $J = \kappa_1 q^2(4d - G(0)(1 - \eta_J))$ is the coefficient of the $k = 0$ component in the uniform phase with free volume condition. The constant $J_v = \kappa_1 q^2 4d$ denotes the coefficient of the volume striction effect. The total LS fraction of the system is given as $\gamma = cx_1 + (1 - c)x_2$, with $c = c_1$.

Figure 13 shows the LS fraction as a function of $\log_{10} Ie_u$. The solid line is the result obtained by using Eq. (28) and the circles are obtained by the MC calculation for 100×100 lattice. They show good agreement in the stronger intensity region, while large deviation in the weaker intensity region. The number of stable solutions decreases to one for very small I cases in Eq. (28). This indicates the uniform phase. But in the intensity range shown in Fig. 13, we have two stable solutions. The HS domain has not appeared for intensity $Ie_u < 1$ during the calculation time for the lattice of the present size, 100×100. Calculations with larger size and also with larger time period seem to be necessary for weaker intensity cases.

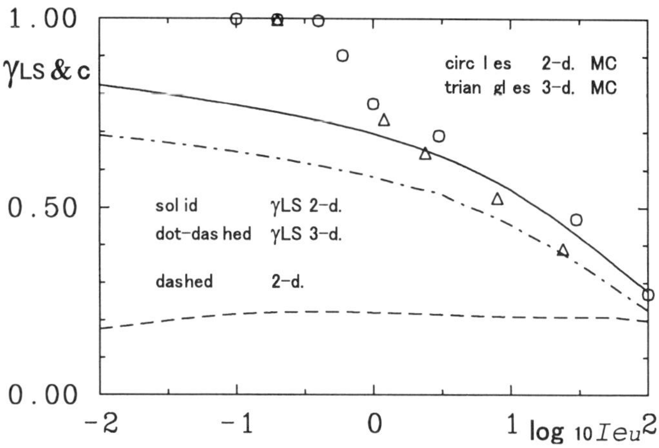

Fig. 13. The LS fraction as a function of the light intensity. The horizontal axis is the logarithm of the intensity, $\log_{10} Ie_u$, with $e_u = 2 \times 10^3$. The solid line is the total LS fraction in the coexistence phase estimated by Eq. (28) for the two-dimensional case, and circles are calculated by MC method for 100×100 lattice. The dashed line is the fraction of HS domain, c, estimated by Eq. (28). The dot-dashed line is the total LS fraction estimated by Eq. (28) for the three-dimensional case, and triangles are calculated by MC method for $50 \times 50 \times 50$ lattice. Parameters for the two-dimensional case are the same as those for the solid line in Fig. 12. Parameters for the three-dimensional case are $q^2 = 20$, $\kappa' = 0.1$, $\varepsilon_0 = 60$, $\eta_G = 2$, and $\eta_J = 1$.

With the dashed line, we show the fraction c of the HS domain. It almost does not change when the intensity increases. We have already noted that the HS concentration in the HS domain is always nearly unity. Therefore the decrease of the LS fraction is caused by the increase of HS concentration in the LS domain. This result is consistent with the snapshot in Fig. 12. The sizes of HS domains in Figs. 12b and 12d are roughly equal, and HS concentration in the LS phase is very dense in Fig. 12d. The fraction of HS domain in Fig. 12c is also not so small even though the light intensity is very weak. The change of the domain pattern is very slow in this case because the effective temperature $T_{\text{eff}} = \frac{\omega I}{\omega + I}$ becomes low.

9. De-pinning of Volume Striction

9.1. *Model of de-pinning*

As demonstrated in the previous section, we have obtained the macroscopic phase separation of HS and LS domains when the volume of the system is fixed. The size of domains is comparable to that of the system, but their growing process is gradual. At the intermediate time period, the individual domain changes its size discontinuously. But when the change of LS fraction is averaged over the whole sample, it becomes continuous in the large system. The step-like decrease (v) has not been reproduced.

The abrupt decrease of the LS fraction will be caused by the abrupt phase conversion of macroscopic size. In the macroscopic phase separation regime, the lattice stretching in the HS domain is relieved by the contraction in the LS domains. Therefore large local volume strain is accumulated in the sample. Usually, the smooth volume change of the crystal is prevented by various pinning centers such as crystal defects. The pinning will be released when the accumulated volume strain force overcomes the pinning force, and abrupt change of volume will appear. At present we expect the step-like decrease to be ascribed to the de-pinning of the volume striction. In fact, the step-like jump in the experiment occurs when the LS fraction value decreases to some values which are common and not dependent on the intensity of excitation light.[4] This may indicate the habit of crystal.

Now we make a simplified calculation taking into account the de-pinning effect. Let us divide the system into two regions, I ($0 \le \ell_x \le (L-2)/2$) and II ($(L-2)/2 < \ell_x \le L-2$), then set an artificial critical fraction n_{cr}. The energy of HS state in region I is chosen to be a little lower $\varepsilon_{0\mathrm{I}} = \varepsilon_0 - \Delta\varepsilon_0$ with $\Delta\varepsilon_0 = \kappa_1 q^2 G^v(0) \times 0.05$. When the LS fraction of I region n_{I} decreases to the critical value n_{cr} after illumination of light, we promptly decrease the energy of I region to $\varepsilon_{0\mathrm{I}} = \varepsilon_0 - \Delta\varepsilon_0 - \kappa_1 q^2 G^v(0)$ by stretching the lattice to mimic the de-pinning of the volume striction. This local stretching of the lattice is introduced by controlling the stress field. The light is further irradiated, and if the LS fraction of II region n_{II} decreases to the critical value, the energy of the HS in the region II is also decreased to $\varepsilon_{0\mathrm{II}} = \varepsilon_0 - \kappa_1 q^2 G^v(0)$. At the same time the energy of the region I is decreased as $\varepsilon_{0\mathrm{I}} = \varepsilon_0 - \Delta\varepsilon_0 - \kappa_1 q^2 G^v(0) \times (1 + n_{\mathrm{II}})$. The term including n_{II} is added to avoid the energy increase caused by the HS state increase in region II.

In the calculation, the stress field f_{L-1} is adjusted to fulfill the condition $\sum_{\ell_x} f_{\ell_x} = 0$. In expression (25) for the ion-stress field interaction, the second term in the brace expresses that the HS energy on the ℓ_x layer decreases when the lattice spacing of its sides is stretched by the stress field. But the first term $\frac{2}{L}\sum_{\ell_x} f_{\ell_x}$ appears because we have used the periodic boundary condition. If the uniform stress $f_{\ell_x} = f$ is applied, the energy of the HS is not changed, since the lattice spacing cannot be changed under the periodic boundary condition. In actual case the uniform stress field can stretch the length of the sample and decrease the HS energy. We need to use the free boundary condition to describe such an effect. However the ion-ion interaction has some complicated form due to the surface effect under the free boundary condition. This makes the calculation tedious. Therefore we tentatively simulate the free boundary condition by introducing an artificial layer which relieves the stress field. The stress filed f_{L-1} is chosen to satisfy the condition $\sum_{\ell_x} f_{\ell_x} = 0$, as $f_{L-1} = -\sum_{\ell_x \neq L-1} f_{\ell_x}$. The HS energy on the 0- and the $(L-2)$-th layers becomes very high. We assume that the spin state is always LS on these layers.

In Fig. 14 we show the time evolution of LS fraction for the two-dimensional case. The solid line is the calculated result with de-pinning effect. After the initial rapid decrease, it shows a gradual decrease. However at about $t = 1.4$, the number of HS in the region I reaches a critical value n_{cr} which is set to 0.3 in the present calculation. The HS energy is suddenly decreased, then the LS fraction in region I decreases steeply. The intensity of light is strong in this case, then the two regions successively convert to the HS domain while in the dot-dashed line case, only one region converts. The dashed line is the result of the volume of the sample being fixed. The two-dots-dashed line is the result for very weak intensity, probably below the threshold intensity.

The snapshot for the case given by the solid line is shown in Fig. 15. At time $t = 0.1$, which is just after the initial rapid decreasing period of LS fraction, many small nucleation centers of HS appear, and they grow

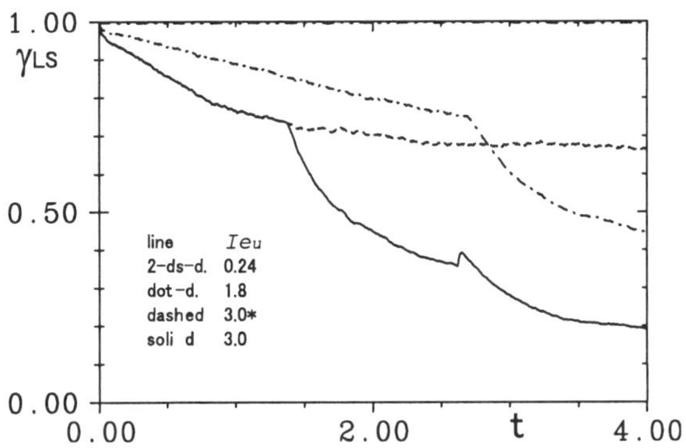

Fig. 14. Time evolution of the low-spin (LS) fraction γ_{LS} for the model with coupling to the volume striction. The two dimensional lattice with 100×100 sites is used. The solid line is calculated for the excitation intensity $Ie_u = 3.0$, the dot-dashed line is for $Ie_u = 1.8$, the two-dots-dashed line is for $Ie_u = 0.24$, and the dashed line for $Ie_u = 3.0$ without de-pinning effect, $e_u = 2 \times 10^3$. Parameters are $\kappa' = 0.1$, $q^2 = 20$, $\varepsilon_0 = 50$, $\eta_J = 1$, $\eta_G = 2$ and $n_{cr} = 0.3$. The two-dots-dashed line which has about 0.997 almost overlaps on the upper axis. The nucleation center has not appeared within the calculation time in this case.

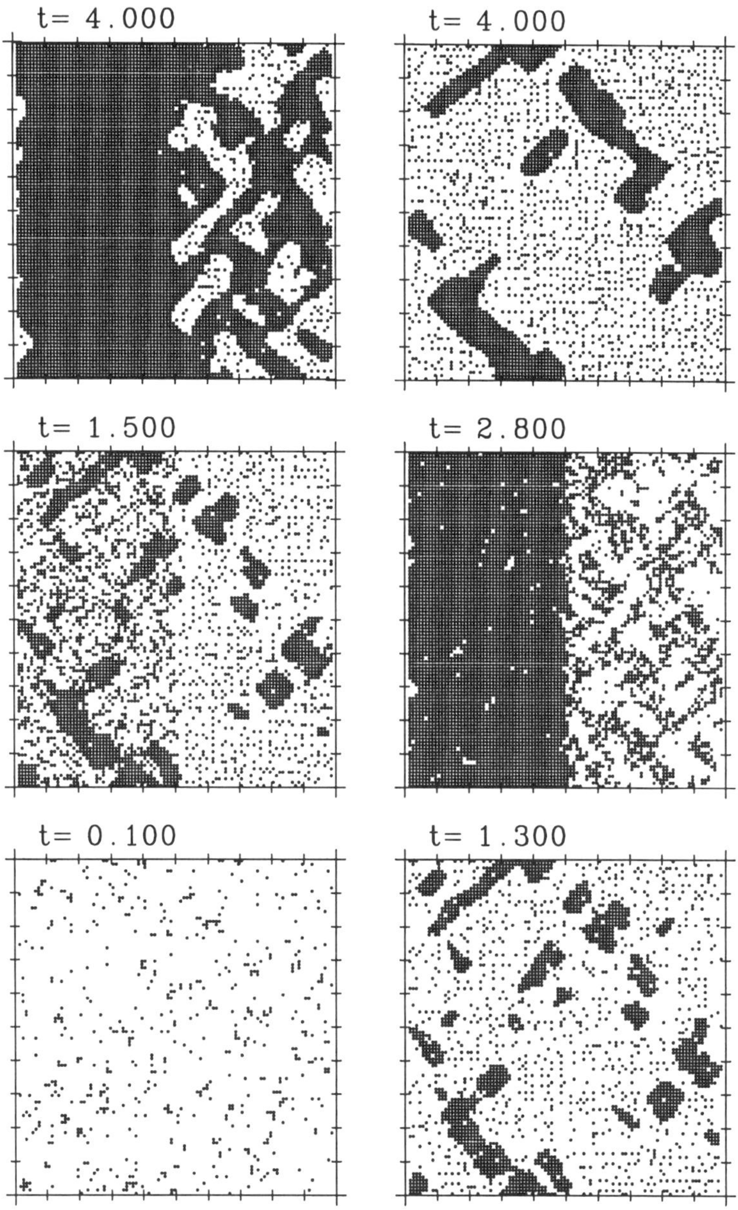

Fig. 15. The snapshot of the ion configuration at various time. Parameters are the same as those of the solid line in Fig. 14. Numbers in the figure indicate the time, whose unit is the same as that of Fig. 14. The right of the top panel is the result without the de-pinning effect (dashed line in Fig. 14).

to domains at $t = 1.3$. The growing process of domains is very similar to that of Koshino–Ogawa's model in the early stage. But their growing speed becomes slow and the total area of HS domains gradually approaches to a saturated value when the volume of the sample is fixed. At about $t = 1.4$, the HS energy in region I is suddenly decreased as the LS fraction reaches the critical value. Just after the time, at $t = 1.5$ the HS fraction steeply increases in region I. The region becomes almost full HS state. The HS fraction in region II gradually increases and the de-pinning of region II takes place at about $t = 2.6$. HS domains in region II gradually grow and become a connected one after a long period. In other parameter cases, we sometimes find that HS domains disappear in region II due to the energy increase caused by the volume strain induced by a HS increase in region I.

At the top of the right panel, we also show the snapshot at $t = 4$ for the fixed volume case without the de-pinning effect. A domain pattern which is grown from one at $t = 1.3$ is observed. The large but isolated domains gradually aggregate and grow to the macroscopic striped pattern noted in Fig. 12 after a very long time, for example, at about $t \sim 40$ for the present parameter case.

At the end of this section, we must point out the time scale of the calculation. The time scale of the microscopic process, which will be characterized by the initial rapid decrease of LS fraction, is chosen to be very small compared with the one for later processes. However in actual experiments, the time scale of the microscopic process will be in the order of 10^{-12} sec, while the domain growing process has a time scale of 1 sec. When we consider this time ratio, the ratio used in the present MC calculation is much smaller than the actual value. However the ratio of both process is already large in the calculation. Hence, we do not expect essential features to be modified even when the ratio is huge.

9.2. *Three-dimensional case*

In Fig. 16 we show the time evolution of LS fraction for the three-dimensional case. The overall features are similar to those of the two-dimensional case. We have assumed that the short range ferro-type coupling is shared by pairs within the second nearest shell.

The dashed line is the calculated result when the volume of the sample is fixed. The growing process of the individual domain is not continuous, but the change of LS fraction becomes gradual when it is calculated in the large sample as already noted for the two-dimensional case. The abrupt

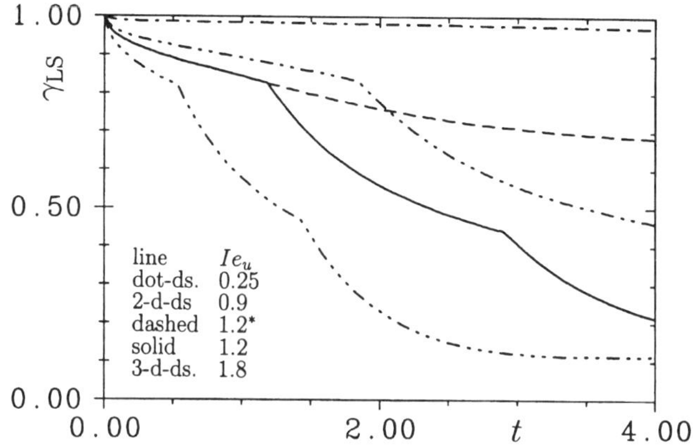

Fig. 16. Time evolution of the low-spin (LS) fraction $\gamma_{\rm LS}$ for three-dimensional lattice with $50 \times 50 \times 50$ sites. The solid line is calculated for the excitation intensity $Ie_u = 1.2$, the dot-dashed line is for $Ie_u = 0.25$, the two-dots-dashed line is for $Ie_u = 0.9$, the three-dots-dashed line for $Ie_u = 1.8$ and the dashed line for $Ie_u = 1.2$ without de-pinning effect, $e_u = 2 \times 10^3$. Parameters are $\kappa' = 0.1$, $q^2 = 20$, and $\varepsilon^{\rm allHS} = 60$. $\eta_J = 1$, $\eta_G = 2$ and $n_{\rm cr} = 0.2$. The short range ferro-type coupling is shared by pairs within the n.n.n. shell.

jump is also not reproduced in the three-dimensional case, when we do not consider the de-pinning of the volume striction. The macroscopic domain pattern has the (111) layer type structure to avoid the repulsion on far neighbor sites in the (1,0,0) direction.

The solid line is the result with de-pinning effect. The intensity of light is strong in this case, then the two regions successively convert to the HS domain while in the two-dots-dashed line case, only one region converts. The dot dashed line is the very weak intensity case. Small domains begin to appear at about $t \sim 4$ even in this intensity case, so the correct threshold value will be smaller than $Ie_u = 0.24$.

The LS fractions obtained after a large period of time are shown by the triangles in Fig. 13. They are reproduced by the dot-dashed line calculated by Eq. (28) in the stronger intensity region, but are not reproduced in the weaker intensity region.

10. Summary and Discussion

10.1. *Cooperative LS to HS transition*

In this report we review the MC method to simulate the LS $\rightarrow$ HS conversion process, in which the ion-ion interaction is mediated by lattice distortion. When the time unit of 1 MC step is made small compared with the time scale characteristic of the elementary process of the original equation, $\tau_{\rm rap}$ in the present system, we can obtain a reliable result for the equation given by the differential form with respect to the time.

The model proposed by Koshino and Ogawa in Ref. 14 was examined in the two- and three-dimensional cases. The characteristic features observed in experiments; (i) the existence of the threshold intensity, (ii) the existence of the incubation period and (iii) steep decrease of the LS fraction after the incubation period were reproduced in the case of large Huang–Rhys factor. Therefore the conclusions obtained in Ref. 14 were not changed in a qualitative sense even though the fluctuation effects are important. The features (i)–(iii) can be observed rather in the wide intensity range in the present calculation because the incubation time corresponds to the nucleation time of the HS domain. However, the features are expected only for the restricted narrow range of excitation intensity when the theory neglecting the fluctuation effect is adopted.

Formal similarity with the usual domain formation which is driven by the energy fluctuation due to the temperature effect was followed up by deriving the Fokker–Planck equation. It was pointed out that the repetition of the excitation and de-excitation processes acts as the driving power to surmount the barrier of the domain nucleation. Following the usual domain formation theory, the nucleation rate of the HS domain was estimated. It grossly accounts for the characteristic time scales of the present cooperative LS $\rightarrow$ HS conversion in the three-dimensional case. The agreement is not so good in the two-dimensional case.

Some miscellaneous problems are considered. Application of the Zeeman field decreases the incubation time. Intermittent excitation also decreases the incubation time when the excitation interval is larger than the time scale of the initial rapid decrease of LS fraction, $\tau_{\rm rap}$. When the excitation light is interrupted after the HS conversion sufficiently proceeds, the relaxation rate to HS is very small compared with that observed in early stage interruption.

In Koshino–Ogawa's model, the coexistence of the LS and HS regions was observed during the conversion process, but macroscopic phase separation (iv) was not reproduced. The step-like decrease of LS fraction (v) was also not reproduced.

We present a model in which the interaction between ions and the acoustic phonons is included to study the volume striction effect. The effective ion-ion interaction is derived for the model and the MC method is developed to simulate the cooperative LS $\rightarrow$ HS conversion by light irradiation.

The model leads the macroscopic phase separation of HS and LS domains and reproduces the characteristic feature (iv) observed in experiment. A quantity which describes the coexistence condition of the HS and LS phases in the fixed volume case is introduced. It can account for the ratio of both phases obtained by the MC simulation. The HS concentration in the HS domain is expected to be almost unity, while that in the LS domain increases gradually as the light intensity increases. The step-like conversion to HS, (v) observed in the experiment was not reproduced as the intrinsic behavior of the model. It is ascribed to the de-pinning effect of the volume striction.

The model shows as an intrinsic properties the features; (i) the existence of the threshold intensity (ii) the existence of the incubation period and (iii) steep decrease of the LS fraction after the incubation period. However we should note that the sigmoid character indicated by these facts is obtained in the experiment after smoothening the abrupt jumps. So we should include in the calculation the detailed process of de-pinning to compare with the experiment. Such a study is left for the future.

Recently, Ogawa *et al.* have observed the photoinduced LS to HS transition in fine particles. The conversion process becomes gentle when it is compared to one in the bulk crystal case.[6] Abrupt volume change of individual particles will be averaged. We think that the interpretation of the abrupt conversion by the de-pinning effect is not so unreasonable.

As noted at the end of Sec. 7, we need to modify the relaxation factor and the ion-ion interaction term so that the cooperative conversion takes place in the parameter region where the LS is stable in equilibrium. This indicates that optical modes will also contribute to the ion-ion interaction in the actual process. Such calculations are also left for future papers.

Tayagaki *et al.* have suggested that the photoinduced HS state is accompanied by the JT distortion.[7] The model based on the JT effect has been presented for the temperature induced transition at $T_C \sim 120\,\mathrm{K}$ by Kambara *et al.*,[13,33,34] though the symmetry change is not observed in the high temperature HS phase.[1] The JT effect in the photoinduced HS state seems to be an interesting issue to clarify the microscopic origin of the cooperative LS $\rightarrow$ HS conversion.

In Sec. 8, we have proposed a quantity which describes the phase coexistence condition in the photoinduced conversion process. The quantity has not direct relation to a potential term obtained in Sec. 5. In the single phase state, the potential term has a character; it should decrease with time, $\dot{v} < 0$. The quantity given in Sec. 8 does not have such property, but has only the stationary property. In fact it can take both the local minimum and the local maximum in the non-interacting case. Further studies are needed whether we can find a function which describes the global stability in more general way in the photoinduced conversion process.

In this review we have restricted the calculation to the $T = 0$ case since experiments are carried out at very low temperatures. But studies at $T \neq 0$ will be also necessary because the experiments are done actually at finite temperatures. It may appear a quantity which describes generally the phase coexistence condition under the thermal and the photoinduced transitions.

10.2. *Characteristics of photoinduced phase transitions (PIPTs)*

Before finishing the present paper, we briefly review general and universal features and remarks on PIPTs[42,43] to stress the special characteristics of the photoinduced LS $\rightarrow$ HS transitions.

- "Phase transitions" are originally defined under the thermally equilibrium condition. On the other hand, the PIPT is a phenomenon in a nonequilibrium situation. In this sense, the PIPT is also called "nonequilibrium phase transitions." Since the term "phase" means a global, macroscopic (not local, microscopic) state of a matter, photoinduced changes in one (or a few) molecule(s) are out of the category of the PIPT.
- In order to yield the PIPTs, *multistability* should be inherently built in the relevant materials. In other words, several local minima are necessary in their free-energy functional.
- In the study of the PIPTs, spatiotemporal dynamics *during* the phase transition is of special importance. How does a local, microscopic change induced by photoexcitation develop to mesoscopic domains and to a global, macroscopic phase? This is a main problem to be solved.
- In the PIPT phenomena, *cooperativity* should always play an essential role. This cooperativity results from nonlocal interactions between molecules (sites) and results in huge *nonlinearity* as a function of external stimulations or time. This nonlinearity also has a great potentiality in applications.
- In usual PIPT phenomena, there coexist many different time scales. Competition among several time scales results in complex and novel dynamics of PIPTs.

- The PIPT dynamics are strongly sensitive to the characters of a reservoir. This is a consequence of nonequilibrium of the PIPTs.

The above items are universal characteristics of the PIPT phenomena. Another important purpose of the PIPT study is to search new phases which cannot be accessed by thermal fluctuations; we should make the best use of the higher energy of photons in comparison to the thermal energy. In other words, we are trying to find the new ground state ("true vacuum") of matters starting from a "false vacuum" (which has been considered as the true vacuum) via photoexcited high-energy states. From the application viewpoint, we have another advantage in the use of photon; the *remote* control of dramatic change of the state of matters is possible in the PIPTs.

We shall mention here several types of the PIPTs. There are two directions of phase change in the PIPTs.

(1) Under a suitable condition, the PIPT is possible from a metastable state (false vacuum) to the absolutely-stable one (true vacuum) induced by pulse-light irradiation. In this case, the phase change can proceed *automatically*, dissipating the energy to an environmental reservoir even after the light irradiation is turned off.
(2) On the other hand, a *photoinduced domain injection* is expected from the absolutely-stable state to a metastable one under the continuous (cw) irradiation (energy injection).

In this article, we concentrate on the latter case (2), that is, the photoinduced LS $\rightarrow$ HS state conversion takes place only when the cw light is irradiated to materials. After the light irradiation ceases, the created HS state automatically changes back to the original LS state. In this sense, the photoinduced LS $\rightarrow$ HS state conversion is one of typical nonequilibrium, cooperative phenomena. The former case (1) has been extensively studied in terms of the photoinduced domino processes, whose dynamics differ greatly from that of case (2). See and compare with Refs. (44–52).

Acknowledgments

The authors would like to thank K. Nasu for stimulating discussions, and also thank Y. Ogawa, S. Koshihara, T. Tayagaki, K. Tanaka and M. Kamata for enlightening discussions about experimental facts, and to K. Boukheddaden for valuable comments. The present work started as collaboration with K. Koshino, to whom we thank. This work was partly supported by the Grant-in-Aids of Japanese Ministry of Education, Science,

Sports and Culture, the priority area (B) [Photoinduced Phase Transition and Their Dynamics]. The numerical computation was partly performed in the Computer Center of Tohoku University and the Supercomputer Center of Institute for Solid State Physics (University of Tokyo).

Appendix: Phase Coexistence Condition

As discussed in previous papers, the LS fraction x for the steady state satisfies the equation,

$$\omega(\varepsilon)(1-x) - Ix = 0, \tag{29}$$

where $\varepsilon = \varepsilon_0^* + J(x-1)$ is the effective energy level of HS state. The change of the HS state energy due to the volume striction term should be included in ε_0^*. We assume that the effect is given by the following expression: $\varepsilon_0^* = \varepsilon_0 - J_v(\gamma - 1)$. Here γ is the LS fraction of the total system, which is determined self-consistently later.

Let us define a function

$$W(\varepsilon) = \int^{\varepsilon} \mathrm{d}\varepsilon' \frac{\omega(\varepsilon')}{I + \omega(\varepsilon')}. \tag{30}$$

Next we define a function,

$$F(\varepsilon, x : \varepsilon_0^*) = W(\varepsilon) - \varepsilon x + \varepsilon_0^* x + \frac{J}{2}(x-1)^2. \tag{31}$$

The condition $\left(\frac{\partial F(\varepsilon,x)}{\partial \varepsilon}\right)_x = 0$ gives Eq. (29) and the condition $\left(\frac{\partial F(\varepsilon,x)}{\partial x}\right)_\varepsilon = 0$ gives the relation $\varepsilon = \varepsilon_0^* + J(x-1)$. Therefore the steady state solution is determined by the stationary condition of $F(\varepsilon, x : \varepsilon_0^*)$ with respect to the changes of ε and x. Usually equations have three solutions for a given ε_0^*. Two of them are stable solutions, and are denoted as x_1 and x_2. But we note that equations have only one solution when ε_0^* is high and/or I is extremely weak or strong cases.

We consider a case in which the sample has two coexisting phases. The local LS fraction in the phase is denoted by x_i, and the fraction of the phase in the sample is denoted by c_i, where $i = 1,\ 2$. We put $c_1 = c$, and then $c_2 = 1 - c$.

Here, let us define a quantity

$$G(\varepsilon_1, \varepsilon_2, x_1, x_2, \gamma, c) = \sum_i c_i[F(\varepsilon_i, x_i : \varepsilon_0 - J_v(\gamma - 1))] + \frac{J_v \gamma^2}{2}. \tag{32}$$

The stationary condition with respect to the change of ε_i and x_i gives Eq. (29) and the relation $\varepsilon_i = \varepsilon_0 + J(x_i - 1) - J_v(\gamma - 1)$, respectively.

The condition $\frac{\partial G}{\partial \gamma} = 0$ leads the relation $\gamma = c_1 x_1 + c_2 x_2$, which gives the LS fraction of the total system. The condition $\frac{\partial G}{\partial c} = 0$ gives the relation $F(\varepsilon_1, x_1 : \varepsilon_0 - J_v(\gamma - 1)) = F(\varepsilon_2, x_2 : \varepsilon_0 - J_v(\gamma - 1))$. We may expect the condition for the coexistence of two phases to be determined by the stationary condition of the function G.

References

1. For comprehensive review see for example, P. Gütlich, A. Hauser and H. Spiering, *Angew. Chem. Int. Ed. Engl.* **33**, 2024 (1994).
2. S. Decurtins, P. Gütlich, K. M. Hasselbach, A. Hauser and H. Spiering, *Inorg. Chem.* **24**, 2174 (1985).
3. A. Mino, Y. Ogawa, S. Koshihara, C. Urano and H. Takagi, *Mol. Cryst. Liq. Cryst.* **314**, 107 (1998).
4. Y. Ogawa, S. Koshihara, K. Koshino, T. Ogawa, C. Urano and H. Takagi, *Phys. Rev. Lett.* **84**, 3181 (2000).
5. Y. Ogawa, A. Oiwa, S. Koshihara and H. Munekata, *Phase Transitions* **74**, 35 (2001).
6. Y. Ogawa, S. Koshihara, M. Takesada and T. Ishikawa, *Phase Transitions* **75**, 683 (2002).
7. T. Tayagaki and K. Tanaka, *Phys. Rev. Lett.* **86**, 2886 (2001).
8. K. Tanaka and T. Tayagaki, *Phase Transitions* **75**, 689 (2002).
9. R. Jacobi, H. Spiering and P. Gütlich, *J. Phys. Chem. Solid* **53**, 267 (1992).
10. For comprehensive review of photoinduced phase transition, see for example, H. Mizouchi, H. Ping and K. Nasu, *Phase Transitions* **74**, 1 (2001).
11. E. Buhks, G. Navon, M. Bixon and J. Jortner, *J. Am. Chem. Soc.* **102**, 2918 (1980).
12. A. Hauser, A. Vef and P. Adler, *J. Chem. Phys.* **95**, 8710 (1991).
13. See for example, T. Kambara, *J. Phys. Soc. Jpn.* **49**, 1806 (1980).
14. K. Koshino and T. Ogawa, *J. Phys. Soc. Jpn.* **68**, 2164 (1999).
15. K. Binder and D. Stauffer, *Adv. Phys.* **25**, 343 (1976).
16. O. Sakai, T. Ogawa and K. Koshino, *J. Phys. Soc. Jpn.* **71**, 978 (2002).
17. O. Sakai, M. Ishii, T. Ogawa and K. Koshino, *J. Phys. Soc. Jpn.* **71**, 2052 (2002).
18. O. Sakai, T. Ogawa and K. Koshino, *Phase Transitions* **75**, 743 (2002).
19. M. Sorai and S. Seki, *J. Phys. Chem. Solids* **35**, 555 (1974).
20. See for example, J. S. Langer: *Solids Far From Equilibrium*, ed. C. Godrèche (Cambridge University Press, Cambridge, 1992), p. 297.
21. H. Romstedt, A. Hauser and H. Spiering, *J. Phys. Chem. Solids* **59**, 265 (1998).
22. T. Kawamoto and S. Abe, *Phase Transitions* **75**, 753 (2002).
23. X. J. Liu, Y. Moritomo, K. Kawamoto, A. Nakamoto and N. Kojima, *Phys. Rev.* **B67**, 012102 (2003).
24. X. J. Liu, Y. Moritomo, T. Kawamoto, A. Nakamoto and N. Kojima, *J. Phys. Soc. Jpn.* **72**, 1615 (2003).

25. H. Romstedt, H. Spiering and P. Gütlich, *J. Phys. Chem. Solids* **59**, 1353 (1998).
26. B. Hoo, K. Boukheddaden and F. Varret, *Eur. Phys. J.* **B17**, 449 (2000).
27. K. Boukheddaden, F. Varret, S. Salinke, J. Linares and E. Codjovi, *Phase Transitions* **75**, 733 (2002).
28. A. Ishikawa, T. Ogawa and V. I. Sugakov, *Phys. Rev.* **B64**, 144301 (2001).
29. M. Nishino, K. Yamaguchi and S. Miyashita, *Phys. Rev.* **B58**, 9303 (1998).
30. M. Nishino and S. Miyashita, *Phys. Rev.* **B63**, 174404 (2001).
31. See for example, A. J. Leggett, S. Chakravarty, A. T. Dorsey, M. P. A. Fisher, A. Garg and W. Zwerger, *Rev. Mod. Phys.* **59**, 1 (1987).
32. Y. Kayanuma, *J. Phys. Soc. Jpn.* **53**, 108 and 118 (1984).
33. T. Kambara, *J. Chem. Phys.* **70**, 4199 (1979).
34. N. Sasaki and T. Kambara, *Phys. Rev.* **B40**, 2442 (1989).
35. In a strict sense, the energy of HS state should be given as $\varepsilon_0 - \kappa q^2 (1-G(0))(1-x)$ as noted in Sec. 2. However, in this place we use the simplified form following Koshino–Ogawa by not removing the self-interaction part.
36. N. Nagaosa and T. Ogawa, *Phys. Rev.* **B39**, 4472 (1989).
37. T. Ogawa and N. Nagaosa, in *Relaxations of Photoinduced Structural Phase Transitions*, ed. K. Nasu (Springer, 1997), p. 45.
38. Strictly speaking, we have assumed that the Da^2 is so large that x_ℓ can be uniform in space.
39. Honestly, the changed range of excitation intensity is very small in the calculation. One will need calculations with incubation time of different orders of magnitude to lead to a correct conclusion. To study the very long incubation time cases, calculations with very large lattice size are necessary because the critical size of droplet becomes large.
40. This will support the present MC simulation method to gain the time evolution of the master equation (1), which is given primarily by the time differential form.
41. We can have cooperative conversion in the stable LS cases even when we put $\eta_G = 1$, if we use extremely large q, $q^2 \sim 100$ and choose parameters within a very limited range.
42. T. Ogawa, *Mol. Cryst. Liq. Cryst.* **361**, 199 (2001).
43. T. Ogawa, *Phase Transitions* **74**, 93 (2001).
44. K. Koshino and T. Ogawa, *J. Phys. Soc. Jpn.* **67**, 2174 (1998).
45. K. Koshino, and T. Ogawa, *Phys. Rev.* **B58**, 14804 (1998).
46. K. Koshino and T. Ogawa, *J. Korean Phys. Soc.* **34**, S21 (1999).
47. K. Koshino and T. Ogawa, *J. Phys. Chem. Solids* **60**, 1915 (1999).
48. V. V. Mykhaylovskyy, V. I. Sugakov, K. Koshino and T. Ogawa, *Solid State Commun.* **113**, 321 (1999).
49. K. Koshino and T. Ogawa, *Phys. Rev.* **B61**, 12101 (2000).
50. V. V. Mykhaylovskyy, T. Ogawa and V. I. Sugakov, *Phys. Stat. Sol. (b)* **231**, 222 (2002).
51. T. Ogawa, *Phase Transitions* **75**, 673 (2002).
52. V. V. Mykhaylovskyy, T. Ogawa, and V. I. Sugakov, *Phys. Stat. Sol. (b)* **238**, 149 (2003).

CHAPTER 4

FEMTOSECOND DYNAMICS OF THE PHOTO-INDUCED LATTICE REARRANGEMENTS IN QUASI-ONE-DIMENSIONAL HALOGEN-BRIDGED PLATINUM COMPLEXES

Tohru Suemoto

Institute for Solid State Physics, The University of Tokyo
Kashiwanoha 5-1-5, Kashiwa-shi, 277-8581, Japan
E-mail: suemoto@issp.u-tokyo.ac.jp

Shinichi Tomimoto

Correlated Electron Research Center (CERC)
National Institute of Advanced Industrial Science and Technology (AIST)
Tsukuba, Ibaraki 305-8562, Japan

Taira Matsuoka

Ricoh Company, Ltd. Omori Office, 1-3-6 Nakamagome
Ota-ku, Tokyo 143-8555, Japan

Recent developments in femtosecond dynamics of the photoexcited state in quasi-one-dimensional platinum complexes $[Pt(en)_2][Pt(en)_2X_2](ClO_4)_4$ with X = Cl, Br and I are reviewed. The experimental results of time-resolved luminescence spectroscopy based on up-conversion technique are presented and analyzed in terms of a theory of wave-packet motion. An attempt to make a movie of wave-packet motion is mentioned. In Sec. 1, a brief introduction to the dynamics of the excited states in quasi-one-dimensional platinum complexes is given. It is stressed that this system can be a good model system for investigating the photo-induced structural phase transition. In order to describe a one-dimensional chain consisting of metal ions and halogen ions, the extended Peierls–Hubbard model is introduced in Sec. 2. The theoretical model of the relaxation dynamics in the excited states with a strong electron-lattice coupling is given in Sec. 3. The model is based on the interaction mode, which is appropriate for understanding the vibrational relaxation of localized centers in solids. Experimental backgrounds with some historical survey are given in Sec. 4. The recent experimental results of time-resolved luminescence for Pt-Cl, Pt-Br and Pt-I systems are presented in Secs. 5 to 8. The main result contains the direct observation of the wave-packet oscillation in the self-trapped excitons. The relaxation process observed in experiments has been successfully interpreted in terms of the model based on the interaction mode and the dynamical aspects are compared with the transient absorption measurements. The lifetime of the STE is shorter in Pt-X with heavier halogen ions. This behavior is discussed in relation with the non-radiative process leading to lattice rearrangements. In Secs. 9 and 10, visualization of the wave-packet form is presented. The basic behavior of the wave-packet is well understood in terms of a harmonic oscillator model. A non-exponential decay profiles are revealed from the center of gravity motion of the wave-packets. The exciton localization process is also discussed in the last section.

Contents

1. Introduction

The lattice rearrangements in solids including change of chemical bonds or production of metastable states from electronic excited states can be initiated by photo-excitation in visible and ultraviolet photon energy regions. Sometimes the lattice rearrangements evolve into semi-macroscopic new orders or new crystal phases. Recently, this phenomenon called "photo-induced phase transition" is receiving much attention as a new concept of photo-induced phenomena in solids. These phenomena are important also for realizing new ordered states, which are not accessible through quasi-thermal-equilibrium paths.

In the study of the photo-induced lattice rearrangements, one-dimensional system has a special meaning. Due to the lack of the mechanism keeping a long range order, the system has a large fluctuation and instability. This gives us a chance for finding various metastable states and the photo-induced inter-conversion processes. The halogen-bridged platinum complex is a prototypical system of quasi-one-dimensional system suitable for studying the properties and dynamics of the metastable states such as solitons, polarons and self-trapped excitons (STE). The series of materials are receiving renewed interest as model systems for investigating the photo-induced phase transitions.

The phase transitions or other lattice rearrangements of atoms in solids will proceed at a speed in the same order as lattice vibration, whose time scale falls in the region between 1 ps and 10 fs. In order to trace the processes of the lattice rearrangements, we need an observation technique in this time range. Due to the recent development of the femtosecond spectroscopy, the dynamical aspects of these phenomena can be observed directly.

In this article, we present recent developments in the study of the dynamics of the excited states in a series of quasi-one-dimensional platinum complexes with various bridging halogen ions. A special emphasis will be placed on the formation and the decay dynamics of the self-trapped exciton, which is the starting point of the formation process of various metastable states accompanying lattice rearrangements. In the last sections, the nuclear wave-packet dynamics during the formation and relaxation processes of the STE is discussed, based on a real-time visualization of the wave-packet form.

2. Excited States in Platinum Complexes

The halogen-bridged platinum complex discussed in this article has a formula of $[Pt(en)_2][Pt(en)_2X_2](ClO_4)_4$ where en is ethylene diamine and X is the halogen ion Cl, Br or I.[1] Hereafter, these materials will be denoted as Pt-Cl, Pt-Br and Pt-I, respectively. The system consists of one-dimensional chains Pt — X — Pt — X — which are isolated each other by ligand molecules, as schematically shown in Fig. 1. As the p orbitals of the halogen ions are deep in energy, we can expect that the system is well described by a one-dimensional Peierls–Hubbard model considering the d_z^2 orbitals of the metal ions. The ground state and the metastable states in this system have been theoretically investigated by Iwano.[2] In the following, the outline of the model will be shown.

The Hamiltonian is given as

$$\begin{aligned} H = & -t_0 \sum_{l\sigma} \left(C_{l\sigma}^{+} C_{l+1\sigma} + C_{l+1\sigma}^{+} C_{l\sigma} \right) \\ & + U \sum_{l} n_{l\alpha} n_{l\beta} + V \sum_{l\sigma\sigma'} n_{l\sigma} n_{l+1\sigma'} \\ & - S \sum_{l\sigma} \left(Q_{l+1} - Q_l \right) n_{l\sigma} + \frac{S}{2} \sum_{l} Q_l^2 \end{aligned} \qquad (1)$$

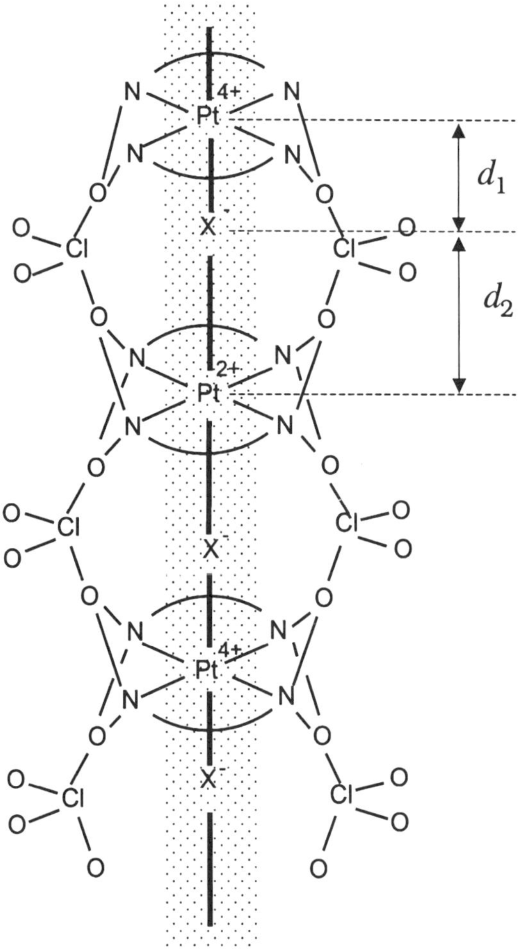

Fig. 1. Structure of the Pt-X chain in the halogen bridged platinum complex $[Pt(en)_2][Pt(en)_2X_2](ClO_4)_4$ with the surrounding ligand molecules. The values of d_1 and d_2 are listed in Table 1.

Here, $C^+_{l\sigma}$ and $C_{l\sigma}$ are the creation and annihilation operators of electron with a spin σ on site l, $n_{l\sigma}$ are the number operators of electrons with spin σ on site l. Q_l is the displacement of the l-th halogen ion on the chain axis. The meanings of the four important parameters are:

U: Coulomb repulsion energy between the electrons on the same site.

V: Coulomb repulsion energy between the electrons on the nearest neighbor sites.

S: Linear coupling constant of electron with the lattice.
t_0: Transfer between the adjacent sites.

The relative magnitude of these parameters determines the stability of the ground state and the metastable states. Figure 2 shows the phase diagram of the Peierls–Hubbard system drawn on the S-U plane in the localized limit ($t_0 \to 0$). In the region defined by $\mathrm{U} > 4\mathrm{S} + 2\mathrm{V}$, the spin density wave (SDW) is realized in the ground state, and in the region of $\mathrm{U} < 4\mathrm{S} + 2\mathrm{V}$, charge density wave (CDW) is the ground state. If the on-site Coulomb energy U is relatively large, the electrons tend to keep distance each other resulting in the SDW state as the ground state. On the other hand, if the contribution from the electron-lattice interaction is large, the CDW is favorable, because the system can lower the total energy by distorting the lattice. As the Pt-Cl system has relatively small on-site Coulomb energy U and a large electron-lattice interaction S, i.e., $U = 2.0$, $V = 1.2$, $S = 0.27$ and $t_0 = 1.3\,\mathrm{eV}$, the ground state is CDW.

Figure 3 schematically shows the configuration of electrons in d_{z2} orbitals of Pt atom in the localized limit. Figure 3(a) shows the SDW configuration, where each metal ion has one electron (M^{3+}) with alternate

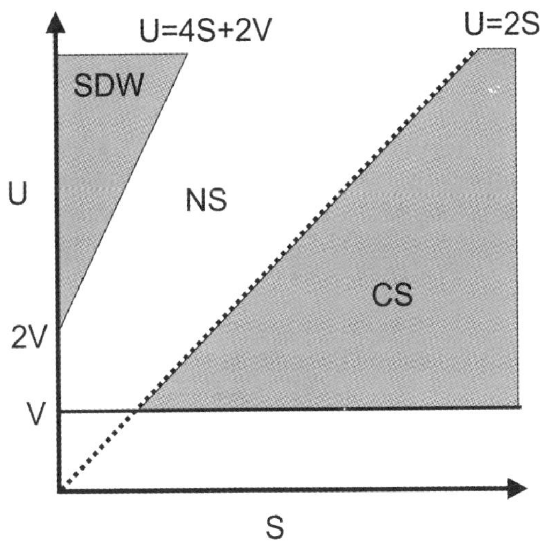

Fig. 2. Phase diagram of the one-dimensional Peierls–Hubbard system in the localized limit after Iwano.[2] In the ground state, SDW state is stable within the region labeled SDW, while CDW is stable in the NS and CS regions. The energy of the neutral soliton is lower in NS region and that of the charged soliton is lower in CS region (reproduced from Fig. 3 of Ref. 2).

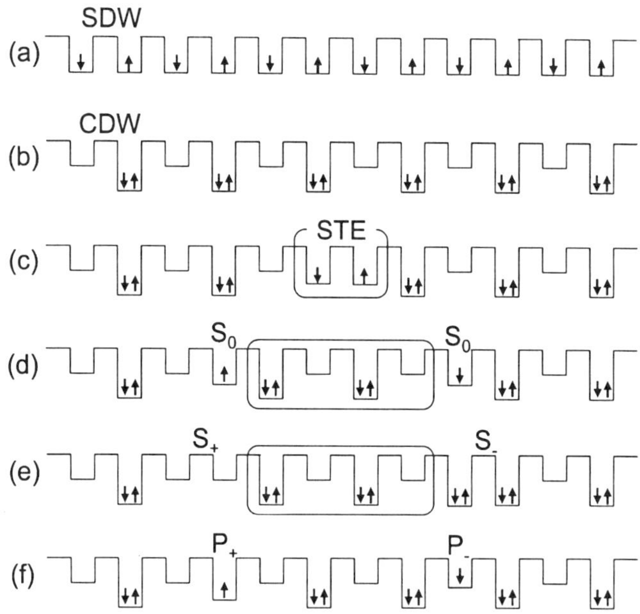

Fig. 3. Electron configurations in the d_{z^2} orbitals in the localized limit (after Iwano[2]; partially modified). (a) SDW ground state, (b) CDW ground state, (c) Self-trapped exciton (STE), (d) Neutral soliton pair, (e) Charged soliton pair, (f) Positive and negative polarons. In (d) and (e), the sequence of empty and occupied sites are inverted in the regions surrounded by rectangles.

spin directions. The configuration (b) is the CDW ground state, where the d_{z^2} orbital is occupied by two electrons every two sites resulting in M^{2+}. The rest is empty, giving M^{4+}. The CDW region is divided into two parts according to the stability of the metastable states. In the configurations shown in (d) and (e), the Pt^{2+}-Pt^{4+} sequences are inverted from the CDW ground state (b) in the regions surrounded by squares. In these cases, we have to assume solitons on the boundaries of the inverted region to connect the sequence inside and outside. We have two choices: One is the soliton denoted as S_0 in (d), which is a site with one electron in the d_{z^2} orbital. Since this configuration does not modify the charge distribution on average and has a spin 1/2, it is called a neutral soliton or a spin soliton. The others are the solitons denoted as S_+ and S_- in (e). These are the empty sites and the doubly occupied sites, respectively. These are called charged solitons, because they are positively or negatively charged. In Fig. 2 neutral solitons or charged solitons are stable in NS or CS regions, respectively. The configuration (f) shows the polarons P_+ and P_-, which are in other words, isolated electrons and holes accompanied by polarization of the lattice. The configuration shown in (c) is the self-trapped exciton (STE), which consists of two M^{3+} ions.

Figure 4 shows the Pt-X chain with lattice relaxations. In the realistic system, the Pt ions do not have valences +2 and +4 but $3 - \rho$ and $3 + \rho$ ($0 < \rho < 1$), because of the finite transfer. The Pt ions are tightly bound to the surrounding nitrogen atoms, while the halogen ions (X^-) are relatively loosely bound between two adjacent Pt ions. The halogen ions are located closer to the Pt^{4+} ions due to stronger Coulomb attraction force.From the initial state shown in (a), the light irradiation will generate free excitons as a Bloch state in the CT absorption band or electron-hole pairs (polaron pairs) when the photon energy is in the continuum above CT band. The theory tells us that the free excitons fall into self-trapped states immediately, because there is no potential barrier between the free state and the self-trapped state in one-dimensional systems.[3] As shown in Fig. 4(b), the STE consisting of two Pt^{3+} ions is accompanied by a lattice distortion. The electronic wavefunction of the exciton is localized on the deformed part of the lattice where the potential energy is lowered.

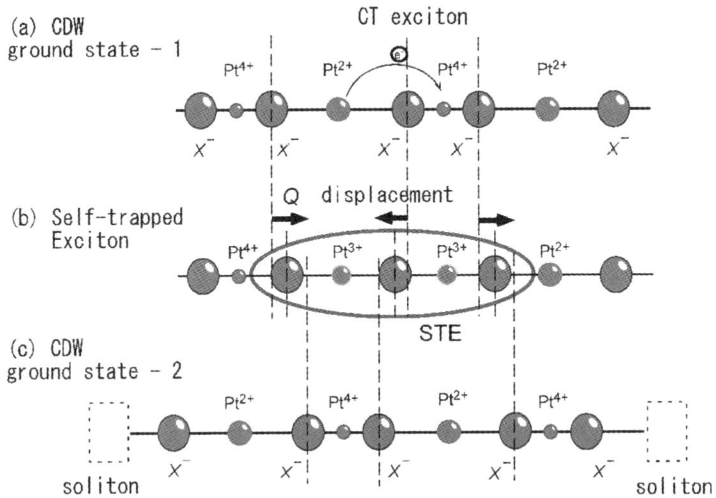

Fig. 4. Schematic drawing of the linear chain of the halogen-bridged platinum complex. Both (a) and (c) are the ground states, but with inverted valence sequences, (b) is the self-trapped exciton consisting of a pair of Pt^{3+} ions. The displacements of the ions in the self-trapped exciton are shown by horizontal arrows.

This is the starting point of creating various metastable states in the Pt-X system. In returning to the ground state, we have two choices. One is the configuration (a) with the electron going back from the Pt^{3+} on the right to the Pt^{3+} on the left. The other is the configuration (c) with one more electron transferred to the right. In this case, the Pt sequence has an inverted valence in the region surrounded by a soliton pair as already shown in Figs. 3(d) and (e).

It is believed that the size of the inverted region (the distance of the solitons) can become an appreciable size, when they are created from the STE.[2,4] Development of semi-macroscopic domains through aggregation of the excited species under high density excitation is predicted.[5] Apart from the excess energy due to the solitons, these two ground states are degenerate in energy and can be mutually converted via the excited state. Since two ground states are crystallographically equivalent, this process is not a real photo-induced phase transition. However, this can be regarded as a model of structural photo-induced phase transition, because the equilibrium positions of the atoms are rearranged in a semi-macroscopic region by light irradiation. If the ligand molecules for the odd and even sites are different (it is at least conceptually possible), this could be a real martensite phase transition.

The structural and optical properties of Pt-X are summarized in Table 1. The valence disproportion ρ is larger for lighter halogen ions. The ratio of Pt-X distances d_2/d_1 becomes smaller with increasing the halogen ion

Table 1. Fundamental structural and optical properties of Pt-X. d_1 and d_2 are distances of $Pt^{3+\rho}X^-$ and $Pt^{3-\rho}X^-$, respectively (Refs. 6, and 7), ρ represents the deviation in valency of Pt ions from 3 (Ref. 8). E_{CT} and E_{lm} are the peak energies of the CT exciton absorption band at room temperature and the STE luminescence band at 2 K, respectively (Ref. 8). ω_R is the frequency of the Raman-active symmetric stretching mode (Refs. 9–11). T_{vib} is the vibration period of the mode.

Pt-X	Pt-Cl	Pt-Br	Pt-I
d_1 (Å)	2.3274	2.4843	2.7918
d_2 (Å)	3.1014	3.0013	3.0368
d_2/d_1	1.33	1.21	1.09
ρ	0.91	0.64	0.36
E_{CT} (eV)	2.72	1.95	1.37
E_{lm} (eV)	1.22	0.76	0.6
ω_R (cm^{-1})	308	166	113
T_{vib} (fs)	108	201	295

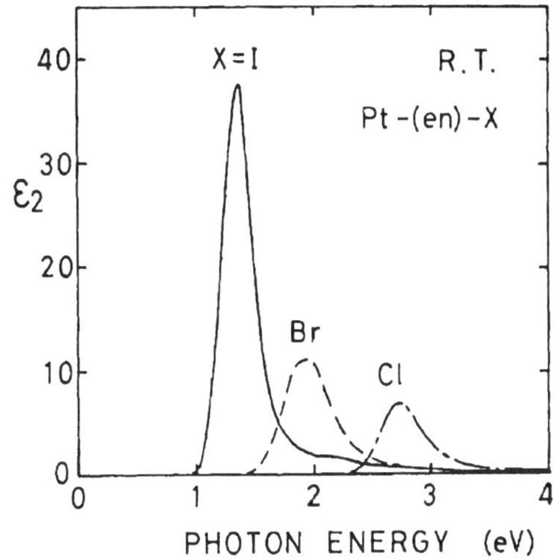

Fig. 5. Optical absorption spectra of Pt-X systems at room temperature. (Wada *et al.*[8])

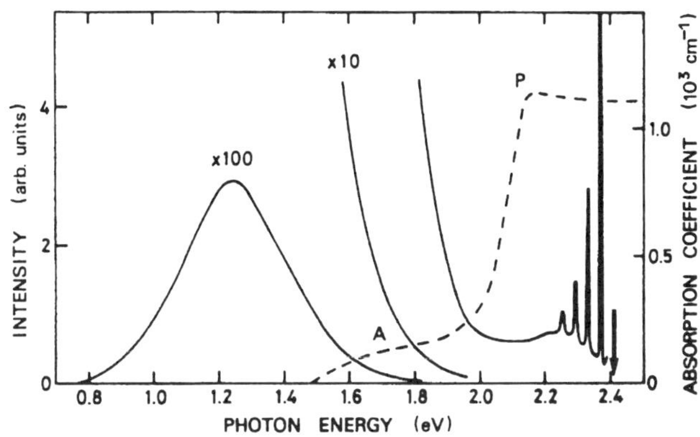

Fig. 6. Solid lines show the resonant Raman and Luminescence spectra of Pt-Cl at room temperature under excitation of 2.41 eV photons. Dashed line shows the absorption spectrum reflecting the defects in the polarization parallel to the chain axis (Tanaka *et al.*[12]).

radius, because the larger ion radius hinders the ions to approach the $Pt^{3+\rho}$ ions. The single crystals of Pt-X are known to exhibit an intense optical absorption, polarized parallel to the chain-axis from near-infrared to the visible wavelength range as shown in Fig. 5.[8] These absorption bands have been assigned to the charge-transfer (CT) excitation, corresponding to an electron transition from $Pt^{3-\rho}$ to $Pt^{3+\rho}$. The energy of the absorption peak E_{CT} and the luminescence peak E_{lm} become larger for lighter halogen ions.

The Stokes shifts also become larger for lighter halogen ions, because the stabilization effect due to lattice relaxation increases. A typical luminescence spectrum for Pt-Cl is shown in Fig. 6.[12] The main part of the luminescence is a Gaussian band around 1.25 eV characteristic of the STE.

3. Theoretical Background of the Relaxation Dynamics

The main interest of this article is the dynamics of the lattice rearrangement phenomena. The process is best described in terms of the trajectory of the lattice system during the relaxation from the initial excited state to the rearranged lattice configuration. The dynamics of the lattice deformation are classically expressed by the motion of individual atoms, but in terms of quantum mechanics, it should be expressed as a propagation of a wave-packet on a multi-dimensional adiabatic potential surfaces. The process may include also diabatic jump between the adiabatic potential surfaces. Although the wave-packet propagation has been relatively well investigated for small molecules both experimentally and theoretically,[13,14] the situation is not simple and less understood in solids.

The localized electronic system with a strong electron-lattice interaction will have adiabatic potential curves shown in Fig. 7. Although the local configuration of STE is similar to a molecule, the relaxation process of the vibrational excitation is very different from isolated small molecules, because the system is strongly coupled to the bulk system having an infinite degree of freedom. In the standard theory of self-trapping, the

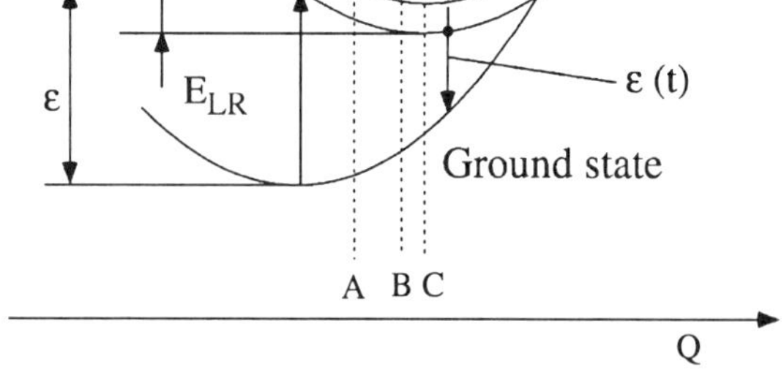

Fig. 7. Adiabatic potential curves of STE and the definition of parameters.

lattice deformation associated with the exciton localization is expressed as a superposition of many bulk phonon modes. The mode is not a "normal mode" but is called as "interaction mode". The abscissa of the configuration coordinate shown in Fig. 7 is the deformation corresponding to this interaction mode. If the linear coupling is assumed, the two potential curves will have the same curvature as shown here. As it has been expected that the wave-packet motion will be reflected in the time-resolved luminescence spectra, the theory of "hot luminescence" is useful in understanding how the behavior of the wave-packets is observed in experiments.

The hot luminescence during the vibrational relaxation of localized center in solids has been theoretically discussed by Kayanuma.[15] Although this theory does not involve the relaxation from free state to localized state, this relaxation of the STE *after localization* will be well described in terms of this model.

The hot luminescence intensity at photon energy Ω_2 at time t is given as

$$F_e\left(\Omega_2, \Omega_1; t\right) = \frac{\sqrt{2\pi}}{\tilde{D}(t)} \exp\left[-\frac{\{\Omega_2 - \tilde{\varepsilon}(t)\}^2}{2\tilde{D}(t)^2}\right]. \tag{2}$$

Here, Ω_1 and Ω_2 are incident and scattered photon energies, respectively. The time-dependent Franck–Condon energy is given by

$$\tilde{\varepsilon}(t) = \varepsilon - 2E_{LR} + 2\int_{-\infty}^{\infty}\left\{\omega^{-1} + \frac{\Omega_1 - \varepsilon}{2D^2}\right\} J(\omega)\cos\omega t\, d\omega. \tag{3}$$

Here, ε is the electronic excitation energy without lattice relaxation and E_{LR} is the lattice relaxation energy, as defined in Fig. 7. The function $J(\omega)$ represents the spectral density of the phonons coupled to the electronic state. D represents the spectral width of the absorption band and $\tilde{D}(t)$ is the time dependent spectral width of the hot luminescence given by

$$\tilde{D}(t)^2 = D^2 - D^{-2}\left\{\int_{-\infty}^{\infty} J(\omega)\cos\omega\, t\, d\omega\right\}^2. \tag{4}$$

In order to realize a local distortion of the lattice from a linear combination of the bulk phonon modes, we have to use those with wavelengths in the same order as the lattice constant. Thus $J(\omega)$ will consist of continuously distributed phonon frequencies near the edge of the Brillouin zone as shown in Fig. 8. This figure shows that the localized deformation of the lattice is Fourier synthesized from the superposition of the phonon modes with different wavelengths. Strictly speaking, this picture is valid within the linear response of the lattice system, because the concept of phonon is based

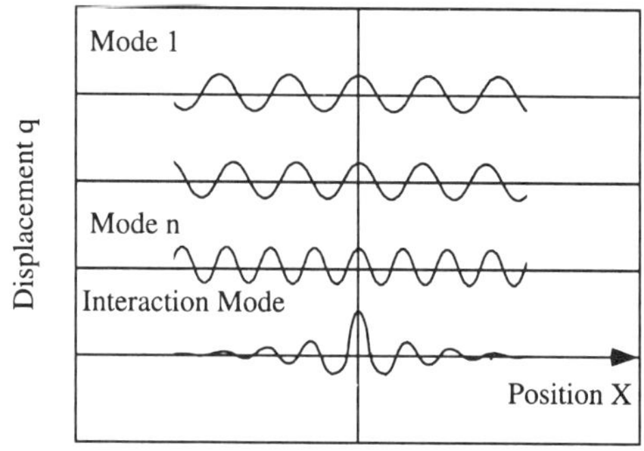

Fig. 8. Schematic illustration of the spatial lattice distortion corresponding to the interaction mode synthesized from those of the bulk phonon modes.

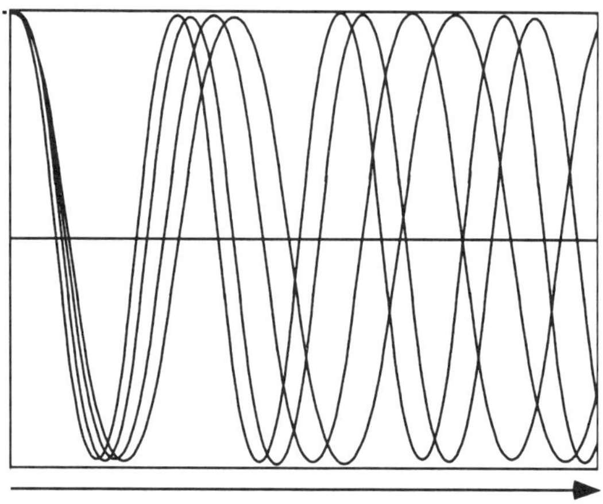

Fig. 9. Schematic illustration of the time evolution of the normal modes involved in an interaction mode.

on the generalized Huck law. This assumption might not be fully justified for the STE in Pt-X, because the change of atomic distance exceeds 10% and the change in the valence of Pt ions is close to one. However, we continue our discussion in this framework.

As far as the oscillations of all the phonon modes are coherent, the wave-packet motion can be observed. However, when the time proceeds, dephasing takes place, because the frequencies of the modes included in the interaction mode are not the same (Fig. 9). Then the wave-packet oscillation becomes faint. This situation is analogous to the dephasing process of inhomogeneously broadened oscillators. The remarkable feature of this treatment is that the dissipation of vibrational energy to bulk crystal is automatically included in this dephasing process.

4. Time-Resolved Measurements and Dynamics

Since the very early stage of the study in the Pt-X systems, the dynamical aspect of the metastable states has attracted attention, because of the diversity of the metastable species characteristic to the one-dimensional system. One important prediction was the non-existence of a potential barrier between the free and the self-trapped states of excitons, in contrast to the three-dimensional system where free excitons are metastable and observed through luminescence.

A pioneering work in 10 picosecond region has been done by Tanino[16] and by Wada[17] for the STE luminescence in Pt-Cl system by using streak cameras. Although the slow decay components were observed, the time resolutions were not enough to resolve the formation and initial relaxation processes of STE. To realize a higher time resolution, it is necessary to use a measurement technique whose time resolution is not limited by electronics. One is the transient absorption/reflection measurement and the other is the up-conversion luminescence measurement. In both cases, the samples are excited by a very short pump pulse and the subsequent response is detected by using probe or gate laser pulses. In the case of transient absorption measurements, one observes the induced absorption from the lowest excited state to the higher excited states. Therefore, in interpreting the spectra of transient absorption, we have to have knowledge on the higher excited states, which is usually very hard to get experimentally or theoretically. On the other hand, the interpretation of the luminescence is more straightforward, because the transition takes place from the lowest excited state to the ground state. Both are well known from steady state luminescence and linear absorption measurements in many cases. Due to this advantage, we can get reliable information from experimental data. This fact is especially important in observing the wave-packet motion of the atomic vibration at arbitrary position on the configuration coordinate axis.

A schematic diagram of the experimental setup for time-resolved luminescence measurements is shown in Fig. 10.[18] The femtosecond pulses from a laser source are divided into two beams by a beam splitter and one beam is directed to the sample as a pumping pulse. The infrared luminescence photons are collected by a pair of paraboloidal mirrors and focused onto a non-linear crystal. The other beam is used as a gating

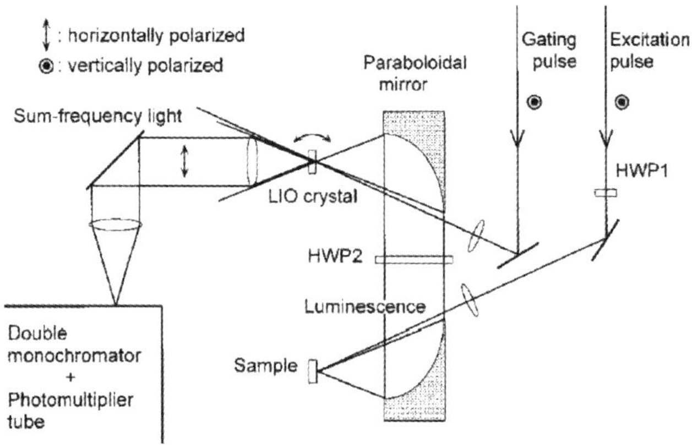

Fig. 10. Diagram of the experimental setup for up-conversion luminescence measurements (Ref. 18). The half-wave plates (HWP1, HWP2) are used for measuring the polarization dependence.

pulse on the non-linear crystal and the sum frequency photons in visible region are directed into a double-grating monochromator and detected by a photon counting system. The time response of the luminescence signal is taken by scanning the optical delay. The fundamental mode-locked pulses from a Ti sapphire oscillator (1.6 eV), second harmonic of the same laser (3.2 eV), or amplified pulses (1.6 eV) from a 100 kHz regenerative amplifier are used. The time resolution ranges from 40 to 170 fs depending on the light sources and the optics design. All the measurements were made at room temperature.

5. Pt-Br System

The luminescence spectrum of STE in Pt-Br has a broad Gaussian shape with a peak around 0.8 eV,[19] reflecting the strong electron-lattice interaction. The luminescence is strongly polarized parallel to the chain axis. Figure 11 shows the decay profiles of the luminescence intensity.[20] Between 0.8 and 1.2 eV, the decay has the same time constant 5.5 ± 0.1 ps. which corresponds to the lifetime of the STE. In addition to this slow component, we can see a fast component at 1.2 eV. Figure 12 shows the fast component measured with a relatively low time resolution (90 fs).[18,20] The most striking feature is the oscillatory structure due to the wave-packet motion between 1.3 and 0.9 eV. This behavior is qualitatively understood in the following way. The wave-packet starts oscillation from the left side of the upper potential curve in Fig. 7. Suppose that we monitor the luminescence intensity

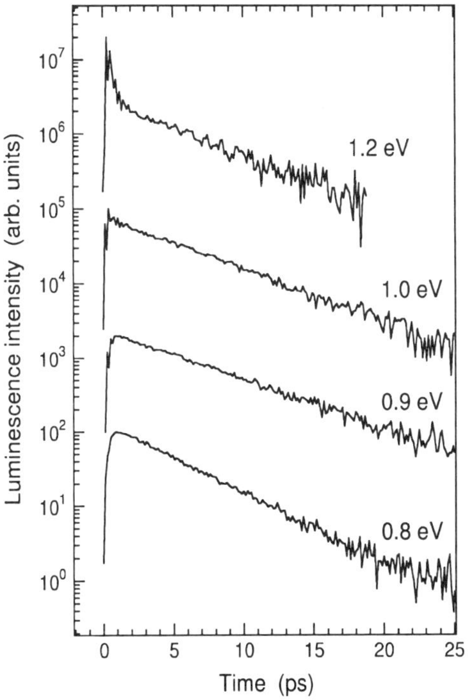

Fig. 11. Time dependence of luminescence intensity in Pt-Br measured at room temperature. The observed photon energy is indicated for each curve. The intensity is plotted in a logarithmic scale and the curves are vertically shifted for a convenient display.[20]

at the photon energy corresponding to the point B. At every moment of passage of the wave-packet, the luminescence intensity increases and an oscillatory feature is observed. From its period, we can directly obtain the frequency of the local vibration in the excited state. The obtained frequency $115\,\mathrm{cm}^{-1}$ (290 fs) is significantly lower than the frequency of the symmetric stretching mode ($175\,\mathrm{cm}^{-1}$) of the Br-Pt-Br unit in the ground state. The oscillation is not clear at 0.8 eV, but becomes clear again at even lower energy 0.7 eV. Recently, a similar oscillation has been observed in $[Pt(en)_2][Pt(en)_2Br_2](PF_6)_4$ system by Dexheimer *et al.*,[21] using transient absorption technique. As shown in Fig. 13, the Fourier transform of the oscillatory part of the signal measured at four different wavelengths (830, 855, 905, 940 nm) had a pronounced peak at $175\,\mathrm{cm}^{-1}$ corresponding to the symmetric stretching mode in the ground state. In addition, a second peak appeared at $110\,\mathrm{cm}^{-1}$ for 830 and 855 nm. This value is in a good agreement with ours.

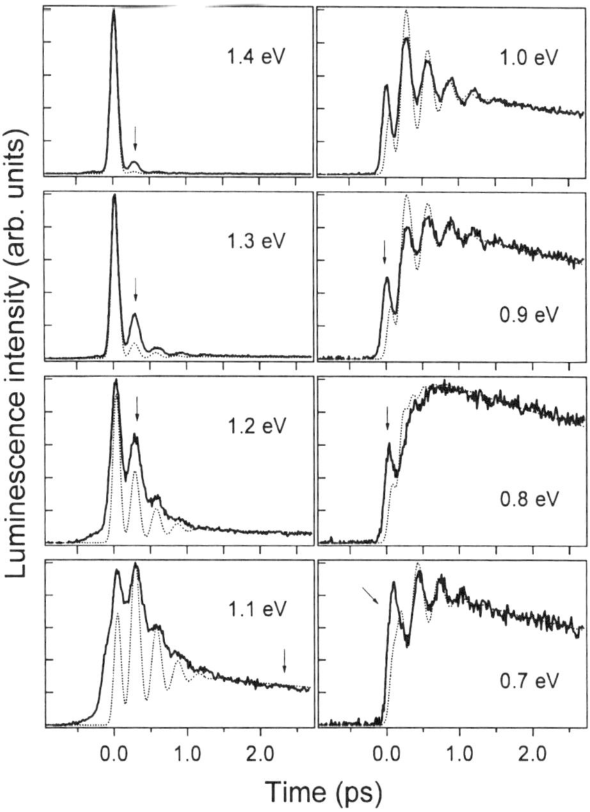

Fig. 12. Time evolutions of the luminescence intensity in Pt-Br at indicated photon energies. The experimental and calculated results are shown by solid and dotted lines, respectively (Ref. 18).

We analyzed the data quantitatively in terms of the theory described in the preceding section.[18] For fitting to the experimental results, a slightly modified formulation is introduced in the following. Firstly, the right side of Eq. (2) should be multiplied by a factor

$$\exp\left(\frac{-t}{\tau}\right) \tag{5}$$

to include the finite lifetime τ of STE. In Eq. (3) we neglect the term $(\Omega_1 - \varepsilon)/2D^2$, because the incident photon energy is always in the CT absorption band, i.e. $\Omega_1 \approx \varepsilon$, then

$$\tilde{\varepsilon}(t) = \varepsilon - 2E_{LR} + 2E_{LR} \int_{-\infty}^{\infty} J'(\omega) \cos \omega t \, d\omega. \tag{6}$$

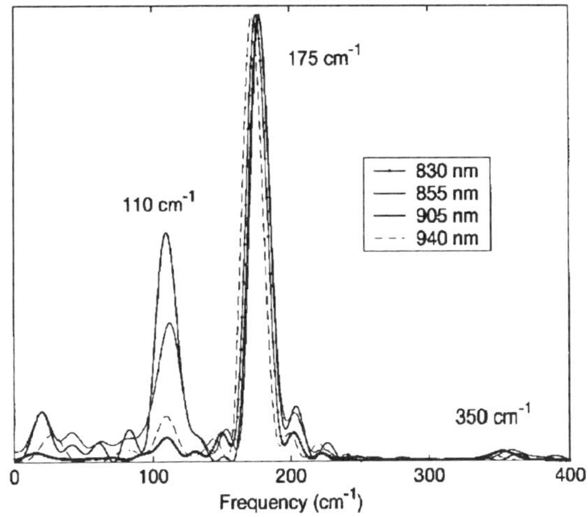

Fig. 13. Fourier power spectra of the oscillatory part of the time evolution data of transmission at given wavelengths. (Dexheimer *et al.*[21]) The $175\,\mathrm{cm}^{-1}$ peak and the $110\,\mathrm{cm}^{-1}$ peak correspond to the symmetric stretching mode in the ground state and the local vibration mode in STE, respectively.

Here, $J'(\omega) = J(\omega)/\omega$ is a dimensionless spectral density function normalized to unity as

$$\int_{-\infty}^{\infty} J'(\omega) d\omega = 1. \tag{7}$$

The time dependent spectral width is modified as

$$\tilde{D}(t) = D_0 - \sqrt{1 - D_1^{-2} \left\{ \int_{-\infty}^{\infty} J'(\omega) \cos \omega t \, d\omega \right\}^2}, \tag{8}$$

where a constant term D_0 is added to include the inevitable broadening determined by Fourier transform limit of the laser pulse.

As can be seen from Eq. (8), the spectral width is equal to D_0 at $t = 0$ and approaches $D_0 + D_1$ at $t = \infty$. The function $J'(\omega)$ is the normalized phonon density spectrum involved in the interaction mode. These modes are excited simultaneously and coherently on start of the lattice deformation toward the self-trapped state. For determining the function $J'(\omega)$, we have to have knowledge on the bulk phonon spectrum, as well as the spectrum of the coupling constants with the localized electronic states. However, it is very difficult to predict these parameters theoretically. Therefore, we assume a Gaussian type spectra,

$$J'(\omega) = \frac{a_i}{\sqrt{\pi} W_i} \exp\left[-\frac{(\omega - \omega_c)^2}{W_i^2}\right]. \tag{9}$$

We used six parameters for calculating the time evolution. The electronic excitation energy ε and the lattice relaxation energy E_{LR} are known from the reported absorption and luminescence data. The oscillation period T was 290 fs (see above) and the lifetime τ is found to be 5.5 ps from the decay curve at longer time scale. Thus four parameters are unambiguously determined. Finally we need information about the spectral density of the phonons involved in the interaction mode. After some try and error processes, we found that a simple Gaussian shape could not reproduce the experimental data and that superposition of two Gaussians centered at the same frequency gave a good fitting result. The widths W_1, W_2 and the weighting ratio a of the two Gaussian components along with D_0 and D_1 are adjusted to reproduce all the data taken for different photon energies between 0.7 and 1.4 eV. The best values are shown in Table 2 and the phonon spectrum is shown in Fig. 14. The calculated time evolution curves are shown in Fig. 12 by dotted lines. We obtained an excellent agreement. Although we cannot decide the shape of the phonon density spectrum uniquely, it is evident that a relatively sharp component and a broad wing are necessary.

Table 2. Values of fitting parameters used for Pt-Br and Pt-Cl in the calculation. ε is the electronic excitation energy. $E_{\rm LR}$ is the lattice relaxation energy. D_0 and $D_0 + D_1$ are the initial and the final spectral widths of hot luminescence, respectively. τ is the lifetime of STE. T is the oscillation period at the central phonon frequency. W_1 and W_2 are the frequency widths of the first and second Gaussian components. a is the weight of the second component.

Parameter	Pt-Br	Pt-Cl
e (eV)	1.4	2.6
$E_{\rm LR}$ (eV)	0.3	0.66
D_0 (meV)	15	15
$D_0 + D_1$ (eV)	0.155	0.235
t (ps)	6	25
$T = 2\pi/\omega_c$ (fs)	290	140
W_1	$2\omega_c/3$	$2\omega_c/9$
W_2	$2\omega_c/14$	$2\omega_c/42$
a	0.3	0.5

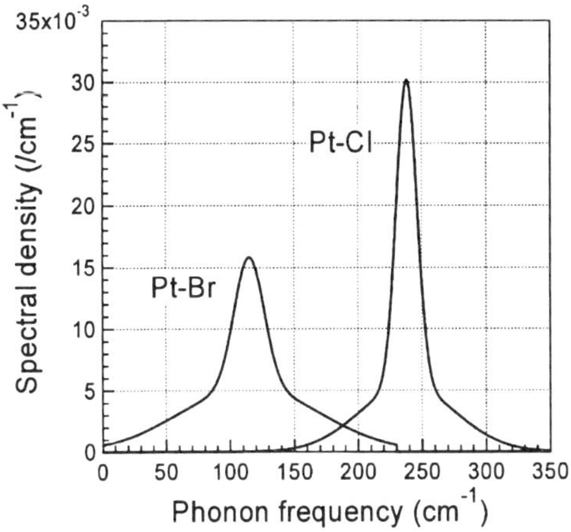

Fig. 14. Spectral density function, $J'(\omega)$, assumed in the calculation of the time-resolved luminescence in Pt-Br and Pt-Cl (Ref. 18).

6. Pt-Cl System

As the CT gap of Pt-Cl system is larger than the photon energy of the titanium sapphire laser, we have to use second harmonic for excitation. The time evolution of the luminescence for Pt-Cl system is shown in Fig. 15.[18,22] The basic behavior is very similar to Pt-Br system, i.e., the fast decay at higher energy and a slow rise and a slow decay are observed near the luminescence peak. In these curves, the oscillation has not been seen, because the time resolution was not sufficient at this stage. We can understand this behavior in the same framework as Pt-Br. The dotted curves in Fig. 15 are the fitting results with the parameter values shown in Table 2. In order to reproduce the experimental curves, the results are convoluted with the response function of the measurement system, and the oscillating feature is smoothed out. The agreement is again very good. The phonon density spectrum of Pt-Cl shown in Fig. 5 has a smaller width and smaller wings compared to Pt-Br.

The wave-packet oscillation has been observed by improving the time-resolution.[23] Figure 16 shows the data taken with a time-resolution of 75 fs under 3.2 eV excitation. The oscillation period is 180 fs, which is longer than the parameter assumed in the fitting in the preceding section. This value is quite close to the oscillation period 185 fs found by Pelt *et al.*[24] who used a transient absorption method in a similar material. The oscillating feature is also similar to that in Pt-Br system, that is, the damping of

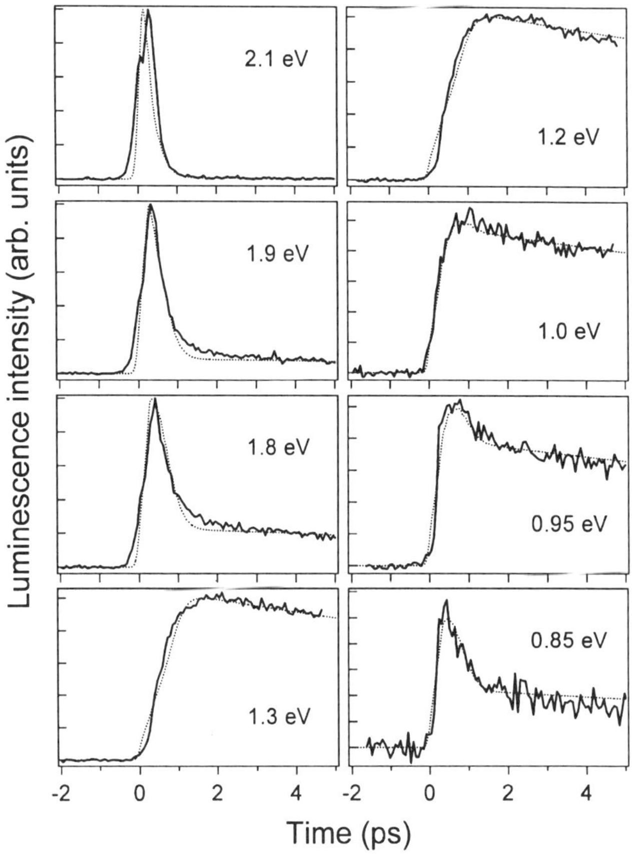

Fig. 15. Time evolution of luminescence in Pt-Cl at indicated luminescence photon energies. The experimental and calculated results are shown as solid and dotted lines, respectively. The luminescence in the energy range between 1.4 and 1.8 eV could not be observed because of the strong stray light of SH generated at the nonlinear crystal.[18]

the oscillating amplitude is large at higher energies (2.3–2.1 eV) and the modulation amplitude is relatively small at 1.2 eV, which is close to the peak energy of the luminescence spectrum (Fig. 6). It is interesting to note that the coherent oscillation is observed even under excitation at the higher energy tail of the CT band (the excitation photon energy is located 0.8 eV above the bottom of the CT band).

The envelope of the decay profile can alternatively be understood in terms of a macroscopic model. As shown in our earlier paper,[22] the rapid decay of the luminescence at 2.1 eV can be interpreted as cooling in the

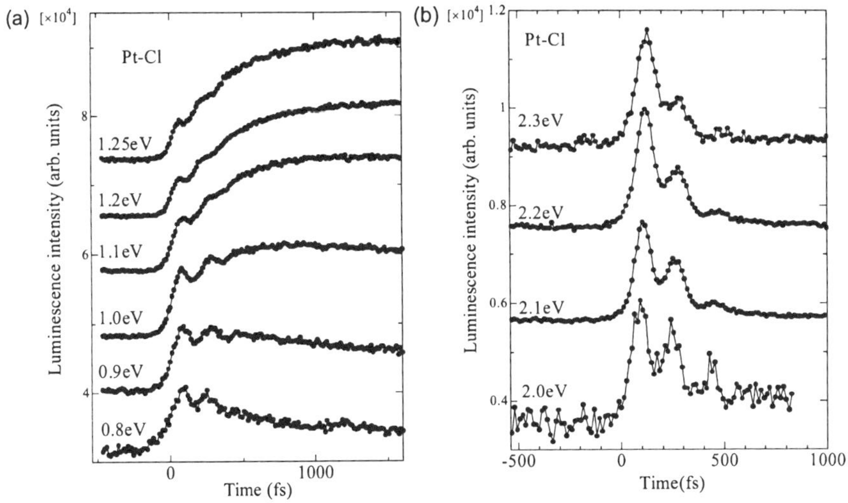

Fig. 16. Time evolution of the luminescence in Pt-Cl with a high time resolution at lower energies (a), and at higher energies (b). The time origins and the intensity scales at different photon energies are not calibrated.[23]

adiabatic potential well. The slow rise near the center of luminescence band then corresponds to the convergence of the population probability towards the potential minimum corresponding to the cooling. The decrease in the oscillation amplitude is directly related to the dephasing among the normal modes involved in the interaction mode model. Thus the energy dissipation termed as "cooling" is nothing more than the dephasing, which is governed by the phonon density spectrum involved in the interaction mode.

7. Pt-I System

The last member of the Pt-X series is Pt-I, which has the heaviest halogen ion. In this system, the ionicity is lower and the CT gap is smaller. A very weak luminescence was observed at liquid He temperature.[19,27] Since the wavelength of luminescence is longer than Pt-Br and the intensity is very weak, the steady state luminescence spectrum at room temperature has not been reported. However, it is possible to measure it by up-conversion techniquc, because the instantaneous intensity is not weak owing to reasonably large oscillator strength. Figure 17 shows the time evolution of the luminescence excited by amplified pulses at 1.6 eV.[18] The luminescence is strongly polarized along the chain axis and the excitation efficiency becomes highest when the polarization of the excitation light

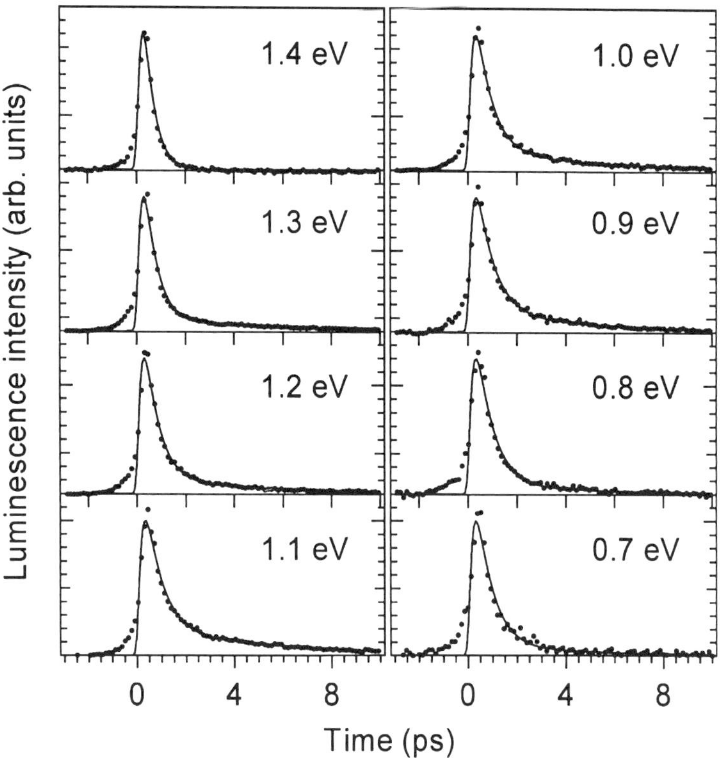

Fig. 17. Time evolution of the luminescence in Pt-I at the indicated photon energies. The dots represent the experimental data and the solid curves show the fitting results. Double exponential decay curves convoluted with the instrumental response function are used in the fitting. The polarizations of the excitation and luminescence light are parallel to the chain axis.[18]

is parallel to the chain axis. From these observations, the luminescing entity is identified as the STE lying along the chain axis similar to those in Pt-Br and Pt-Cl systems. However, there are several major differences compared to the systems with lighter halogen ions. The time evolution is almost the same for all the luminescence photon energies from the high energy tail to the peak of the luminescence band. The slow rise which is commonly observed around the luminescence peak energy in other Pt-X is missing.

A Fourier component with a frequency of $107\,\mathrm{cm}^{-1}$ (oscillation period = 311 fs) has been reported by Sugita *et al.*[25] in the response of the transient absorption for Pt-I system. However, there is no indication of an oscillating feature at any photon energy in our luminescence measurements. This might be due to the lower time resolution or the very short lifetime of the STE.

8. Halogen Ion Dependence

The dynamical properties of the STE at room temperature are summarized in Table 3. The observed lifetimes are ascribed to the non-radiative process, because the radiative lifetime of the STE in Pt-X is estimated to be much longer (4–6 ns).[26] The non-radiative process in Pt-Cl is proved to be thermal activation type with an activation energy of 15 meV from the temperature dependence of the luminescence lifetime.[17] From the table we can see that the non-radiative transition rate increases as the halogen ion becomes heavier. This tendency is in accordance with the fact that the luminescence intensity becomes weaker for heavier halogen ions.[27]

In describing the wave-packet, we assumed a configuration coordinate model shown in Fig. 7, where the abscissa represents the interaction mode leading to the self-trapped state. However, the dynamics of the system should be described on a multi-dimensional space. One realistic model assumes two degrees of freedom explicitly, that is, the displacement towards the STE and the separation between two solitons which define the new domain with an inverted $Pt^{+3+\rho}$-$Pt^{+3-\rho}$ sequence.[2] The adiabatic potential surfaces defined on this two-dimensional space are illustrated schematically

Table 3. Dynamical parameters of STE in Pt-X complexes. The values refer to the high temperature phase for Pt-Cl and to the low temperature phase for Pt-Br. The lifetimes are determined from the decay of the luminescence[18].

		Pt-Cl	Pt-Br	Pt-I
Lifetime (ps)		30 ± 10	5.5 ± 1	0.65 ± 0.05
Oscillation period (ps)		0.180[a]	0.290[b]	
		0.185[c]	0.303[d]	0.311[e]
Dephasing time (ps)	fast[f]	0.63	0.435	
	slow[f]	2.94	2.03	
	fast[g]		0.22	
	slow[g]		0.78	

[a]Oscillation of luminescence intensity[23].
[b]Oscillation of luminescence intensity[20].
[c]Transient absorption by Pelt *et al.*[24].
[d]Transient absorption by Dexheimer *et al.*[21].
[e]Transient absorption by Sugita *et al.*[25].
[f]This work ($= 2\pi/W_i$). [g]This work (center of gravity analysis).

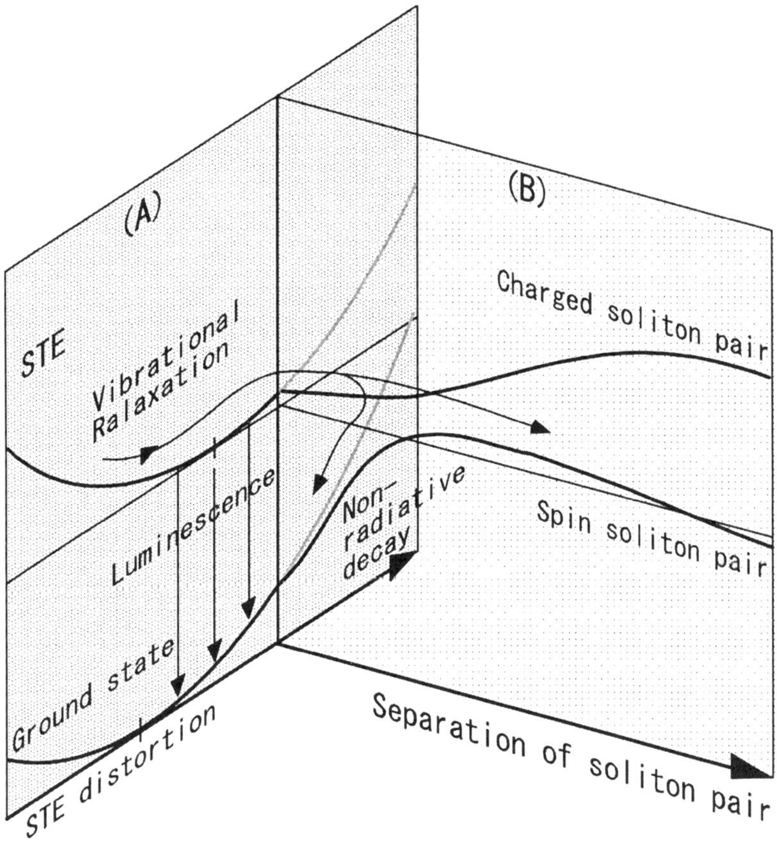

Fig. 18. Cross-sections of the adiabatic potential surfaces of the excited states in Pt-X (schematic). The abscissa on plane (A) and that on plane (B) correspond to the STE distortion and to soliton pair distance, respectively. The NS (spin soliton) is assumed to be lower than CS. The downward arrows show the radiative transitions, and the curved arrows show the possible non-radiative channels leading to the ground state or the soliton pair state.

in Fig. 18. The plane (A) is parallel to the interaction mode axis shown in Fig. 7 and the plane (B) is perpendicular to this axis. The axis lying on the plane (B) represents the separation of the soliton pair. A theoretical investigation based on an extended Peierls–Hubbard model[2] shows that the charged soliton pair state connects to the STE state and that the spin soliton pair state connects to the ground state with moderate values of material parameters.

The ground state crystal is at the minimum point of the lower surface on plane (A). By photo-excitation in the CT band, the free exciton states (not shown in this figure) with well-defined wave vectors in momentum space will be produced. Then accompanying the shrinkage of the electronic wave function, the system will come into the upper potential surface on plane (A). Then the lattice starts to deform towards the self-trapped state. The wave-packet starts to oscillate from the left to the right.

In Pt-Br, the oscillation decays in several hundreds femtoseconds and appreciable part returns non-radiatively to the ground state in several picoseconds. The non-radiative process will occur through a part where the upper and lower surfaces come closer. This is located on plane (B) in this figure. The system jumped to the lower surface will go either to the left (ground state-1 in Fig. 4) or to the right (spin soliton pair or ground state-2) with a certain branching ratio. The process of generating the spin soliton pairs from the STE has been shown by Okamoto *et al.*[28,29] for Pt-X-Pt and Pt-X-Pd families.

In Pt-Cl, the vibrational relaxation is mostly completed within one picosecond, and the luminescence transition from the minimum of the upper surface continues about 30 ps, which reflects mainly the non-radiative lifetime. This non-radiative relaxation is supposed to be caused by thermal activation through the potential barrier towards the initial ground state.[17]

In Pt-I, most of the excited states will go down immediately to the lower potential surface experiencing only one or two oscillations on plane (A). This is probably due to very low potential barrier between the curve on plane (A) and the jumping point on plane (B) or large jumping rate between the potential surfaces on plane (B). We can conclude that the light induced structural change is most easily realized in the systems with heavier halogen ions. By changing the material parameters only slightly, we can expect disappearance of the self-trapped exciton state and more efficient non-radiative process.

9. Visualization of the Wave-Packet

As already discussed in Sec. 5, the relaxation dynamics including the lattice distortion should be described in terms of the wave-packet propagation on the adiabatic potential surface. As for the shape of the wave-packets or the hot luminescence, there exist a few arguments from both experimental and theoretical sides. For example the theory of Kayanuma[15,30] or Toyozawa[31,32] discusses the shape of the hot luminescence reflecting the wave-packet, However, it was treated within an approximation which can represent only a Gaussian spectral shape.

The problem from the experimental side is the difficulty in observing the shape of the wave-packet. The oscillations of wave-packet in solids were observed in F-centers[33] and in self-trapped excitons (STE) of NaCl,[34] Pt-Br,[21] PtI[25] and $Pt(dmg)_2$.[35] In most of the experiments, however, the oscillatory behavior has been observed only at one or few points on the adiabatic potential surface. In addition, it has not been possible to correlate the observed data with the displacement of the atoms and it has been hard to elucidate the shape of the wave-packets. The shape of the wave-packet had to be estimated from the temporal behavior of the signal observed at certain point on the configuration coordinate axis, assuming a model. The comparison with theory has often been done only as a reproduction of the temporal behavior of the signal at a fixed point. One of the advantages of the luminescence up-conversion method for observing the wave-packet is

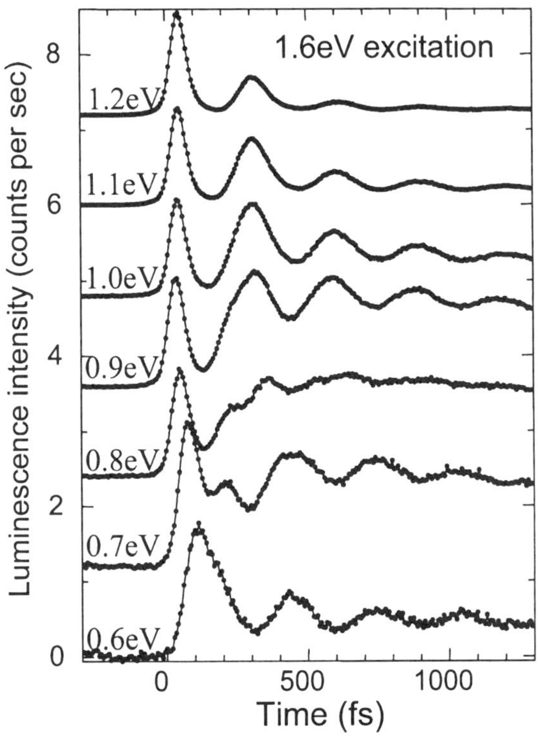

Fig. 19. Time evolutions of the luminescence intensity in Pt-Br under 1.6 eV excitation. Average power of the gate beam was 100 mW and of the pump beam was 1 mW. The observed photon energies are indicated near the corresponding curves. The ordinate shows the calibrated luminescence intensity (photons/unit energy interval). (Ref. 38)

that we can monitor the wave-packet at arbitrary point on the adiabatic potential surface in a very wide range.

Figure 19 shows the time evolution of the luminescence intensity measured with a greater time-resolution (50 fs) than in Sec. 5. The data were taken at every 0.1 eV between 0.6 and 1.2 eV with a time interval of 6.67 fs. At the photon energy of 0.8 eV (peak of luminescence spectrum), we can see the overtone oscillation with a period of about 150 fs, which is one half of the period of the wave-packet oscillation.[36] This corresponds to the passage of the wave-packet through the potential minimum every 150 fs in alternate directions. Moreover, we can observe the anti-phasing behavior between the oscillations at 0.7 and 0.9 eV. The luminescence below and above 0.8 eV probes the wave-packet at the points on the opposite sides of the potential minimum on the adiabatic potential curve. This clearly shows that the wave-packet is moving back and forth around the potential minimum. After a careful correction against the spectral response of the measurement system, we plotted the 7×240 data on a plane of the photon energy and the time delay in a gray scale as shown in Fig. 20.[36–38] The data between the points were estimated by linear interpolation. We can see clearly the sinusoidal motion of the wave-packet around 0.8 eV from 60 fs till 1300 fs, The amplitude of the first downward swing is about 0.2 eV, corresponding to about 0.087A displacement of the Br^- ion.

Fig. 20. Two-dimensional gray scale display of the time evolution of the luminescence intensity for Pt-Br. The ordinate is the luminescence photon energy, which approximately corresponds to the spatial position of the Br^- ion. The brighter part has larger amplitude.[23]

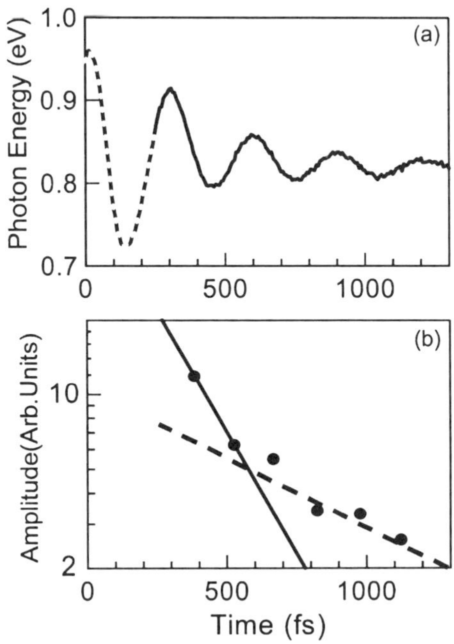

Fig. 21. (a) Time evolution of the 1st moment of the luminescence spectra for Pt-Br calculated for the data shown in Fig. 20. The dotted part of the curve does not correctly reflect the center of gravity, because large part of the spectrum is outside the observation range. The solid part of curve is slightly shifted upward due to the same effect. (b) Oscillation amplitude of the 1st moment plotted on a semi-logarithmic scale. The solid and dashed straight lines correspond to time constants of 220 fs and 780 fs, respectively (Ref. 23).

To see the damping behavior of the "center of gravity" of the wave-packet, we calculated the first moments of the spectra between 0.6 and 1.2 eV, and plotted them in Fig. 21(a). Since a large portion of the luminescence spectrum is outside the observed energy range in the first downward swing between 100 and 200 fs, the dashed part of the curve is shifted upward and does not correctly reflect the center of gravity. However, the error will be relatively small after 250 fs (solid part of the curve) for estimating the motion of the Br^- ion, because the area outside the energy range is expected to be less than 15% from the calculation shown later. The amplitude for each half cycle is shown in Fig. 21(b) in a logarithmic scale. It is clearly seen that the decay is highly non-exponential, showing the existence of multiple dephasing mechanisms. The large distribution in the relaxation time is in accordance with the existence of two Gaussian components in the spectrum of the phonon involved in the interaction mode (see Sec. 5). It is worth noting that the apparent decay time seen in the raw data in Fig. 19 strongly depends on the energy of observation. It is very fast at 1.2 eV while it is rather slow at 0.9 eV. This fact shows that the observation of the total shape of the wave-packet is inevitable even for deducing a correct decay time constant.

It is possible to make a movie showing a vivid motion of the wave-packet from the snapshots cut from this figure. Figure 22(a) shows some impressive snapshots. We can see not only the swing of the wave-packet, but also the change of its shape. The edge of the wave-packet is quite sharp in the first downward swing (87 fs). At 127 and 167 fs, the peak of wave-packet seems to locate outside the measured energy range and it comes back in sight at 207 fs. Then the peak moves to the left turning point around 287 fs. One of the most interesting features is that the shape becomes asymmetric and the amplitude goes higher

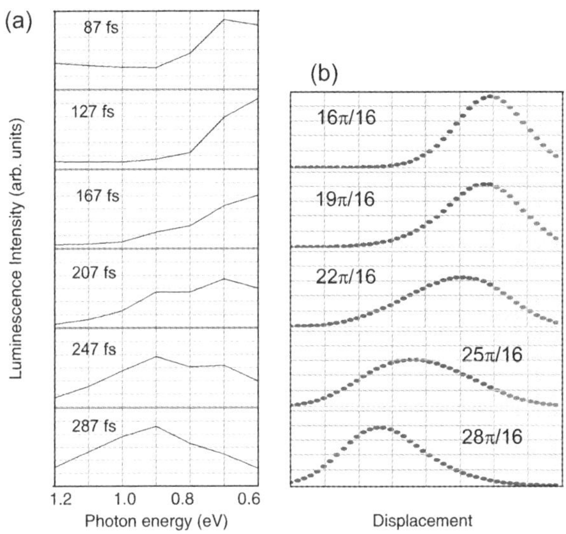

Fig. 22. Snapshots of the wave-packets till the end of the first cycle in Pt-Br. The curves in the column (a) are the vertical cross-sections cut from the 2D-plot shown in Figure 20. The frames shown are taken every 40 fs from 87 fs after excitation. The plots in the column (b) are the calculation based on the harmonic oscillator model without damping. The abscissa represents the configuration coordinate and the scale is adjusted to match the experimental data in the column (a). The area covered by gray shading is not observed in the experiment (Ref. 23).

at the turning points (127 and 287 fs) than at the center of oscillation (247 fs).

This behavior can be understood with a help of a simple calculation assuming a harmonic oscillator. In Secs. 5 and 6, we analyzed the dynamics of the wave-packet in terms of an interaction mode model. However the model assumed the Gaussian shape for the wave-packet on all the way of relaxation, which is evidently inadequate in understanding the time dependence of the wave-packet shape. Therefore, we use a model that expresses correct wave-packet shape but does not include the damping effect. In the following, we calculate the time-dependent wave-packet shape $\Psi(t,Q)$ by superposing the well-known wave functions of the harmonic oscillator.

$$\Psi(t,Q) = \sum_n c_n \psi_n(Q) \exp\left[i\hbar\omega nt\right] \tag{10}$$

$$\psi_n(Q) = \sqrt{\frac{1}{(2^n n! \sqrt{\pi})b}} H_n(Q/b) \exp\left(-\frac{1}{2}(Q/b)^2\right) \tag{11}$$

Here n is the index of vibrational levels, b is a constant having a dimension of length, $H_n(x)$ are the Hermite functions. The instantaneous luminescence intensity at configuration coordinate Q is given as

$$I(t,Q) = I_0 \left|\Psi(t,Q)\right|^2 .$$

For calculating the wave-packet, we have to estimate the vibrational quanta to be included. From Fig. 20, the amplitude of the first downward swing is estimated to be 0.2 eV from the center (0.8 eV). As the CT band edge locates at 1.4 eV, the Stokes shift is estimated to be 0.6 eV, and the lattice relaxation energy to be 0.3 eV. Then the vibrational energy corresponding to the amplitude of 0.2 eV is $0.3/(0.6/0.2)^2 = 0.033$ eV. As the vibrational quantum obtained from the oscillation period (300 fs) is 12.4 meV, this corresponds to 2.7 quanta. Subtracting 0.5 quantum for zero point vibration, the excited vibration quanta is about 2. Therefore, five eigenfunctions for $n = 0$ to 4 vibrational levels of a harmonic oscillator are superposed with an appropriate Gaussian weighting (1 : 4 : 6 : 4 : 1). An additional broadening of 0.2 eV (FWHM) was assumed in calculating the spectra. Figure 22(b) shows the calculated squared wave-packet amplitudes for one period of oscillation. The horizontal scales are adjusted to the plots of the experimental data. Since the damping is as large as factor two in a half period, the harmonic oscillator model without damping is not very appropriate. Therefore, we had to artificially adjust the parameters to reproduce the wave-packet

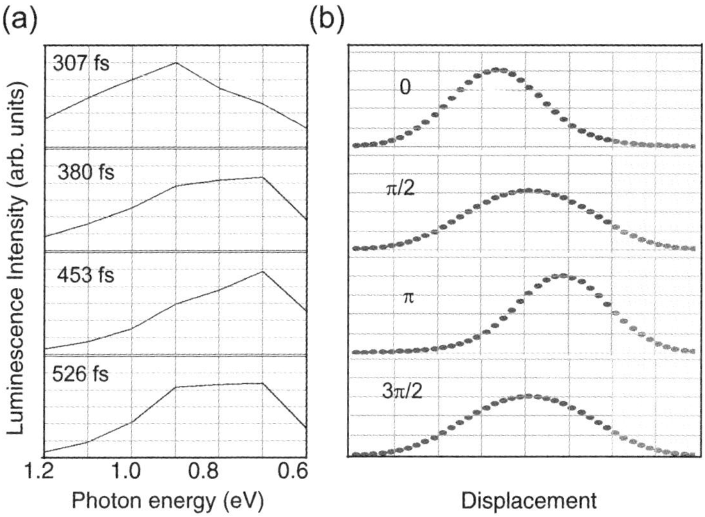

Fig. 23. Snapshots of the wave packet in the second cycle for Pt-Br. The frames shown are taken every 73 fs. The plots in the column (b) are the calculation for a harmonic oscillator. The area covered by gray shading is not observed in the experiment (Ref.38).

shape. The oscillation period in the calculation is taken as 426 fs, which gave the best fit. At a glance, we can notice a very close resemblance between the experiment and the calculation. At the turning points, the wave goes higher and becomes asymmetric, while it becomes low and flat near the center of oscillation, in accordance with the experiment. This is because the amplitude of the wave-packet becomes higher near the turning points due to nearly zero velocity there.

Another interesting feature is that the intensity at the center (0.8 eV) stays almost constant after 380 fs (Fig. 23(a)). This means that the levels higher than $n = 2$ are not excited at this stage. If the wave-packet had contained the components of symmetric wave functions ($n =$ even) other than $n = 0$, the amplitude of the wave-packet at the center would be modulated. Actually, the wave-packet shapes in this time range are well reproduced by a superposition of the wave-functions for $n = 0$ and $n = 1$ with the same weight as shown in Fig. 23(b). Here, the period is taken as 292 fs in the calculation.

10. Exciton Localization Process

The most intriguing and challenging problem in the exciton dynamics is the process from the extended free exciton state to the localized STE state.

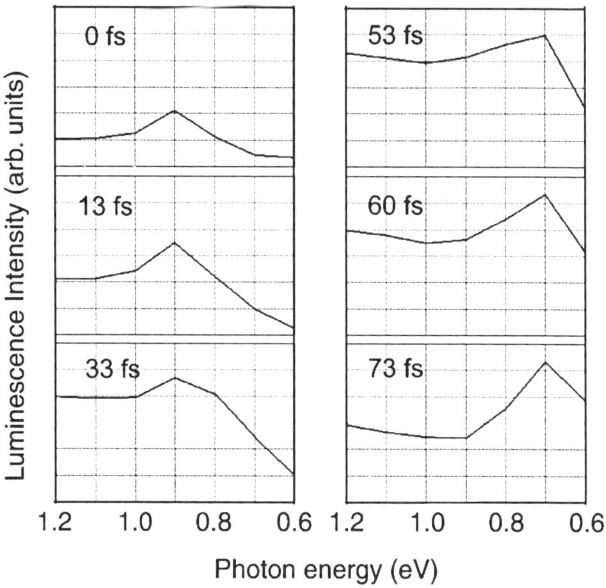

Fig. 24. Snapshots of the luminescence spectra at the early stages form 0 to 73 fs in Pt-Br.

An attempt to describe this process by molecular dynamic method has been reported by Ishida.[39] He has shown that the shrinkage of the electronic wavefunction will start after some waiting time corresponding to several periods of oscillations of the bulk phonon. Tanaka *et al.*[40] has shown that the hot luminescence peak stands still around the excitation photon energy for a while and then it starts oscillating in lower energy region, when the transfer energy is sufficiently large.

In Pt-Cl system, we estimated the upper limit of the formation time of STE as 140 fs from a shape analysis of the rising edge of the luminescence at 0.95 eV shown in Fig. 15.[22] A similar analysis for the Pt-Br system (Fig. 19) would give even smaller value comparable with the instrumental response for any observation photon energy. However, this argument is not appropriate for deducing "formation time" of STE, because it is not guaranteed that the onset time of luminescence defines the true time origin.

The formation time can be estimated more properly from the experimental data with precise determination of the time origin. The time origin of the data shown in Fig. 19 was defined as the peak position of the cross-correlation of the elastic scattering from the sample surface and the gate pulse. The overlapping of the beams at the nonlinear crystal might bring a timing error, because it depends on the phase-matching angle of the nonlinear crystal. However, this was avoided by setting the crystal so as to generate the sum frequency at the position very close to the front surface of the crystal. The second source of the timing error will be the finite penetration depth of the exciting light in the Pt-X sample. If there is an appreciable size of this effect, the shape of each peak in the time evolution curves in Fig. 19 will be asymmetric (tailing to the right). Judging from the quite symmetric shape of the first peak (e.g. at 0.9 eV) in this figure, the error is estimated to be smaller than one third of the time resolution (17 fs), if it exists. Therefore, we can believe the finite delay (35 fs) of the first peak at 0.9 eV in Fig. 19 is not an artifact. This delay should correspond to the lower limit of the formation time of the STE.

Another curious fact is that the wave-packet does not appear from the highest point (1.2 eV) in Fig. 20, but it starts from around 0.9 eV. If the wave-packet were produced on the adiabatic potential curve at the point vertically above the equilibrium position of the ground state as in the case of isolated molecule, the wave-packet should come into sight from the top of this figure at $t = 0$ with a finite downward slope. In contradiction to this expectation, the luminescence appears from 0.9 eV and spreads almost simultaneously in the range between 0.9 and 1.2 eV (see the vertical bright line at 50 fs in Fig. 20). If we extrapolate the sinusoidal wave from 250 to 1300 fs to the negative direction on the time axis in Fig. 21(a) keeping the same period, the center of gravity position at $t = 0$ is estimated to be at the highest position in the luminescence energy (cosine like behavior). These facts suggest that the lattice distortion starts from the initial equilibrium position just on the light excitation but the luminescence is not observed due to some reason. If we define the formation time of STE as the time required for the definite oscillation to appear, it will be about 70 fs from Fig. 20, corresponding to a quarter of oscillation period. Thus we can estimate the STE formation time to fall between 35 and 70 fs. Even if the lattice distortion starts at $t = 0$, it takes at least a quarter of oscillation period to reach the center of oscillation (0.8 eV). Therefore, we can conclude that the waiting time predicted in Refs. 39 and 40 is negligible.

The self-trapping process of the exciton will start from the free exciton band with delocalized electron and hole wave functions. The shrinkage of the electronic wavefunctions and the lattice deformation towards the STE equilibrium will proceed cooperatively. Before establishment of the fully localized state, the overlapping of the electron and hole wavefunctions is small, and the oscillator strength will be small. This can be the reason why the luminescence at early stage is weak.

The origin of the broad band luminescence around 50 fs is not clear now. Luminescence from FE is unlikely to give a very broad spectrum in such a low energy region far from the band edge. In order to describe the dynamics of the self-trapping, we have to consider an adiabatic potential surface in a two-dimensional space with the axes corresponding to the STE distortion and the size of the deformed region. The broad luminescence spectrum might occur during this localization process, due to the complicated shape of the 2D potential surfaces connecting the free state and the STE minimum.

Acknowledgments

The author is indebted to Dr. S. Saito, Dr. K. Sakata, Prof. J. Takeda and Prof. S. Kurita for collaboration. This work has been supported by Grant-in-Aid for Scientific Research (A), (B) and Grant-in-Aid on priority areas, "Photo-induced Phase Transition and Their Dynamics", from the Ministry of Education, Culture, Sports, Science and Technology of Japan.

References

1. K. Toriumi, M. Yamashita, S. Kurita, I. Murase and T. Ito, *Acta Cryst.* **B49** 497 (1993).
2. K. Iwano, *J. Phys. Soc. Jpn.* **66** 1088 (1997).
3. K. S. Song and R. T. Williams, *Self-Trapped Excitons*, Vol. 105 of *Solid-State Sciences* (Springer, Berlin, 1993).
4. H. Okamoto, Y. Oka, T. Mitani and M. Yamashita, *Phys. Rev.* **B55** 6330 (1997).
5. K. Iwano, *Phys. Rev.* **B61** 279 (2000).
6. S. C. Huckett, B. Scott, S. P. Love, R. J. Donohoe, C. J. Burns, E. Garcia, T. Frankcom and B. I. Swanson, *Inorg. Chem.* **32** 2137 (1993).
7. H. Endres, H. J. Keller, R. Martin, H. N. Gung, and U. Traeger, *Acta Crystallogr, Sect. B, Struct. Crystallogr. Cryst. Chem.* **35** 1885 (1979).
8. Y. Wada, T. Mitani, M. Yamashita and T. Koda, *J. Phys. Soc. Jpn.* **54** 3143 (1985).
9. J. Takeda, M. Okada, S. Kurita, K. Tanaka and T. Suemoto, *Phys. Rev.* **B52** 14441 (1995).
10. R. J. Donohoe, L. A. Worl, C. A. Arrington, A. Bulou and B. I. Swanson, *Phys. Rev.* **B45** 13185 (1992).
11. G. F. Strouse, B. Scott, M. Berkey, S. Johnson, T. Darling, A. Migliori and B. I. Swanson, *Synth. Met.* **86** 1919 (1997).
12. M. Tanaka, S. Kurita and Y. Okada, T. Kojima and Y. Yamada, *Chem. Phys.* **96** 343 (1985).
13. Todd S. Rose, Mark J. Rosker and Ahmed Zewail, *J. Chem. Phys.* **91** 7415 (1989).

14. A. H. Zewail, *J. Phys. Chem.* **A104** 5660 (2000).
15. Y. Kayanuma, *J. Phys. Soc. Jpn.* **57** 292 (1988).
16. H. Tanino, W. W. Rühle and K. Takahashi, *Phys. Rev.* **B38** 12716 (1988).
17. Y. Wada, U. Lemmer, E. O. Göbel, M. Yamashita and K. Toriumi, *Phys. Rev.* **B52** 8276 (1995).
18. S. Tomimoto, S. Saito, T. Suemoto, Jun Takeda and Susumu Kurita, *Phys. Rev.* **B66** 155112 (2002).
19. H. Okamoto, private communications.
20. S. Tomimoto, S. Saito, T. Suemoto, K. Sakata, J. Takeda and S.Kurita, *Phys. Rev.* **B60** 7961 (1999).
21. S. L. Dexheimer, A. D. V. Pelt, J. A. Brozik and B. I. Swanson, *Phys. Rev. Lett.* **84** 4425 (2000).
22. S. Tomimoto, H. Nansei, S. Saito, T. Suemoto, J. Takeda and S. Kurita, *Phys. Rev. Lett.* **81** 417 (1998).
23. T. Matsuoka, T. Suemoto, J. Takeda and S. Kurita, to appear in *Procs. of International Conference on Dynamical Processes in Excited States of Solids*, Christchurch, 2003 (Elsevier).
24. A. D. Van Pelt and S. L. Dexheimer, *Ultrafast Phenomena XII* (Springer-Verlag, Heidelberg, 2001) p. 393.
25. A. Sugita, T. Saito, H. Kano, M. Yamashita and T. Kobayashi, *Phys. Rev. Lett.* **86** 2158 (2001).
26. Y. Wada, K. Era and M. Yamashita, *Solid State Commun.* **67** 953 (1988).
27. H. Okamoto, T. Mitani, K. Toriumi and M. Yamashita, *Materials Science and Engineering* **B13** L9 (1992).
28. H. Okamoto and M. Yamashita, *Bull. Chem. Soc. Jpn.* **71** 2023 (1998).
29. H. Okamoto, Y. Kaga, Y. Shimada, Y.Oka, Y. Iwasa, T. Mitani and M. Yamashita, *Phys. Rev. Lett.* **80** 861 (1998).
30. Y. Kayanuma, *Phys. Rev.* **B41** 3360 (1990).
31. Yutaka Toyozawa, Akio Kotani and Atsuko Sumi, *J. Phys. Soc. Jpn.* **42** 1495 (1977).
32. Y. Toyozawa, *Optical Processes in Solids* (Cambridge University Press, Cambridge, 2003) Ch. 11.
33. M. Nisoli, S. D. Silvestri, O. Svelto, R. Scholz, R. Fanciulli, V. Pellegrini, F. Beltram and F. Bassani, *Phys. Rev. Lett.* **77** 3463 (1996).
34. T. Tokizaki, T. Makimura, H. Akiyama, A. Nakamura, K. Tanimura and N. Itoh, *Phys. Rev. Lett.* **67** 2701 (1991).
35. S. Iwai, T. Kamata, S. Murata, K. Yamamoto and T. Ohta, *J. Luminescence* **87** 629 (2000).
36. T. Matsuoka, S. Saito, J. Takeda, S. Kurita and T. Suemoto, *Nonlinear Optics* **29**(10–12) 587 (2002).
37. T. Matsuoka, T. Suemoto, J. Takeda and S. Kurita, *Technical digest of CLEO/QELS 2003 (Postdeadline papers book)* (Baltimore, June 2003).
38. T. Matsuoka, T. Suemoto, J. Takeda and S. Kurita, *Phys. Rev. Lett.* **91**, 247402 (2003).
39. K. Ishida, *Solid State Commun.* **90** 89 (1994).
40. S. Tanaka and Y. Kayanuma, *J. Luminescence* **87** 936 (2000).

CHAPTER 5

MONTE CARLO SIMULATIONS ON ISING-LIKE MODELS FOR PHOTOINDUCED PHASE TRANSITIONS

Tohru Kawamoto and Shuji Abe
Nanotechnology Research Institute
National Institute of Advanced Industrial Science and Technology (AIST)
1-1-1 Umezono, Tsukuba 305-8568, Japan
E-mail: tohru.kawamoto@aist.go.jp

Monte Carlo simulations on Ising-like models are an important theoretical approach to the problem of photoinduced phase transitions. The purpose of this chapter is threefold: (1) to give a basic description of the Ising-like models and the Monte Carlo simulation method for non-specialists; (2) to demonstrate typical behaviors of the photoinduced phase transition with examples of calculated results; (3) to discuss a theoretical proposal on the structural design for efficient photoinduced phase transitions.

Among the materials exhibiting photoinduced phase transitions, there are a class of materials in which the origin of bistability is local, like in spin-crossover complexes. In such a system, each local unit has two (or more) possible states, which can be most simply modeled as spin states. Then the system is regarded as a collection of so-called pseudospins interacting with each other. This Ising-like model is characterized by the energies of individual pseudospins and the interaction energies between pseudospins. The physical meanings of the parameters in the model vary, depending on the material. There are also various situations depending on the condition of photoexcitation. We describe in detail how this model with the effect of external excitation is treated by the Monte Carlo simulation method.

We illustrate basic characteristics of photoinduced phase transitions, showing typical results of Monte Carlo simulations. We discuss steady states under continuous excitation as well as transient processes after the sudden application of excitation. The existence of threshold excitation intensity for the transition is demonstrated. We also show results for the dynamical phase transition, which occurs only under a sufficient intensity of continuous excitation, in a system with relatively small interactions between pseudospins.

To achieve bidirectional photoinduced phase transitions with high efficiency, we have proposed a theoretical model of a block superstructure with an appropriate combination of two kinds of bistable units. Our Monte Carlo simulations clearly demonstrate that such a combined system, in comparison with the uniform system, exhibits an accelerated phase transition with reduced threshold excitation intensity, as a result of rapid nucleation of converted domains.

Contents

1. Ising-Like Model for Photoinduced Phase Transitions

The term "photoinduced phase transition" implies two aspects. Firstly, it is a cooperative phenomenon analogous to ordinary phase transitions. Secondly, it is a unique phenomenon induced by light irradiation. This is interpreted here in a strict sense so that light is directly responsible for the phenomenon, in contrast to the case where light is first converted to heat, which then drives an ordinary thermal phase transition. Actually, this latter type of phase transition has been utilized in today's optical storage devices such as DVD-RAM, where laser light melts an optical material to switch between amorphous and crystalline phases. If an efficient photoinduced non-thermal phase transition is realized, it may also be utilized for optical storage devices. From a fundamental point of view, even a new phase that cannot be realized by changing temperature may be created by photoirradiation.[1]

The cooperative nature of the photoinduced phase transition appears in a couple of characteristics observed typically: (i) the existence of a threshold light intensity for the transition;[2–4] (ii) the existence of the initial "incubation period" in the temporal evolution of the transition.[2] Although the crystal of photochromic molecules may change its color by photoirradiation, it may not be regarded as a photoinduced phase transition unless there are strong interactions among molecules.

Photoinduced phase transitions are observed in various materials and hence are described by various theoretical models. However, most of the essential features of the phenomenon are captured by a class of rather simple mathematical models called Ising-like models. The Ising model is a standard model to describe the magnetic properties of systems consisting of localized electron spins as shown schematically in Fig. 1. In this system, electrons cannot travel in the crystal but stay on each atomic site, so that the freedom of each electron is only the direction of spin, up or down. The Ising Hamiltonian represents the internal energy of the system,

$$E = -\sum_i HS_i - \sum_{i,j} J_{ij} S_i S_j, \tag{1}$$

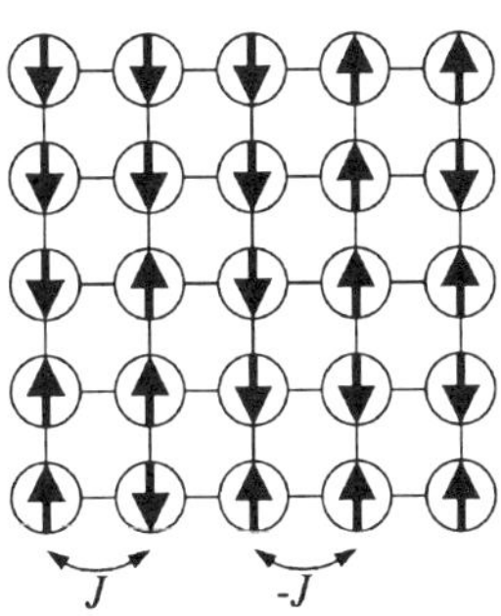

Fig. 1. Lattice of localized spins whose states are represented by the direction of arrows, up or down, in the Ising model. The interaction energies between parallel and anti-parallel spins are $-J$ and J, respectively.

where S_i reprcsents a spin variable at ith site: it is assumed that S_i takes either 1 or -1, corresponding to up or down. The first term of the right hand side represents the Zeeman energy due to the external magnetic field H. If H and S_i are parallel, the system energy decreases. Otherwise it increases. The second term is the interaction energy between spins with exchange parameters J_{ij}. Depending on whether two spins are parallel or antiparallel to each other, the interaction energy changes its sign. The state of the whole system is specified by the complete set of the spin variables,

$$\mathbf{S} = (S_1, S_2, \ldots, S_i, \ldots, S_{N_S}), \tag{2}$$

with N_S being the number of spins (sites) in the system. In the case of all J_{ij} being positive, the ground state of the system is ferromagnetic, that is, all the spins align in the same direction.

The Ising model can be used in a much wider context for a system consisting of a set of local states, not necessarily spin states. An example is bistable local states accompanied by lattice distortions. A schematic adiabatic potential diagram for such bistable states is shown in Fig. 2(a) with a configuration coordinate u_i representing a variation of atomic positions. The two local energy minima separated by an energy barrier correspond to two different states, say, A and B. If the energy of the system is determined mostly by the two states, we can use, instead of the continuous variable u_i, the discrete *pseudospin* variable S_i, which takes the values of 1 and -1 for the states A and B, respectively, as shown in Fig. 2(b). Then the energy

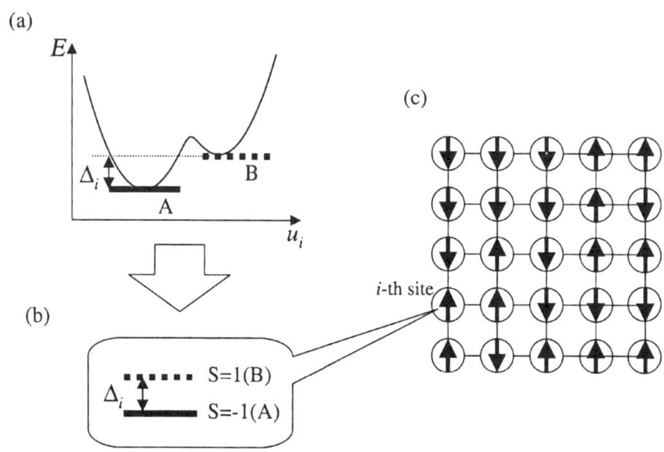

Fig. 2. Physical implications of pseudospins in a Ising-like model. (a) The adiabatic potential of a constituent unit that have bistable states A and B. (b) It is represented by a two-level model, or a pseudospin. (c) The Ising-like model consisting of pseudospins.

of the system may have a form similar to the original Ising model as

$$E = \sum_i \varepsilon_i S_i - \sum_{i,j} J_{ij} S_i S_j. \tag{3}$$

The first term represents the intrinsic energy difference between the state A and B at the i-th constituent unit. Note that ε_i is a material parameter which depends on the unit and usually not zero, in contrast to the corresponding external field H in Eq. (1). The second term represents interactions between constituent units. In many cases the interaction strength J_{ij} is set to be positive (corresponding to a ferromagnetic interaction in the original Ising model), although the microscopic origin of the interaction depends on the system. It should be noted that the pseudospins are usually treated as classical variables in this context.

The Ising-like model is applicable if (i) the system consists of local units, each of which has two distinct configurations, and if (ii) the state of the whole system is determined by the set of the local configurations.

The Ising-like model given by Eq. (3) has been applied to transition metal complexes, where the transition metal ions with ligand molecules can be regarded as constituent units with localized bistable spin states, i.e., low-spin and high-spin states. In particular, the model has been widely used to describe the low-spin to high-spin transition in Fe(II) spin-crossover complexes.[5–8] Another example of bistability is the adsorbed and desorbed states of an atom on the surface. In contrast, Ising-type models are not suitable for the description of photoinduced phase transitions in metallic systems with itinerant electrons, *e.g.*, photoinduced metal-insulator transitions in manganese perovskite.

2. Ising-Like Model for Spin-Crossover Complexes

As an example of real materials described by the Ising-like model, we introduce the spin-crossover complexes in more detail.

2.1. *Experimental facts*

Spin-crossover complexes are a group of Fe(II) coordination compounds exhibiting a thermal spin-state transition.[9] The high- and low-temperature phases correspond to the high spin ($^5T_2$: S = 2) and low spin ($^1A_1$: S = 0) states of Fe(II), as shown in Fig. 3(a). Switching between the two phases was realized also by photoirradiation bidirectionally,[1,2,6,9–11] although some structural differences between the thermally induced phase and the photoinduced one have been reported.[1] The photoinduced transition cannot be

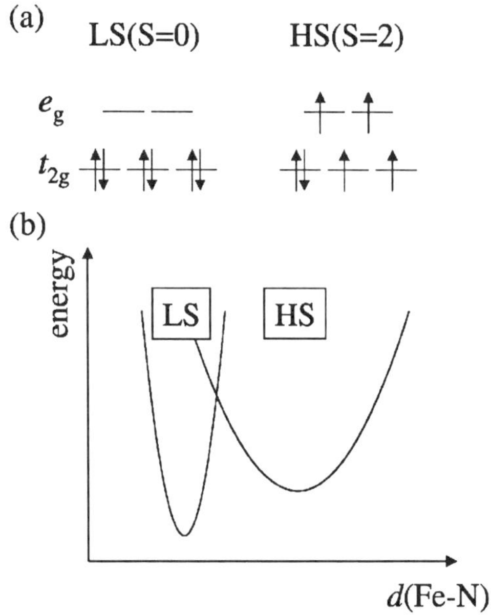

Fig. 3. (a) Spin configurations of Fe(II) in spin crossover complexes in the low temperature phase (low-spin: $S = 0, {}^1A_1$) and in the high temperature phase (high spin: $S = 2, {}^5T_2$). (b) A schematic diagram of adiabatic potentials along the distance between Fe and N, d(Fe-N), in spin crossover complexes (see Fig. 5).

explained by thermal effects, since the transition from the high temperature phase to the low temperature phase has also been realized.

Here we introduce relevant experimental results in spin crossover complexes.[2] Koshihara and his coworkers studied the dynamical processes of photoconversion in single crystals of a typical spin crossover complex, $[Fe(2\text{-pic})_3]Cl_2 \cdot EtOH$ (2-pic = 2-aminomethyl-pyridine) at low temperatures. The main features observed in the photoinduced transition from the low-spin state to the high-spin state are: (a) the existence of a threshold light intensity, (b) the existence of a so-called incubation period before of the phase transition, and (c) step-like temporal behavior associated with phase separation. These features were observed in the evolution of the low spin component under continuous excitations, as shown in Fig. 4. In addition, the photoconversion efficiency was very high: a photon on average can change about tens of Fe sites. These features indicate cooperative interactions between the spin states of Fe ions. The spin-crossover complexes satisfy the requirements for the application of the Ising-like model mentioned in Sec. 1.

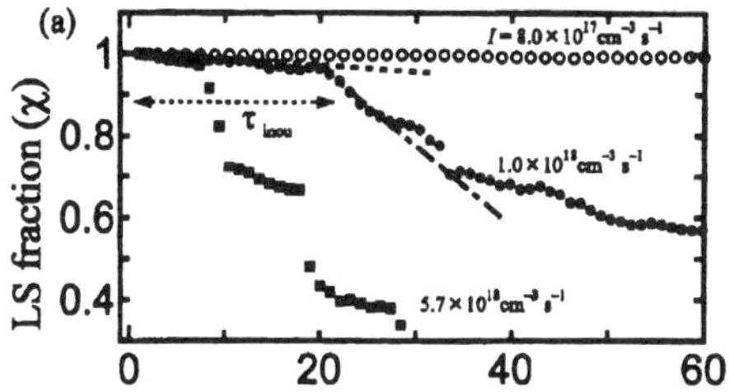

Fig. 4. Time-dependence of photoconversion from the low-spin state to the high-spin state observed in the spin crossover complex $[Fe(2\text{-pic})_3]Cl_2 \cdot EtOH$ (2-pic = 2-aminomethyl-pyridine) at 2.2 K for various excitation photon fluxs (I). $\tau_{\rm incu}$ indicates the incubation period in the case with $I = 1.0 \times 10^{18}\,\mathrm{cm^{-3}\,s^{-1}}$. [Y. Ogawa, S. Koshihara, K. Koshino, T. Ogawa, C. Urano, and H. Takagi, *Phys. Rev. Lett.* **84**, 3181 (2000). Copyright (2003) by the American Physical Society.]

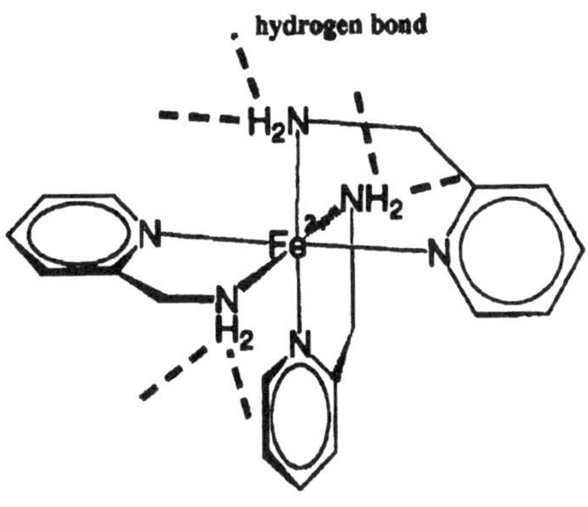

Fig. 5. The local structure around Fe(II) in the $[Fe(2\text{-pic})_3]Cl_2 \cdot EtOH$ (2-pic = 2-aminomethyl-pyridine. [Y. Ogawa, S. Koshihara, K. Koshino, T. Ogawa, C. Urano, and H. Takagi, *Phys. Rev. Lett.* **84**, 3181 (2000). Copyright (2003) by the American Physical Society.]

The structure of the constituent unit of $[Fe(2\text{-pic})_3]Cl_2 \cdot EtOH$ is shown in Fig. 5. Iron atoms are surrounded by six nitrogen atoms octahedrally, resulting in the splitting of Fe 3d orbitals into triply degenerate lower t_{2g} orbitals and doubly degenerate higher e_g orbitals, as shown in Fig. 3(a). At low temperatures, each Fe(II) atom is in its low spin state where all the 3d electrons stay in the t_{2g} orbitals. At temperatures higher than the transition temperature, Fe(II) changes to its high spin state, although there is no magnetic order at the high spin phase because of very weak magnetic interaction between Fe(II) ions in this material.

The transition from the nonmagnetic phase to the paramagnetic phase is accompanied by a structural change. The averaged distance between iron and surrounding nitrogen atoms, d(Fe-N), depends on the spin state. Figure 3(b) shows a schematic view of adiabatic potentials as a function of d(Fe-N). The equilibrium distance of the high-spin state is longer by about 0.2 Å than that of the low-spin state, and the two local minima are separated by an energy barrier. This is the origin of local bistability.

Figure 3(b) also illustrates two important facts. One is the relative energies of the two local minima: The energy minimum of the low-spin state is lower than that of the high-spin state. Consequently, the low-spin state is the ground state at $T = 0$. The other point is the different curvatures of the two adiabatic potentials: The curvature of the low spin state is larger than that of the high-spin state. This difference becomes important in discussing an entropy effect for thermal phase transitions in the next subsection.

2.2. *Internal entropy effect*[5,8]

In order to treat temperature-dependent properties of the spin crossover complexes, the Ising-like model given by Eq. (3) must be slightly modified, because entropy effects associated with the degrees of freedom other than the pseudo-spin variables, *e.g.*, the local vibrational entropy as mentioned above. A modified Ising-like model is constructed as:

$$E = \sum_i \left(\varepsilon_i + \frac{k_B T}{2} \ln g \right) S_i - \sum_{i,j} J_{ij} S_i S_j, \tag{4}$$

where k_B is the Boltzman constant and T is temperature. Here $S_i = -1$ and 1 correspond to the low- and high-spin states, respectively. The relative energy difference ε_i is supplemented by the additional term proportional to kT due to the internal entropy effect, where g represents the degeneracy ratio $g = g_+/g_-$ with g_- and g_+ being the degeneracies of $S_i = -1$ and $+1$ states. This is a phenomenological way of correcting the relative energies of the two states, whose internal degrees of freedom are not explicitly taken into account in the Ising-like model.

In many theoretical studies, the degeneracy ratio g has been treated as a phenomenological parameter. However, it is instructive to discuss the origin of the degeneracy. In the actual material, the main contributions come from electron spins, electronic orbitals, and vibrations.

The electronic parts in the degeneracy are simply obtained. The spin degeneracy $2S + 1$ is 1 and 5 for the low temperature state with $S = 0$ and for the high temperature state with $S = 2$, respectively. The orbital degeneracy is 1 and 3 for the low and high temperature states due to the Jahn–Teller splitting of the 3d-orbitals as shown in Fig. 3(a). Note that the orbital degeneracy is sometimes lifted due to an asymmetric ligand field. As for the vibrational degeneracy, the high temperature state may have a larger degeneracy, *i.e.*, a larger density of states due to smaller vibrational frequencies. In fact, the Raman spectroscopy measurements by Bousseksou *et al.* indicate that the vibrational degeneracy is much larger in the high temperature state, and that this is the major contribution to the full entropy.[12] Thus all the contributions to the degeneracy are larger in the high temperature state than in the low temperature state, implying that the free energy difference between the bistable states is strongly temperature dependent, resulting in the thermally induced phase transition.

2.3. *Cooperative interactions*

The second term in Eq. (4) represents ferromagnetic-like interactions between constituent units. In the spin crossover complexes, the origin of the interunit interaction is homogenization of the Fe(II) atomic radii, as shown schematically in Fig. 6. The atomic radius is larger in the high-spin state than in the low-spin state. The homogeneous states in which all of the units are in the same state (-1 or $+1$) are shown in Figs. 3(a) and (b). There is no elastic stress because the atomic radii are identical. On the other hand, in a coexistent state shown in Fig. 3(c), there is a strain on the boundary between the two states. Therefore, the uniform states are preferable compared to the coexistence state.

In Eq. (4), the interaction term is assumed to be a sum of products of pseudospins, although its microscopic derivation may not be easy. An important question is the dependence of the interaction strength on the interunit distance. The elastic interaction depicted in Fig. 6 is regarded primarily short-ranged, but also a long-range interaction has been considered with the idea of "fictitious pressure".[13] This is especially important

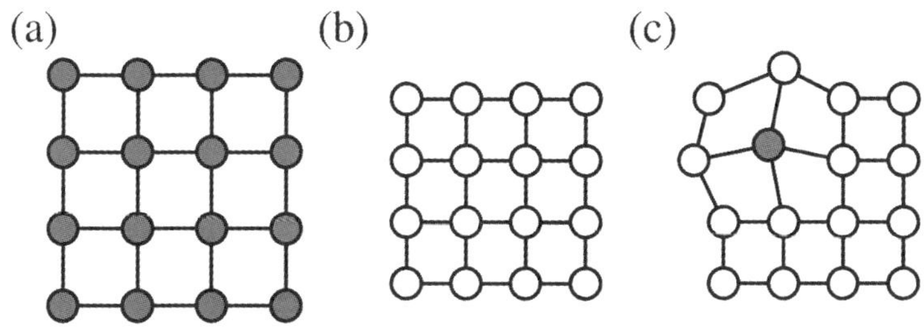

Fig. 6. Cooperative interactions due to correlated interunit lengths. (a) All the units having a large ionic radius. (b) All the units having a small ionic radius. (c) When both the radii coexist, the lattice has a large stress.

in the Monte Carlo method, which takes account of the spatial correlation of pseudospins **S**. Overall experimental results have been simulated with the nearest-neighbor coupling model, although long-range interactions have also been necessary for accurate fitting. In the case of a mean-field approximation, the range of interaction does not matter, because spatial correlation is not taken into account for the averaged pseudospin $\langle S_j \rangle$.

2.4. *Photoinduced vs. thermally induced phases*

A motivation for the study of the photoinduced phase transitions has been search for a new solid phase that does not appear by thermal treatment. However, the Ising-like model cannot be applied to such a system where a photoinduced phase is different from both the low-temperature and high-temperature phases. Even in the spin crossover complex $[Fe(2-pic)_3]Cl_2 \cdot EtOH$, it has been claimed that the photoinduced high-spin phase obtained from the low temperature phase by photoirradiation is different from the high-temperature high-spin phase with respect to the crystal structure, although the difference is not yet very clear.[1] This difference is neglected in the present approach with the Ising-like model, where each constituent unit have only two states. If the difference between the photoinduced phase and the thermally induced phase is quite large, further modification of the model is required.

3. Monte Carlo Simulation Method

This section provides a tutorial overview of the Monte Carlo simulation method used in our calculations. The readers who are not interested in the details of the theoretical procedures can skip this section and go directly to Sec. 4.

3.1. *Monte Carlo simulations for Ising-like models*

The Monte Carlo method is a generic term representing computer simulation techniques using stochastic procedures, although there are a variety of actual procedures in this category. In the case of the original Ising model, various properties in equilibrium as well as in nonequilibrium states have been investigated by this method. For example, magnetization behavior in the oscillating field,[14] relaxation dynamics of spin-glass systems,[15] or effect of a small difference in a initial state on the time evolution process (so-called damage spreading)[16] have been studied.

The Monte Carlo method has been successfully used also in studying a wider class of phase transitions described by the Ising-like model. For the spin-crossover system, Sakai *et al.* reported applications of this method to the photoinduced phase transition.[17,18] Romstedt *et al.* studied the thermal relaxation process of excited state.[6,7] Nishino *et al.* discussed photoinduced magnetization.[19,20] We also applied this method to examine the efficient photoinduced phase transition in nanostructured materials,[21–23] which is discussed in Sec. 5.

Both equilibrium and non-equilibrium states can be investigated with Monte Carlo simulations. In the Ising-like model we define two quantities: the excitation rate W and the state fraction ρ_{X} (X = A or B). W is a given parameter representing a rate of creating an excited state corresponding to photoexcitation. The case of $W = 0$ and $W \neq 0$ indicate the condition without or with photoirradiation, respectively. Actual procedure of the treatment of W in simulations is described later. On the other hand, ρ_{X} is a physical quantity obtained by calculations. It represents the fraction of the units in each state, given by

$$\rho_{\mathrm{B}} = \frac{1}{N_{\mathrm{S}}} \sum_i \delta_{S_i+1}. \tag{5}$$

$$\rho_{\mathrm{A}} = 1 - \rho_{\mathrm{B}} \tag{6}$$

The equilibrium state is defined as $d\rho_{\mathrm{B}}/dt = 0$ under $W = 0$. A nonequilibrium state defined by $d\rho_{\mathrm{B}}/dt \neq 0$ under $W = 0$ can also be investigated by Monte Carlo simulations,[6,7] although this is not the topic of the present article. The main interest here is the steady state defined by $d\rho_{\mathrm{B}}/dt = 0$ under photoexcitation $W \neq 0$, and also the transient state defined by $d\rho_{\mathrm{B}}/dt \neq 0$ under photoexcitation $W \neq 0$.

3.2. *Master equation*

To apply the Monte Carlo method, we assume that the dynamics of the system is described by the probability $P(\mathbf{S}, t)$ of the system being in each state $\mathbf{S}$ at each time t. It is assumed that the probability distribution function $P(\mathbf{S}, t)$ is governed by a master equation. One of the most popular master equations is the Glauber dynamics:

$$\frac{dP(\mathbf{S}, t)}{dt} = \sum_{i=1}^{N} \{-w_i(S_i)P(\mathbf{S}, t) + w_i(-S_i)P(S_1, S_2, \ldots, -S_i, \ldots, S_N, t)\}. \tag{7}$$

Here $w_i(S_i)$ is the rate constant of the pseudospin S_i changing its configuration to $-S_i$ in a unit time. The first and the second terms of the right hand side-represent the rate of changing from $(S_1, S_2, \ldots, S_i, \ldots, S_N)$ to $(S_1, S_2, \ldots, -S_i, \ldots, S_N)$ and that of the inverse process, respectively. Note that this master equation assumes that each state change is local and independent flipping of a pseudospin. For example, it does not include simultaneous flipping of two opposite pseudospins, which is called Kawasaki dynamics.

$P(\boldsymbol{S}, t)$ is obtained by the time integration of Eq. (7) starting from a given initial distribution $P_0(\boldsymbol{S}, t_0)$. To obtain an equilibrium state we should continue the time integration until $P(\boldsymbol{S}, t)$ converges to a certain equilibrium distribution $P_{\text{eq}}(\mathbf{S})$. Then, the expectation value of A is calculated as

$$\langle A \rangle = \sum_{\mathbf{S}} A(\mathbf{S}) P_{\text{eq}}(\mathbf{S}), \tag{8}$$

where $A(\mathbf{S})$ represents the value of A when the system in the state $\mathbf{S}$.

In the Monte Carlo simulation method, $P(\boldsymbol{S}, t)$ is not calculated directly, but the state $\mathbf{S}$ is calculated as a function of t as $\mathbf{S}(t) = (S_1(t), S_2(t), \ldots, S_i(t), \ldots, S_N(t))$ stochastically, so that we obtain a single sample of time evolution of $\boldsymbol{S}(t)$. Various physical properties are obtained with statistical averaging over many runs.

In Monte Carlo simulations, time is discretized with a small step Δt called the Monte Carlo step (MCS), which is usually used as a unit of time, i.e., $\Delta t = 1$. At each MCS, all the sites are scanned for stochastic jumps, as schematically shown in Fig. 7. The state $S_i(t) = s$ of site i at time t is changed to $S_i(t + \Delta t) = -s$ with a probability $w_i(s \to -s)$ or keeps its value with a probability $1 - w_i(s \to -s)$.

The time dependence of a quantity A is calculated as $A(t) = A(\boldsymbol{S}(t))$ for each run of Monte Carlo simulation. The expectation value of $A(t)$ is calculated by performing an average over sufficient large number of Monte Carlo runs:

$$\langle A(t) \rangle = \frac{1}{N_t} \sum_{k=1}^{N_t} A(\mathbf{S}_i(t)), \tag{9}$$

where N_{t} is the number of runs. When the system approaches to a steady state, $\langle A(t) \rangle$ tends to be independent of time.

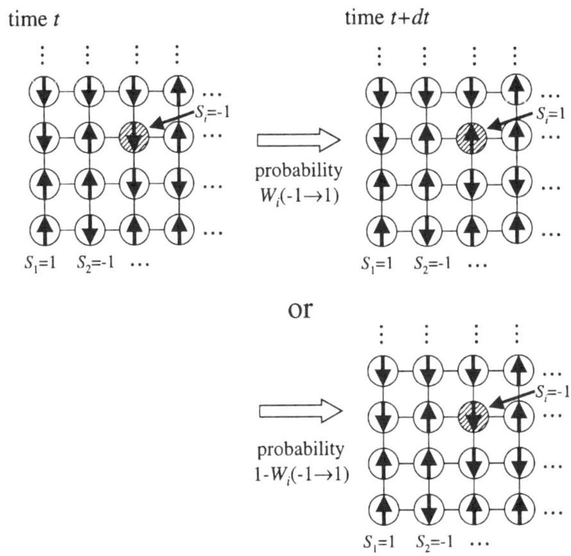

Fig. 7. Individual procedure of a Monte Carlo simulation with Glauber dynamics.

3.3. *Transition rates*

Because the transition rates w_i govern the dynamics of $\mathbf{S}(t)$, they must be determined carefully. In general, w_i is a function of $\mathbf{S}(t)$ and contains various contributions including thermal fluctuation part $w_i^{\mathrm{T}}(S_i)$, quantum mechanical tunneling part $w_i^{\mathrm{Q}}(S_i)$, and photoexcitation part $w_i^{\mathrm{P}}(S_i)$. We introduce these three rates in due order.

3.3.1. *Thermal fluctuations*

In thermal equilibrium, $P(\mathbf{S}, t)$ should be given by a canonical distribution

$$P_{\mathbf{eq}}(\mathbf{S}) = \frac{\exp(-\beta E(\mathbf{S}))}{\sum_{\mathbf{S}} \exp(-\beta E(\mathbf{S}))}, \tag{10}$$

where $E(\mathbf{S})$ represents the system energy calculated by Eq. (4) and $\beta = 1/k_{\mathrm{B}}T$. The detailed balance

$$\begin{aligned} &w_i^{\mathrm{T}}(S_i) \exp(-\beta E(S_1, S_2, \ldots, S_i, \ldots, S_N)) \\ &\quad = w_i^{\mathrm{T}}(-S_i) \exp(-\beta E(S_1, S_2, \ldots, -S_i, \ldots, S_N)), \end{aligned} \tag{11}$$

is a sufficient condition for the thermal equilibrium condition. It is easily found with Eqs. (7) and (10) that $dp_{\mathrm{eq}}(\mathbf{S})/dt = 0$ for the transition rates

w_i that satisfies Eq. (11). We define the local effective field H_i and the local energy E_i as

$$H_i = \varepsilon_i - \sum_j J_{ij} S_j, \tag{12}$$

$$E_i = H_i S_i. \tag{13}$$

Then Eq. (11) is rewritten as

$$w_i^{\mathrm{T}}(S_i)\exp(-\beta E_i) = w_i^{\mathrm{T}}(-S_i)\exp(\beta E_i). \tag{14}$$

As specific forms of $w_i^{\mathrm{T}}(S_i)$ satisfying this condition, the Arrhenius, the heat bath and the Metropolis models are often used.

(a) *The Arrhenius model*
The Arrhenius model uses the form

$$w_i^T(S_i) = A\exp(\beta E_i), \tag{15}$$

which is derived from the Arrhenius law in the chemical reaction theory. It is useful for systems where a pseudospin flip means a structural change in each constituent unit, as in the spin crossover complexes. According to the Arrhenius law, the reaction rate with an energy barrier E_{A} is

$$k = k_0 \exp(-\beta E_A), \tag{16}$$

where k_0 is a temperature independent factor. The relation between the energy diagram in the Ising-like model and adiabatic potential is schematically shown in Fig. 8. In the case of the change $S_i \to -S_i$, the barrier height

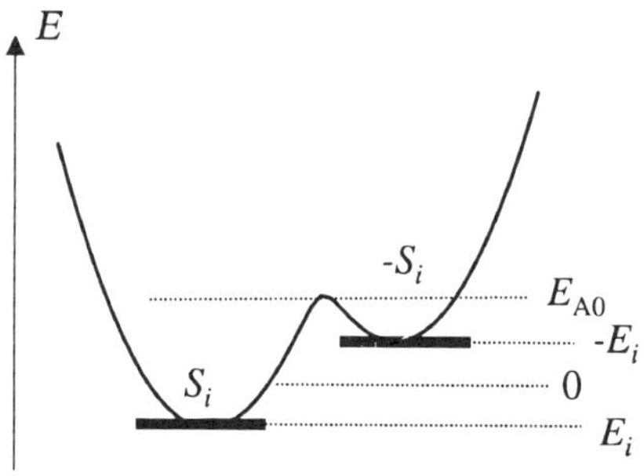

Fig. 8. Relation between the energy diagram in the Ising-like model and the adiabatic potential for the derivation of the Arrhenius model. E_{A0} represents the energy of the transition state and is a barrier height from $E = 0$.

is $E_A = E_{A0} - E_i$. Therefore the transition rate is expressed as

$$w_i^T(S_i) = k_0 \exp\lfloor -\beta(E_{A0} - E_i) \rfloor. \tag{17}$$

By putting $A = k_0 \exp(-\beta E_{A0})$, Eq. (15) is obtained. The Arrehenius model is especially suitable for studying nonequilibrium states at high temperatures.

(b) *The heat bath model*
The heat bath model has been widely used for simulations on the original Ising model for magnetism. It has the form

$$w_i^{\mathrm{T}}(S_i) = \frac{1}{\exp(\beta \Delta E_i) + 1}, \tag{18}$$

where $\Delta E_i = -2E_i$, which is the energy difference between the states after and before the flip of S_i. This corresponds to the canonical distribution of S_i with assuming that the other spins are fixed, i.e.,

$$P_i(S_i) = \frac{\exp(-\beta E(S_1, S_2, \ldots, S_i, \ldots, S_N))}{\exp(-\beta E(S_1, S_2, \ldots, S_i, \ldots, S_N)) + \exp(-\beta E(S_1, S_2, \ldots, -S_i, \ldots, S_N))}. \tag{19}$$

Although this model has been widely used in Monte Carlo simulations, its application to the simulation of photoinduced phase transitions has been limited to general arguments without specifying the material system. This model is advantageous in its calculational cost. An equilibrium state can be reached with this model more rapidly than with the other models, along with qualitatively correct dynamical behavior. Because a Monte Carlo simulation requires many runs to calculate an expectation value, such a fast relaxation speed is beneficial.

(c) *The Metropolis model*
The Metropolis model was the first model satisfying the condition of the detailed balance. Its form is given by

$$w_i^{\mathrm{T}}(S_i) = \min[1, \exp(-\beta \Delta E_i)]. \tag{20}$$

This model has been often used for investigating the equilibrium states of Ising systems.

The transition rates of these three models are compared in Fig. 9 for a system of simple cubic lattice with $\varepsilon_\iota = 0$ for all units and $J_{ij} = J$ for all the nearest neighbors. In this case E_i takes its minimum $-6J$ and maximum

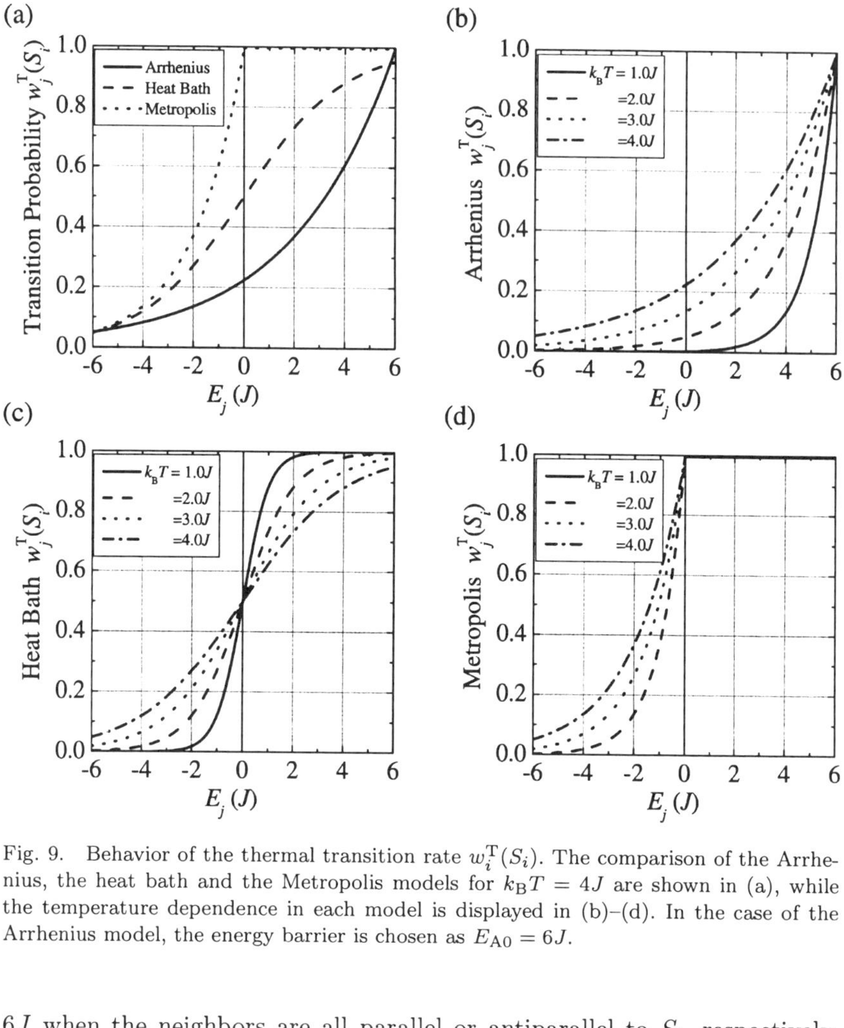

Fig. 9. Behavior of the thermal transition rate $w_i^{\mathrm{T}}(S_i)$. The comparison of the Arrhenius, the heat bath and the Metropolis models for $k_{\mathrm{B}}T = 4J$ are shown in (a), while the temperature dependence in each model is displayed in (b)–(d). In the case of the Arrhenius model, the energy barrier is chosen as $E_{\mathrm{A0}} = 6J$.

$6J$ when the neighbors are all parallel or antiparallel to S_i, respectively. The difference is prominent, especially in the region of $E_i > 0$, where the rate decreases in the Arrhenius model but increases in the heat bath model with temperature, while the Metropolis model gives no temperature dependence. This affects the temperature dependence of the phase transition kinetics very much. Basically, the thermal relaxation of photoexcited states becomes slower in lowering temperature in the Arrhenius model, making a photoinduced phase transition to occur more easily. The situation is opposite in the heat bath model, so that stronger excitation is required for a phase transition at lower temperatures.

3.3.2. *Quantum mechanical tunneling*

If we use the Arrhenius model for thermal fluctuations, jumps between the states rarely occur at temperatures much lower than the barrier height E_{A0}. In such a situation, the effect of quantum mechanical tunneling sometimes plays a key role. For example, temperature-dependent relaxation in spin-crossover complexes at low temperatures could not be explained with the Arrhenius model. Figure 10 shows an Arrhenius plot of the experimental transition rate $k_{HL}(T)$ from the high-temperature state to the low-temperature state in a diluted spin crossover complexes.[24,25] The curves deviate from the Arrhenius behavior at low temperatures. Hauser explained this problem theoretically by considering the tunneling effect as shown in Fig. 11.[24,25]

Here we briefly introduce Hauser's theory based on nonadiabatic multiphonon relaxation.[24,25] A single configuration coordinate Δu of the totally symmetric normal mode is used for the potential curves of the low-spin and high-spin states. The quantized phonon states are considered as shown in Fig. 12. In the Condon approximation, the relaxation rate $k_{HL}(T)$ is written as

$$k_{\mathrm{HL}}(T) = \frac{2\pi}{\hbar^2 \omega} \beta_{\mathrm{HL}}^2 F_p(T), \tag{21}$$

where ω is the phonon frequency and β_{HL} represents the electronic matrix element between the initial and final states, which originates from the

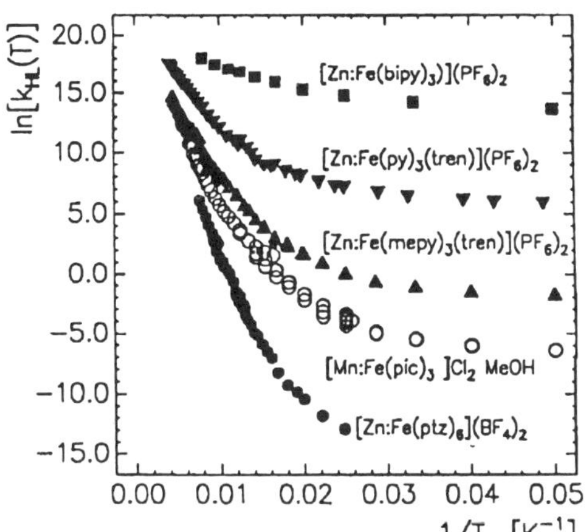

Fig. 10. Arrhenius plot of the experimental transition rate $k_{HL}(T)$ from the high-temperature state to the low temperature state in dilute Fe(II) systems. [A. Hauser, *Coordin. Chem. Rev.* **111**, 275 (1991). Copyright (2004) by Elsevier.]

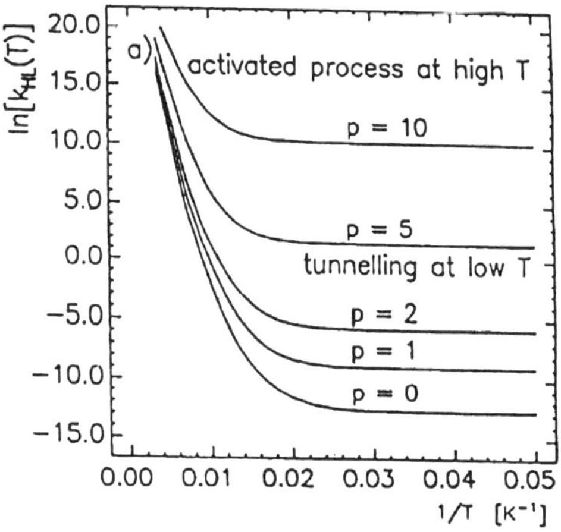

Fig. 11. Logarithm of $k_{\mathrm{HL}}(T)$ against $1/T$, calculated for $S = 45$ and $\mathrm{h}\nu = 250\,\mathrm{cm}^{-1}$ with varying the reduced energy parameter p. [A. Hauser, *Coordin. Chem. Rev.* **111**, 275 (1991). Copyright (2004) by Elsevier.]

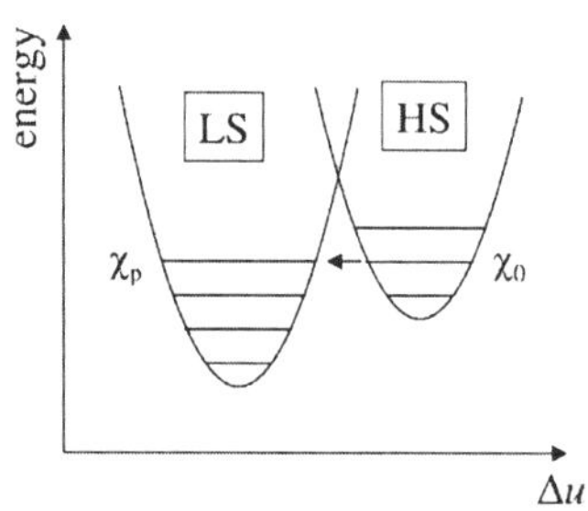

Fig. 12. A schematic energy diagram of phonons in the double well potential for a spin crossover complex. χ_0 and χ_{p} indicates the initial and final state of the intersystem crossing.

spin-orbit coupling. The thermally averaged Frank-Condon factor $F_p(T)$ is given by

$$F_p(T) = \frac{\sum_m |\langle \chi_n | \chi_m \rangle|^2 \exp(-\beta m \hbar\omega)}{\sum_m \exp(-\beta m \hbar\omega)}, \tag{22}$$

where the summation runs over all the phonon number m of the initial state. The quantum number n of the final state is given by $n = m + p$ ($p = \Delta E/\hbar\omega$) because of energy conservation. If the phonon energy $\hbar\omega$ is sufficiently larger than $k_{\mathrm{B}}T$, the phonon state in the initial state is the ground state χ_0, and F_p is approximated by

$$F_p(T \to 0) = |\langle \chi_p | \chi_0 \rangle|^2 = \frac{S^p \exp(-S)}{p!} \tag{23}$$

where the Huang–Rhys factor $\mathrm{S} = f\Delta u^2/2\hbar\omega$.

The transition probability $w_i^{\mathrm{Q}}(S_i)$ in a unit time of Monte Carlo simulation is determined from $k_{\mathrm{HL}}(T)$ in Eq. (21) with a proper factor of time scale. Note that $w_i^{\mathrm{Q}}(S_i)$ and $w_i^{\mathrm{T}}(S_i)$ should not be used additively, because $w_i^{\mathrm{Q}}(S_i)$ contains not only the quantum tunneling effect but also a thermal contribution.

3.3.3. *Photoexcitation*

In Subsec. 3.1, W was defined as the transition rate corresponding photoexcitation. In general, the transition rate can depend on the site and/or its state. It is often assumed that photoexcitation occurs only when the state S_i of the unit i is a particular state S_0, implying that the state changes always in one way from S_0 to $-S_0$, i.e.,

$$w_i^{\mathrm{P}}(S_i) = W_i \delta_{S_i, S_0}. \tag{24}$$

Note that the probability W is defined in the time unit of MCS, i.e., with respect to the time scale of thermal fluctuations. Generally S_0 is the initial state before starting photoexcitation, when we investigate a phase transition from one phase to another. Experimentally such a one-way excitation mechanism works in the case of resonant excitation with a selected photon energy. In fact, the forward and backward photoinduced phase transitions were induced selectively with two different photon energies in spin crossover complexes[26] and Prussian blue analogues.[27]

In addition to the case of single unit excitation given by Eq. (24), we also investigated the case of block excitation, where a cluster of many units change their states simultaneously. It was experimentally reported that many molecules were converted per one photon.[2,28] This situation can be simulated by the block excitation model.

3.4. **Calculation procedure**

We summarize here a typical calculation procedure for the Monte Carlo simulation of the Ising-like model with thermal fluctuation $w_i^{\mathrm{T}}(S_i)$ and photoexcitation $w_i^{\mathrm{P}}(S_i)$.

First, we prepare an initial state $\mathbf{S}_0(t_0)$ at the starting time t_0. Usually the equilibrium state without photoexcitation is prepared for $\mathbf{S}_0(t_0)$ by performing a sufficiently long simulation under no excitation. After the preparation of $\mathbf{S}_0(t_0)$, a stochastic time evolution of the system is carried out under photoexcitation. The state-change trials of photoexcitation and thermal fluctuation are performed alternately, randomly choosing a site every time. Both photoexcitation trials and thermal fluctuation trials are performed N_S times in 1 MCS, where N_S is the number of sites in the system. Thus the procedure in 1 MCS is summarized as:

(1) Select a site i randomly for a trial of photoexcitation.
(2) Prepare a random number p $(0 < p < 1)$.
(3) If $p < w_i^\mathrm{P}(S_i)$, change the state S_i to $-S_i$.
(4) In the block excitation case, the other sites in the block of a given shape containing the site i at its center are put to the same state simultaneously.
(5) Select another site i randomly for a trial of thermal fluctuation.
(6) Calculate the transition probability $w_i^\mathrm{T}(S_i)$.
(7) Prepare another random number p $(0 < p < 1)$.
(8) If $p < w_i^\mathrm{T}(S_i)$, change the state S_i to $-S_i$.
(9) Repeat the steps 1–8 for N_S times.

Steps 1–4 are skipped when we prepare the equilibrium state.

4. Photoinduced Phase Transition in Uniform Systems

In this section, we give an overview of the results of Monte Carlo simulations on the Ising-like model with a uniform structure. As a typical case, we pick up the case of a simple cubic lattice with $\varepsilon_i = \varepsilon$ for all i and with $J_{ij} = J$ for nearest neighbor pairs in Eq. (3). Monte Carlo simulations were performed for a system size $48 \times 48 \times 48$ with the Arrhenius model for thermal jump rate $w_i^\mathrm{T}(S_i)$ and by assuming a photoexcitation rate $W_i = W$ for all i in Eq. (24). Without loss of generality the system is supposed to be in the A $(S_i = -1)$ phase initially, and the number fraction of units in the state B $(S_i = 1)$, ρ_B, is traced under one-way excitation from the A state to the B state, i.e. $w_i^\mathrm{P}(S_i)$ with $S_0 = -1$ in Eq. (24).

4.1. *Stationary state under constant photoexcitation*

First of all, we examine stationary states under photoexcitation W. Figure 13 shows ρ_B as a function of W. Starting from the A phase and

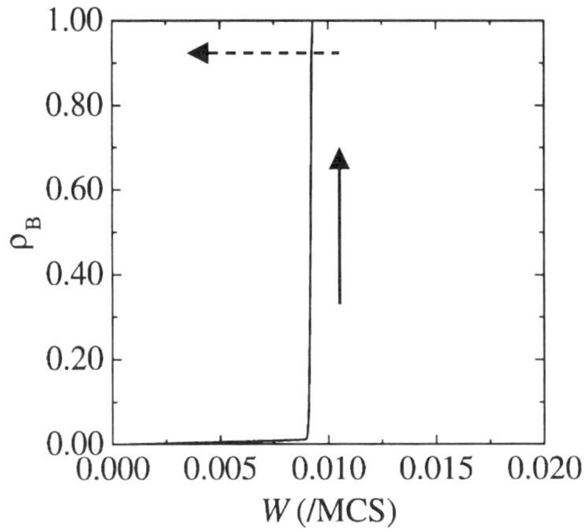

Fig. 13. Hysteresis behavior of the fraction ρ_B against the excitation rate W in the steady state under continuous excitation, calculated at $J = 1.2k_BT$ with the Monte Carlo method. The system is a $48 \times 48 \times 48$ simple cubic lattice with $\varepsilon_i = 0$. The Arrhenius model is applied as the thermal transition rate with $E_{A0} = 15k_BT$.

increasing W sufficiently slowly so that the system is kept in its stationary state at each W, the system remains in the A phase up to a certain critical value $W = W_c$ but then jumps to the B phase. If we then reduce W gradually again, the system remains in the B phase even down to $W = 0$. This hysteresis behavior indicates the global bistability and the photoinduced phase transition inherent in this system. W_c determines the threshold light intensity for the photoinduced phase transition.

Figure 14 shows how the hysteresis curve of ρ_B depends on the energy ε when the A phase is the initial state. In increasing ε, the photoinduced phase transition occurs at larger W. This shows a simple but essential principle: The photoinduced phase transition becomes easier as the energy of the final state becomes lower compared to that of the initial state. If we look for a material exhibiting a *reversible* photoinduced phase transition material, the energies of the two states should be as close to each other as possible.

4.2. *Phase transition kinetics under photoexcitation*

We proceed to the temporal evolution of phase conversion when the excitation is suddenly raised at $t = 0$ from 0 to a constant value W. The fraction ρ_B is plotted as a function of time t in Fig. 15 for the system of $\varepsilon = 0$ and $J = 2T$ with various W. For $t < 0$, the system is in an equilibrium state with $\rho_B \sim 0$. After the start of excitation at $t = 0$, the system remains in the A phase for some time (the incubation period) and then suddenly

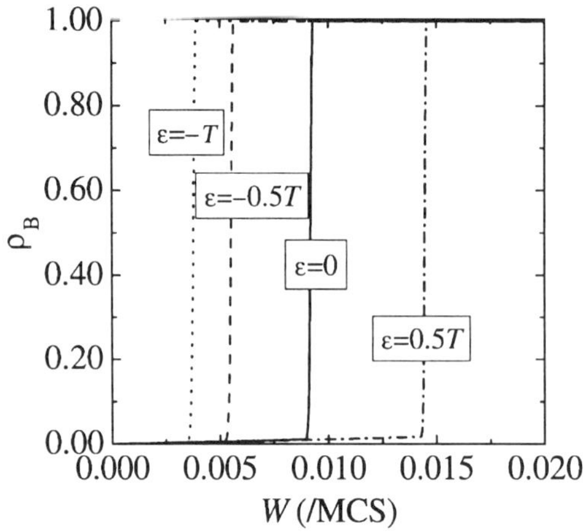

Fig. 14. Hysteresis curves of ρ_B against W calculated for various ε. The other parameters are the same as in Fig. 13.

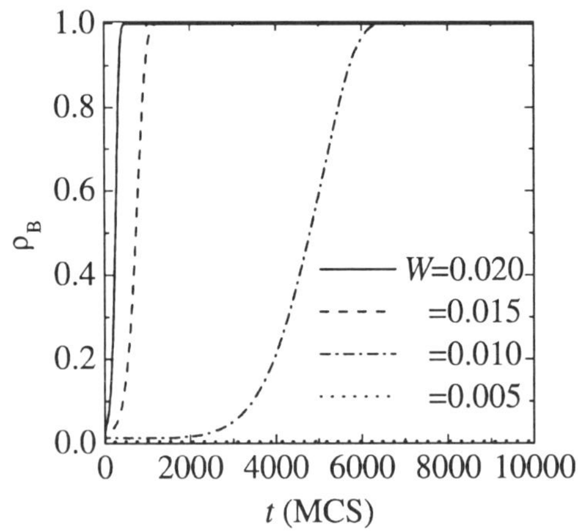

Fig. 15. Temporal evolution of the phase transition calculated in the same system as in Fig. 13 ($J = 1.2k_BT$, $\varepsilon_i = 0$). The fraction ρ_B is plotted as a function of the excitation time t for various excitation rate W per MCS.

increases and approaches almost unity, implying the completion of the phase transition, if W is larger than a threshold $W_c = 0.009$ (see the solid curve in Fig. 14). The phase transition becomes faster in increasing W. As mentioned in Section 2, the incubation period was observed in the photoinduced phase transition of a spin crossover complex.[2] In the case of $W < W_c$, the system does not exhibit a phase switching.

We have seen the basic characteristics of the photoinduced phase transition: the threshold of excitation rate W_c, a sudden change of the steady state at W_c, and the incubation period. These characteristics originate from the transition processes similar to the first order thermal phase transition, involving nucleation and growth. In the present simulation, this mechanism is incorporated in the form of $w_i^T(S_i)$. The lifetime τ of a photoexcited B ($S = 1$) state is the inverse of $w_i^T(1)$. For example, in the case of the Arrhenius model, τ is calculated with Eqs. (12), (13), and (15) as

$$\tau = \frac{1}{w_i^T(1)} = A \exp[\beta\{\varepsilon - (m_- - m_+)J\}], \tag{25}$$

where m_- and m_+ represent the number of neighboring units having $S_j = -1$ and 1, respectively. As m_+ increases, τ grows exponentially, that is, clustered excited states have a longer lifetime. When the lifetime becomes comparable to the average excitation time, the cluster of the excited states can stay stably. This is a nucleation process. As the lifetime becomes further longer, the excited state cluster starts to expand. This is a growth process.

Such nucleation and growth processes can be visualized in the Monte Carlo simulation. Figure 16 shows snapshots of the conversion process from the A phase to the B phase. The time points of the snapshots shown in Fig. 16(b–d) are indicated in Fig. 16(a). At $t = 1000$ MCS, which is in the incubation period, B-state sites are distributed randomly and separately. In this stage, photoexcited states goes back to the initial states immediately. At $t = 3000$ MCS, around which ρ_B gradually starts to increase, nucleation of the B phase is found. After such a nucleus emerges anywhere in the system, it grows quickly as we see in Fig. 16(d).

The behavior shown in Fig. 16 is for $W = 0.01$, which is slightly larger than W_c. In this case, some time is required to form a nucleus, but once it is formed, it grows quickly. This is the origin of the incubation period that appears as an initial plateau in the time evolution of ρ_B shown in Figs. 15 and 16(a). However, for W much larger than W_c, the incubation period cannot be defined, and ρ_B tends to increase linearly with time t, as we see in the casc of $W = 0.02$ in Fig. 15. In this case the nucleation time is much smaller than the excited-state lifetime, and many nuclei emerge quickly and simultaneously just after the start of excitation.

The tendency of excited-state clustering during the phase transition process is demonstrated by examining nearest-neighbor pair correlation.

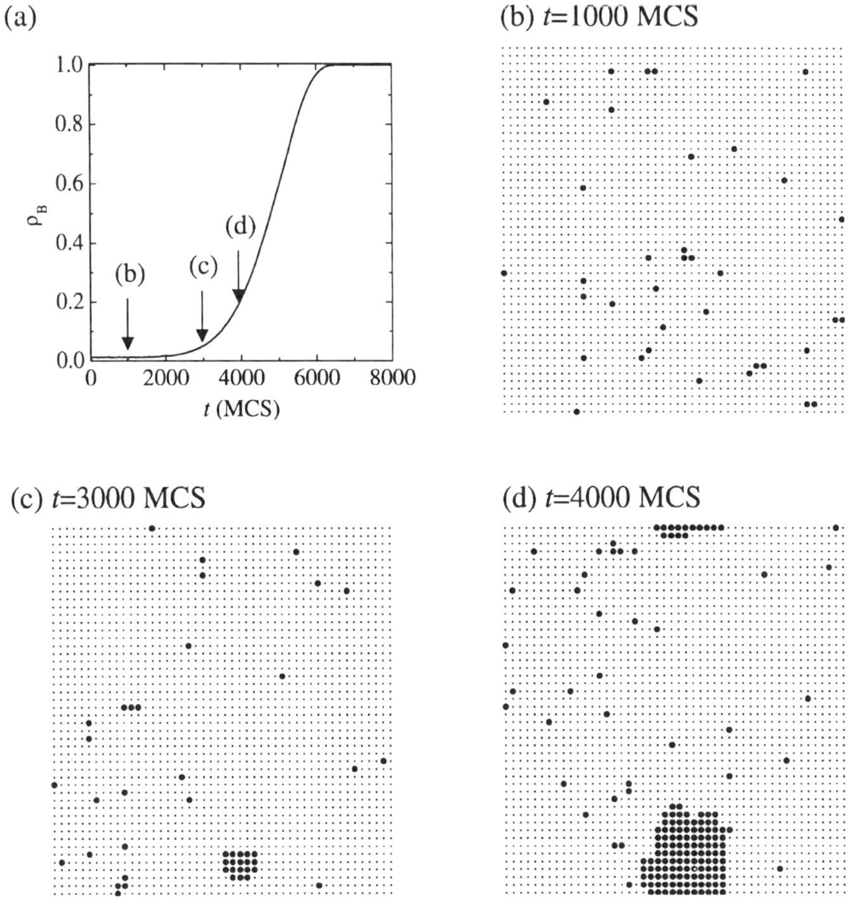

Fig. 16. Snapshots during the conversion process from the A phase to the B phase in the case of $W = 0.01$ in Fig. 15. In (b)–(d), small and large dots represent the A-state sites and the B-state sites.

Figure 17 is a diagram showing the fractions of nearest neighbor A-A, B-B, and A-B pairs at each time t. The A-B component is always very small even during the transition process, implying that the interface between the A phase and B phase is kept as small as possible due to strong interaction J.

4.3. *Dynamical phase transition*

So far we have shown the situation where the photoinduced phase is stable even after turning the excitation off, as indicated by the hysteresis curve in Fig. 13. This is the case of sufficiently strong interaction J. In contrast,

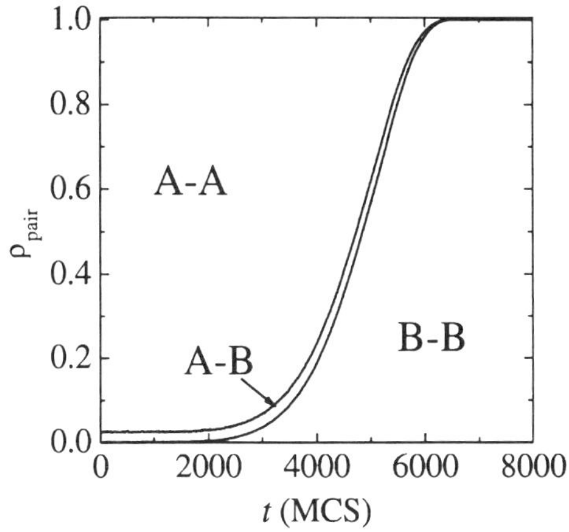

Fig. 17. Diagram of fractions of the three kinds of nearest neighbors, A-A, B-B, and A-B as a function of the excitation time t in a single run of the Monte Carlo simulation as that used in Fig. 15.

in the case of weaker interaction, the bistability of the system tends to disappear because of short lifetime of excited states or because of thermal disorder.

As J decreases relative to ε, the lifetime of the excited state τ given by Eq. (25) converges to a constant independent of m^+. Then the cooperative nature weakens, so that a photoexcited phase cannot be kept without photoirradiation. The J-dependence of the steady state under excitation W is shown in Fig. 18, in which Fig. 18(a) is the same as Fig. 13. With decreasing J, the hysteresis region narrows and shifts to larger W as shown in Fig. 18(b). In this situation the photoinduced phase is a nonequilibrium steady state. It can exist only under continuous excitation, whereas it decays to the initial state when the excitation is switched off. This type of phase transition is called the dynamical phase transition. In further decreasing J, the hysteresis disappears as shown Fig. 18(c) and finally a well-defined phase transition vanishes as shown in Fig. 18(d). This is because the pseudospins behave almost independently with each other.

The dynamical phase transition as shown in Figs. 18(b) and (c) was experimentally observed in the spin crossover complexes. The experimental result in $[Fe(ptz)_6](BF_4)_2$, (ptz = 1-propyltetrazole) is shown in Fig. 19(a).[29] The high spin fraction n_{HS} converted from the low-spin ground state at 77 K was traced by slowly varying the intensity of light with photon energy $h\nu = 2.41\,eV$. The light intensity was increased from 0 to $180\,mW/cm^2$ step by step and then decreased again. The result clearly

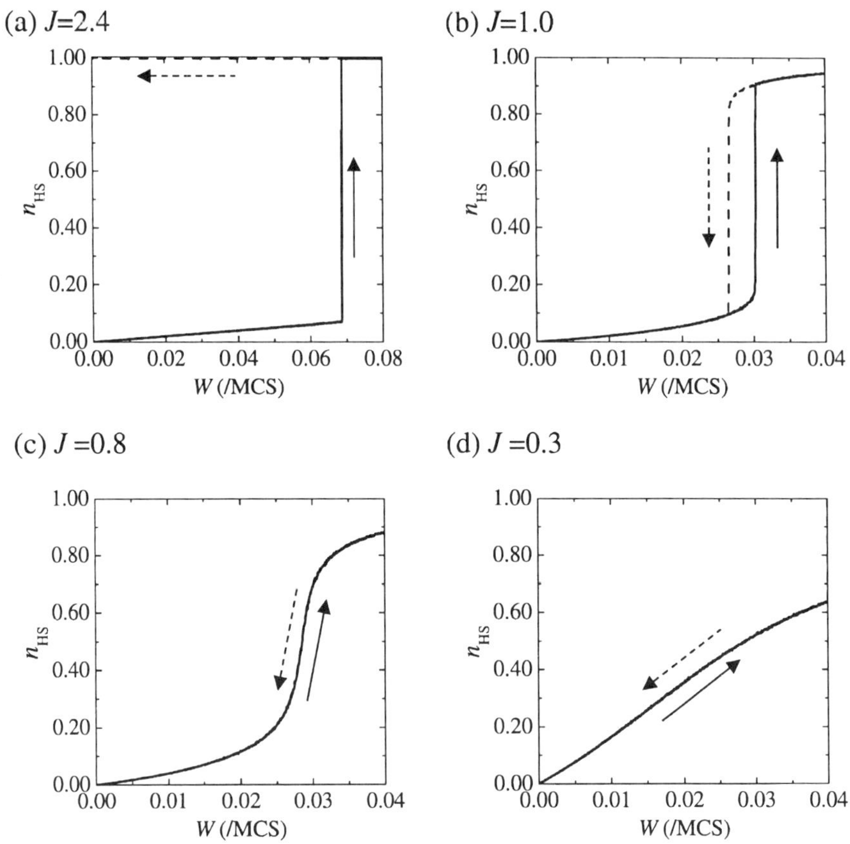

Fig. 18. The fraction ρ_B in a nonequilibrium state calculated by the Monte Carlo method. ρ_B is drawn as a function of the excitation rate W with various interaction strengths J with $\varepsilon = 5.4$ and $k_BT = 2$.

shows the hysteresis of a dynamical phase transition. Figure 19(b) is our theoretical simulation for this experiment, by choosing appropriate parameters in the Ising-like model.[29] The observed fraction n_{HS} saturates much below one, but this may be due to the experimental uncertainty of the absolute scale of n_{HS}.

The case of no hysteresis as shown in Fig. 18(c) was also reported in another spin crossover complex, Nafion-[Fe(Htrz)$_3$] film (Htrz = 1,2,4-4H-triazole), as shown in Fig. 20.[30] The experimental procedure was the same as in Fig. 19, but no hysteresis was observed. The S-shape nonlinear increase of n_{HS} against P is observed below 35 K but disappers at higher

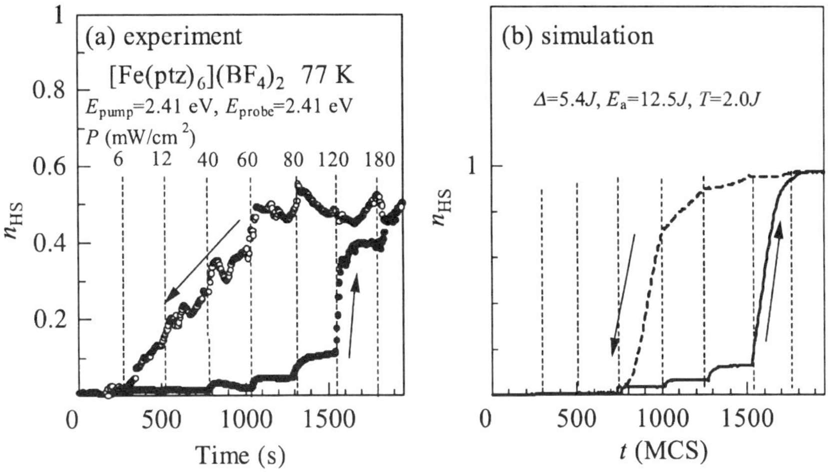

Fig. 19. (a) Dynamical phase transition observed in [Fe(ptz)6](BF4)2 observed at 77 K Temporal evolution of density $n_{\rm HS}$ of the HS molecules under photoexcitation. (b) Hysteresis behavior reproduced by a Monte Carlo simulation. [X. J. Liu, Y. Moritomo, T. Kawamoto, A. Nakamoto, and N. Kojima, *Phys. Rev.* **B 67**, 012102 (2003). Copyright (2004) by the American Physical Society.]

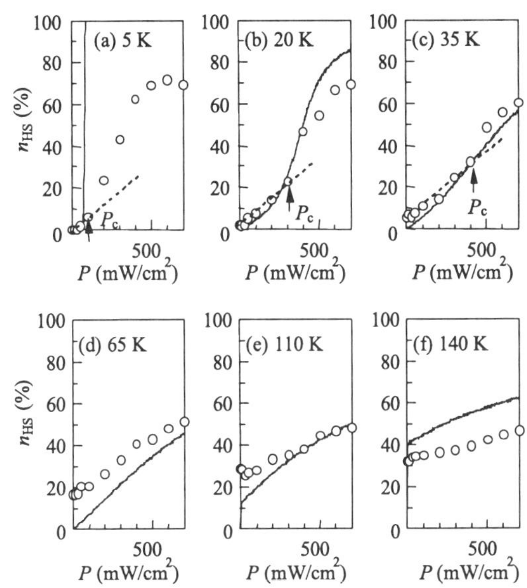

Fig. 20. Density $n_{\rm HS}$ of the excited HS molecules against excitation power density P in the Nafion-[Fe(Htrz)$_3$] film at various temperatures. $P_{\rm c}$ is the critical excitation power density. Thin curves are the results of Monte Carlo simulations for $J = 10\,$K and $\Delta = 300\,$K. [X. J. Liu, Y. Moritomo, T. Kawamoto, A. Nakamoto, and N. Kojima, *J. Phys. Soc. Jpn.* **72**, 1615 (2003). Copyright (2004) by the Physical Society of Japan.]

temperatures. These results can be simulated by using the phenomenological Hamiltonian of Eq. (4), as shown with solid lines in Fig. 20. The experimental result at the lowest temperature $T = 5\,\mathrm{K}$ is not well simulated, presumably due to the effect of quantum-mechanical tunneling. At high temperatures, $T > 65\,\mathrm{K}$, there is no indication of the dynamical phase transition. This is because the thermal equilibrium state is a disordered phase, in which a substantial number of thermally excited high-spin states are randomly distributed, leading to the loss of bistability between the two uniform phases.

5. Photoinduced Phase Transitions in Combined Structures

5.1. *Proposed model*

From a practical point of view, it is desirable to make the photoinduced phase transition bi-directional and reversible, i.e., inducible both in the forward and backward directions as many times as possible. It is also preferable to reduce the light intensity required for the phase transition, because the irradiation of strong light might cause side effects such as thermal deformation and degradation. In practice it is not very easy to find a material that satisfies these conditions. We proposed a novel idea of material design based on nanoscale binary superstructures to overcome this difficulty.

The main idea is as follows: Suppose there is a bistable material α in which the metastable state B is much higher in energy than the stable state A, so that photo-conversion is difficult. Then we prepare a companion material β in which the relative stability is opposite (see Fig. 21). If the two materials are combined in an appropriate fashion, e.g., in a superlattice as shown in Fig. 22, and if there are sufficient cooperative interactions that favor the α and β units being in the same state (either A or B), then the energy of the A phase in the total system can be made degenerate with that of the B phase. This is a rather obvious way of designing new suitable materials in general. However, this approach yields more than that: It turns out that the photo-conversion efficiently is drastically enhanced in such combined structures compared with uniform structures.

The actual model we studied is as follows. We set ε_i in Eq. (3) to be $\Delta_\alpha/2$ for the α unit and $\Delta_\beta/2$ for the β unit to represent the energy difference shown in Fig. 21. The blocks of α and β units are combined to form superstructures as shown in Fig. 22. We investigated mostly symmetric structures shown in Fig. 22(a)–(c): These are symmetric with respect to exchange of α and β. Each unit has a block size n in d-dimensional

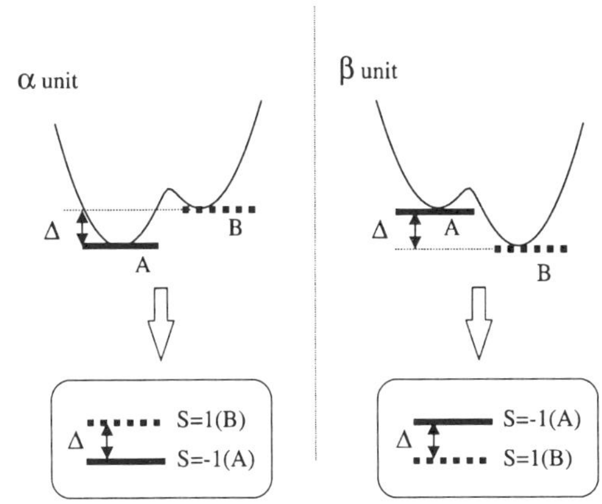

Fig. 21. Schematic illustration of two kinds of a constituent unit, α and β, with different relative stabilities between the bistable states A and B.[23]

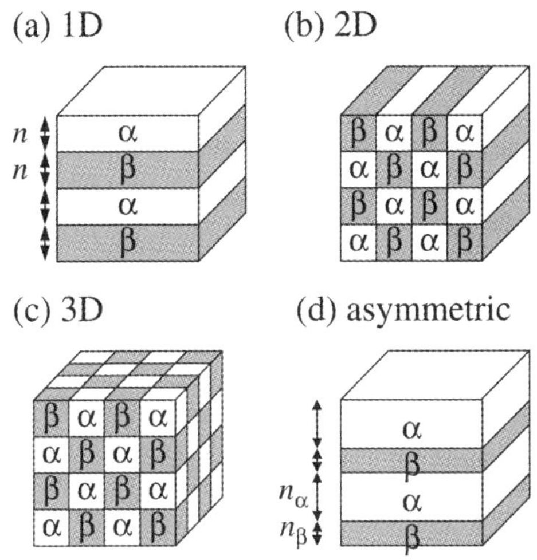

Fig. 22. Examples of superstructures: (a)–(c)1D, 2D, and 3D confined superstructures of two kinds of constituent units, which are symmetric with respect to the inchange of α- and β-units. n represents the confinement size of each block. (d) An asymmetric structure, where the sizes of the blocks of α and of β unit are n_α and n_β, respectively.[23]

($d = 1, 2, 3$) directions. The energy difference between the bistable states is set as $\Delta_\alpha = -\Delta_\beta = \Delta$ (> 0). This means the α and the β blocks have an opposite relative stability between the A and B states as illustrated in Fig. 21. We further studied an asymmetric structure shown in Fig. 22(d), where the block size n is different between the α and β units.

The system still tends to be in a uniform phase without phase separation, if the block thickness is not very large and the interblock interface energy proportional to J outweighs the intrablock energy loss. The critical value Δ_c, at which the ground state changes from the uniform states to a separated phase, is calculated as

$$\Delta_c \equiv 4dJ/n. \tag{26}$$

When $|\Delta| < \Delta_c$, the A and B phases are the doubly degenerate ground states.

5.2. *Simulation results for superstructures*

We have performed Monte Carlo Simulations on the Ising-like model with superstructures in the same way as for the uniform structures. A difference is that we use the heat-bath model for the transition rate $w_i^T(S_i)$ rather than the Arrhenius model, to save computation time.

5.2.1. *Symmetric superstructures*

In the symmetric structures, the forward and backward processes are essentially the same. Figure 23 shows the overall behavior of steady states as functions of W for the uniform system ($\Delta = 0$) and for a 2D-confined superstructure with $\Delta = 0.8\Delta_c$ ($= 6.4J$), for transitions from the A phase to the

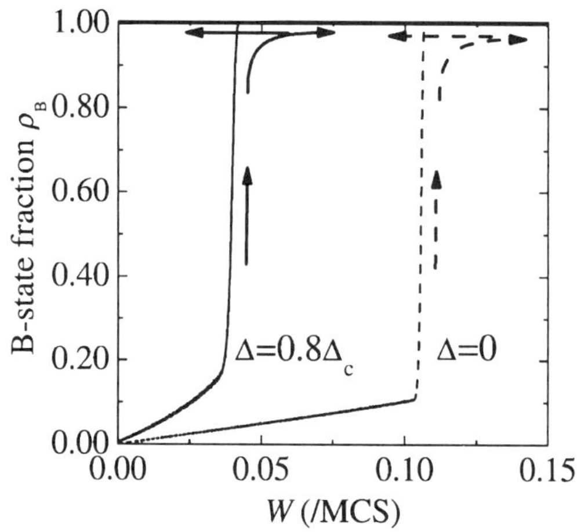

Fig. 23. Hysteresis in the fraction ρ_B as a function of the excitation rate W at $k_B T = J$ for the 2D confined superstructure with $\Delta = 0.8\Delta_c$ ($= 6.4J$), $n = 1$, compared with the uniform case of $\Delta = 0$. The lattice size is $60 \times 60 \times 60$. [T. Kawamoto, and S. Abe, *Phys. Rev.* **B68**, 235112 (2003). Copyright (2003) by the American Physical Society.]

B phase. In both cases there is a critical excitation rate W_c for switching, as discussed in the previous section: it is much smaller in the superstructure than in the uniform system. This means that the phase transition can occur with weaker light intensity in the supersturctures. (The typical values of W in this section are different from those in the previous section, because of difference in the used model of $w_i^T(S_i)$.)

The kinetics of the photoinduced phase transition is displayed in Fig. 24, where ρ_B is plotted as a function of the excitation time t for two cases of W. Figure 24(a) shows that, compared to the uniform case, the phase switching proceeds more quickly in the superstructures than in the uniform structure. For a much weaker excitation case shown in Fig. 24(b), the phase switching does not occur in the uniform system but occurs in the superstructure. This is because W is higher than W_c in the superstructure but lower than that in the uniform system (see Fig. 23).

The excitation time t_p required for reaching an equilibrium state depends on the excitation intensity W, as shown in Fig. 25 for various Δ in the 2D superstructure. (Here t_p is defined as the time of ρ_B exceeding 0.8, and determined as an average over 10 Monte Carlo samples.) It is found that t_p sharply diverges at above W_c. We have confirmed the critical behavior by extending simulations up to 100,000 MCSs. Similar results have been obtained for 1D and 3D superstructures. The threshold W_c at which t_p diverges is reduced significantly with increasing $|\Delta|$. The results in Fig. 25 imply that the required time t_p at each W significantly decreases with increasing $|\Delta|$. For example, at $W = 0.12$, the phase switching in the case of $\Delta = 0.9\Delta_c$ requires only one tenth of the excitation time for a homogeneous structure with the same excitation intensity.

The fractions of B-state sites in the α blocks and in the β blocks are plotted separately in Fig. 26. We see that the fraction ρ_B of the β blocks grows rapidly from the beginning, and then that of the α blocks gradually follows. This suggests that the origin of the rapid phase switching in the superstructure is the accelerated nucleation of converted domains in the β blocks, i.e., the structural blocks relatively close to local instability.

Figure 27 shows snapshots during the simulation from the A phase to the B phase. At $t = 20$ MCS, immediately after the start of photoirradiation, the units in the excited state appear to be randomly distributed except for some small aggregations. More β (square) units are in the excited state than the α (triangle) units. The reason is as follows. When a unit in the B state is alone in the A phase, this is a transient excited state due to interunit interaction. The excited B state returns to the A state after a

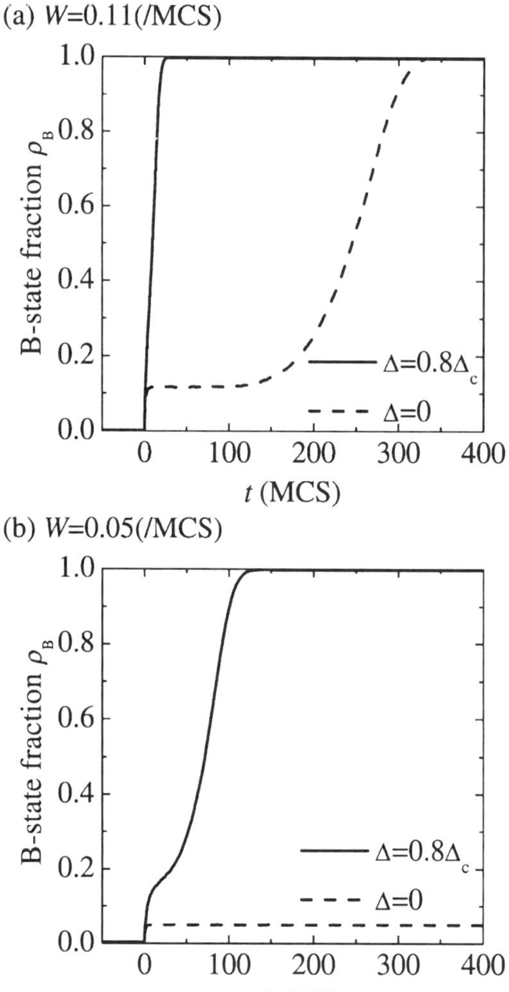

Fig. 24. Temporal evolution of ρ_{B} in the 2D confined superstructure (solid line) compared with the case of the uniform lattice (broken line), under excitation in $t > 0$ with (a) $W = 0.11$ and (b) $W = 0.05$/MCS. The other parameters are the same as in Fig. 23. [T. Kawamoto, and S. Abe, *Phys. Rev.* **B68**, 235112 (2003). Copyright (2003) by the American Physical Society.]

certain lifetime. But the lifetime of the excited state is longer in the β unit than in the α unit because the excitation energy is smaller in the β unit than in the α unit due to positive Δ_β. In Fig. 27(c), the nucleation of the B phase starts. The outer shell of the critical nucleus mainly consists of the β unit. The β units can balance the energy loss at the interface between the initial

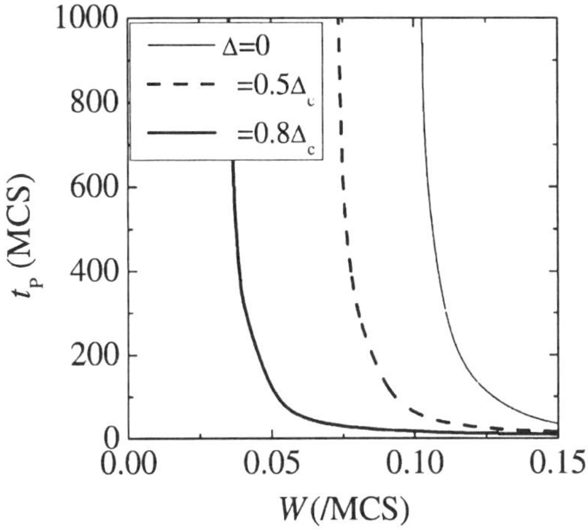

Fig. 25. Dependence of the conversion time $t_{\rm P}$ on the excitation intensity W in the 2D superstructure with $n = 1$ at $k_{\rm B}T = J$ for various Δ with $\Delta_{\rm c} = 8J$. [T. Kawamoto, and S. Abe, *Phys. Rev.* **B68**, 235112 (2003). Copyright (2003) by the American Physical Society.]

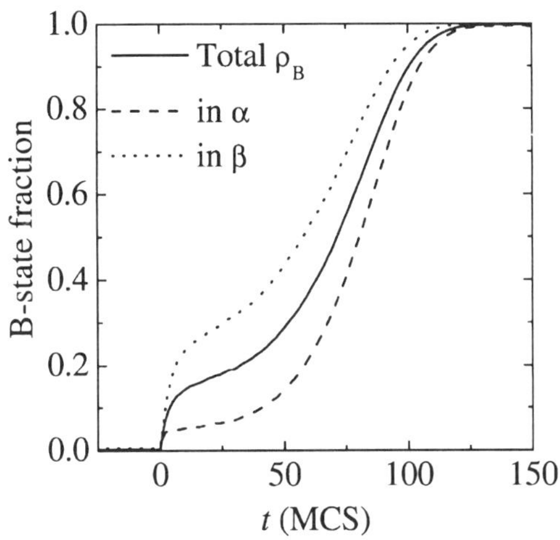

Fig. 26. Temporal evolution of the conversion fraction in the 2D superstructure with $n = 1$ for $W = 0.05$/MCS at $k_{\rm B}T = J$, in the case of $\Delta = 0.8\Delta_{\rm c}$. Solid, broken and dotted lines show the total fraction, and the partial fractions in the α and β blocks, respectively. [T. Kawamoto, and S. Abe, *Phys. Rev.* **B68**, 235112 (2003). Copyright (2003) by the American Physical Society.]

and final states with their intrinsic stability of the final state. The lifetime of the excited states in the α unit surrounded by the excited β unit becomes longer because of the "ferromagnetic" interunit interaction. Consequently, the excited α units remain in the nucleus. This is the mechanism of easier

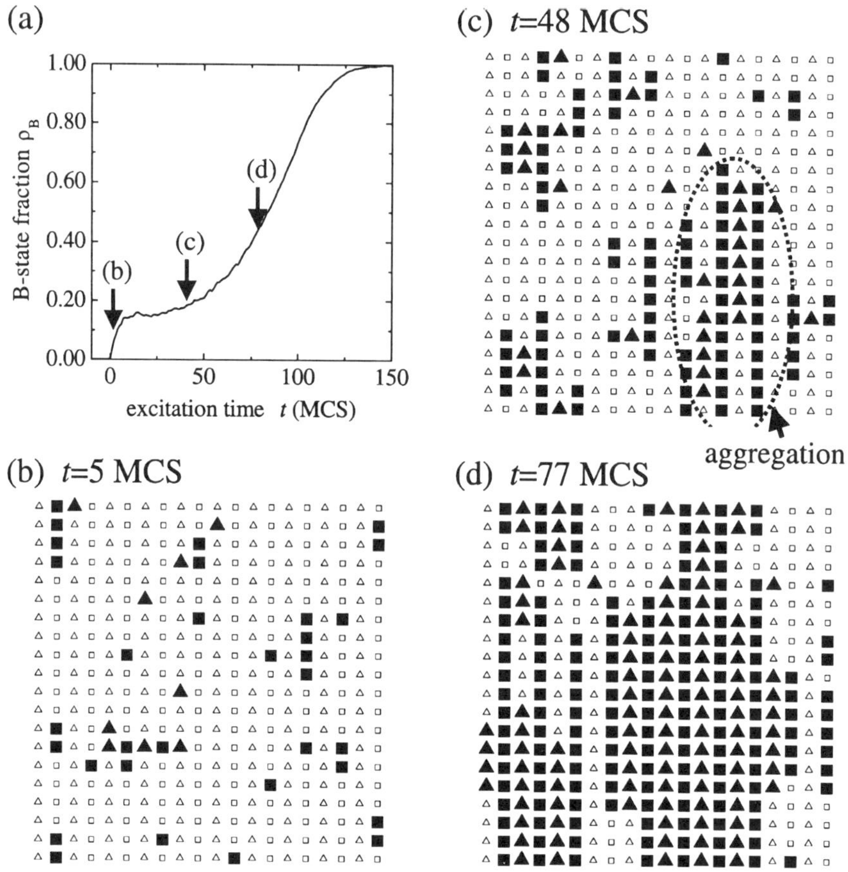

Fig. 27. Snapshots of pseudospin states in the Monte Carlo simulation at $t = 20$, 65, and 105 MCS during conversion from the A phase to the B phase in the 2D superstructure with $n = 1$, $\Delta = 6.4J$, $N_\ell = 20$, $k_BT = J$, and $W = 0.05$/MCS. Triangle and squares represent the α and β units, respectively. Open and filled marks indicate that the units are in the A-state and in the B state, respectively.

nucleation in the superstructure. If such a nucleus arises anywhere in the system, the growth of the final phase easily proceeds.

The dependence of the threshold W_c upon Δ_c is plotted in Fig. 28, not only for the 2D system but also for the 1D and 3D superstructures with $n = 1$. In all cases, W_c decreases with increasing Δ towards Δ_c. The optimum condition is realized near Δ_c in the 3D superstructure, for which actual values of Δ are larger because of the larger $\Delta_c = 4dJ/n$ for $d = 3$. The dependences of W_c on the block size have also been investigated.

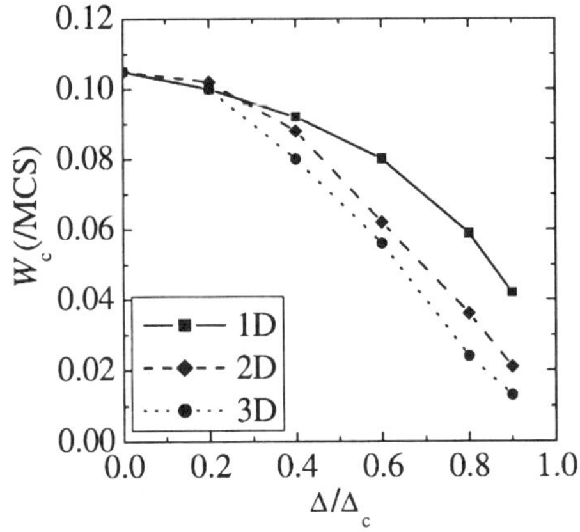

Fig. 28. Dependence of the threshold excitation rate W_c upon Δ in the 1D, 2D, and 3D superstructures with $n = 1$ at $k_BT = J$. The abscissa is normalized by $\Delta_c = 4J$, $8J$, and $12J$ for 1D, 2D, and 3D, respectively. [T. Kawamoto, and S. Abe, *Phys. Rev.* **B68**, 235112 (2003). Copyright (2003) by the American Physical Society.]

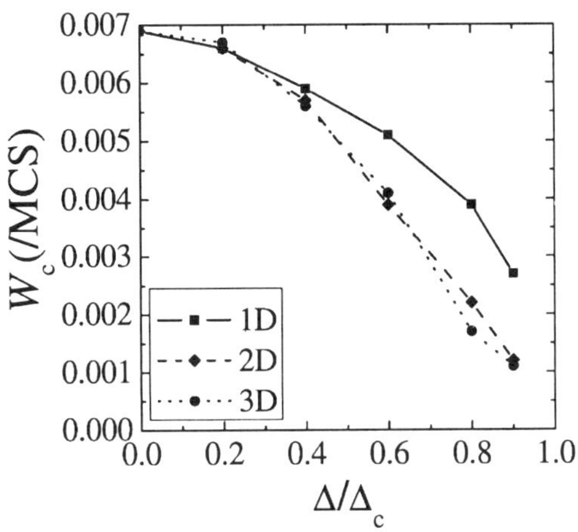

Fig. 29. Dependence of the excitation rate threshold W_c upon Δ in the case of $1 \times 1 \times 10$ block excitation in the same system as in Fig. 28. [T. Kawamoto, and S. Abe, *Phys. Rev.* **B68**, 235112 (2003). Copyright (2003) by the American Physical Society.]

There is a possibility that a photon can excite a cluster of units almost immediately. This situation can be simulated with block excitation. Figure 29 shows W_c as a function of Δ/Δ_c in the case of simultaneous excitation with a block size $1 \times 1 \times 10$. The threshold W_c has the same tendency as the case of the single excitation shown in Fig. 28. Note that the conversion efficiency is higher in this case than in the case of single unit excitation. Threshold W_c is reduced to less than a tenth, although 10 units at most are excited in each time. This is because a nucleus of converted units easily exceeds the critical size for growth in the block excitation case.

5.2.2. *Generalization*

In general, the two kinds of blocks need not have the same size. As an example, we considered an asymmetric 1D superstructure shown in Fig. 22(d), where the width n of the α block is different from that of the β block. Figure 30 displays the threshold of excitation rate W_c calculated for both the directions as a function of n_β with fixed $n_\alpha = 1$. The two sets $(\Delta_\alpha, \Delta_\beta) = (0.9J, -0.9J)$ and $(1.8J, -0.9J)$ have been investigated. In increasing n_β, the threshold W_c of switching from the A phase to the B phase decreases whereas that of the opposite direction increases. This is due to the variation of relative stability between the initial and final phases. The system energies per unit at zero temperature for the A and B phases are obtained from Eq. (4) respectively by

$$E_T(\mathrm{A}) = (-n_\alpha\Delta_\alpha + n_\beta\Delta_\beta)/2(n_\alpha + n_\beta) + E(J) \tag{27}$$

$$E_T(\mathrm{B}) = (+n_\alpha\Delta_\alpha - n_\beta\Delta_\beta)/2(n_\alpha + n_\beta) + E(J), \tag{28}$$

where E(J) is a constant contribution from the interaction term. The energy gain of the A phase compared with that of the B phase is

$$\begin{aligned}\Delta E_T &= E_T(B) - E_T(A)\\ &= (n_\alpha\Delta_\alpha - n_\beta\Delta_\beta)/(n_\alpha + n_\beta).\end{aligned} \tag{29}$$

As ΔE_T increases, switching from the A phase to the B phase becomes increasingly difficult, and the opposite becomes increasingly easy. Therefore, in order to reduce W_c in both the directions simultaneously, ΔE_T should be close to zero. This condition is satisfied for $n_\beta = 1$ in the case of $(\Delta_\alpha, \Delta_\beta) = (0.9J, -0.9J)$, and for $n_\beta = 2$ in the case of $(\Delta_\alpha, \Delta_\beta) = (1.8J, -0.9J)$. These are in fact optimum cases for reducing the threshold W_c in both forward and backward transitions simultaneously, as shown in Fig. 30.

Another direction of extending the model is the range of interaction. As an opposite extreme to the case of nearest neighbor interactions, we examined an artificial model with infinite range interactions, where interaction strength is a constant for any pairs in the system. In this case, the switching process depends on only by the number ratio of

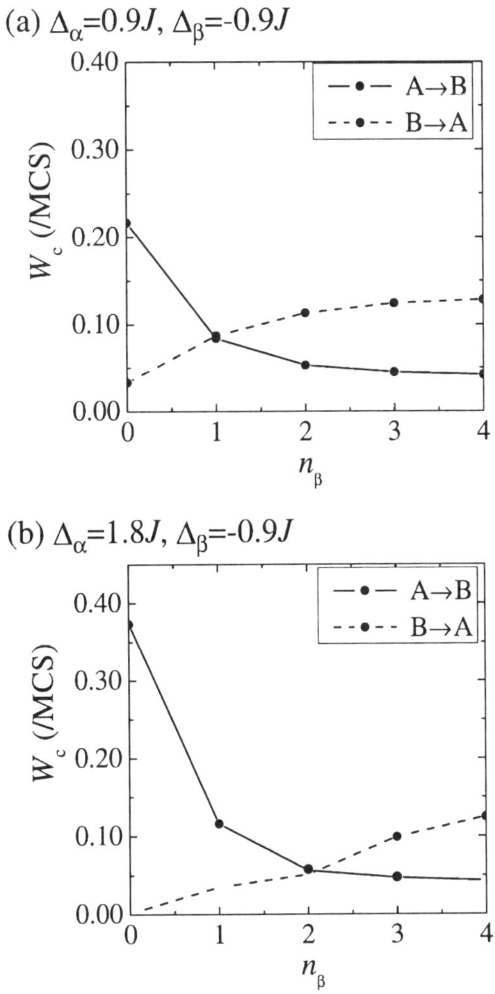

Fig. 30. Threshold excitation rate W_c as a function of the β block size n_β with $n_\alpha = 1$ in the 1D confined asymmetric superstructure shown in Fig. 22(d) with parameters (a) $\Delta_\alpha = 0.9J$, $\Delta_\beta = -0.9J$ and (b) $\Delta_\alpha = 1.8J$, $\Delta_\beta = -0.9J$. The solid and broken lines represent W_c for the transition from the A phase to the B phase and vice versa, respectively. [T. Kawamoto, and S. Abe, *Phys. Rev.* **B68**, 235112 (2003). Copyright (2003) by the American Physical Society.]

the A and B units, unaffected by microscopic structures (e.g., regular or disorder). As shown in Fig. 31, Monte Carlo simulations in this case also indicate similar acceleration of photo-conversion with finite Δ. We also studied the case of intermediate interactions, e.g.,

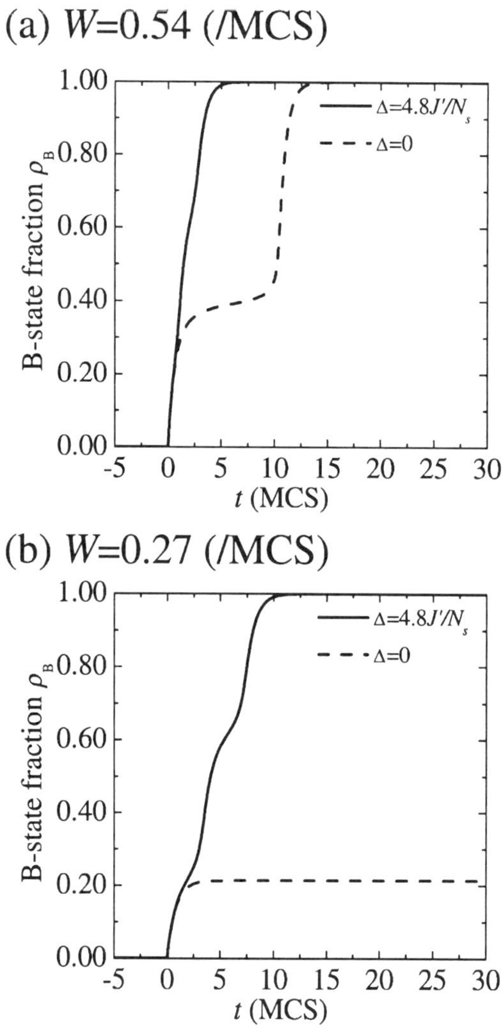

Fig. 31. Temporal evolution of the B-state fraction under excitation in a mixed structure with an infinite range interaction J' at $k_BT = N^3J'/6$ for (a) $W = 0.54$/MCS; (b) $W = 0.27$/MCS. The solid and broken lines represent the mixed structure with $\Delta = 0.8NJ'$, and the uniform lattice with $\Delta = 0$, respectively. [T. Kawamoto, and S. Abe, *Phys. Rev.* **B68**, 235112 (2003). Copyright (2003) by the American Physical Society.]

interactions between second nearest neighbors, and obtained similar results. Therefore, the rapid switching in combined structures is rather a universal phenomenon, independent of the detailed form of interunit interaction.

5.3. *Discussion*

The concept of bidirectional photoinduced phase transition in combined systems discussed in the previous section is related with the photoinduced magnetism in the Co-Fe Prussian blue analogue.[27,31–33] In this material photoinduced charge transfer results in the low-spin to high-spin transition of Co atoms accompanied by a change in its ion radius. Detailed discussions on this material have been given in a review.[31] Here we just mention that this material effectively contains various different local units with different relative stabilities between the two spin states, as a result of the replacement of CN by H_2O surrounding Fe vacancies, which are located randomly in the crystal. Figure 32 displays the adiabatic potential curves of cobalt-centered clusters calculated with an ab initio quantum chemical technique[33] for various numbers of replaced CN. The cluster of Co with full CN coordination has an adiabatic potential like the α unit in Fig. 21, and that of Co with two CN ions replaced by a H_2O molecule has a potential similar to the β unit. We demonstrated that the light frequencies responsible for forward and backward photoinduced transitions can be interpreted with the adiabatic potentials of these clusters including higher excited states.[33]

Another material relevant to our study is manganese oxides, in which the photocontrol of magnetoresistance was reported. An insulator-metal

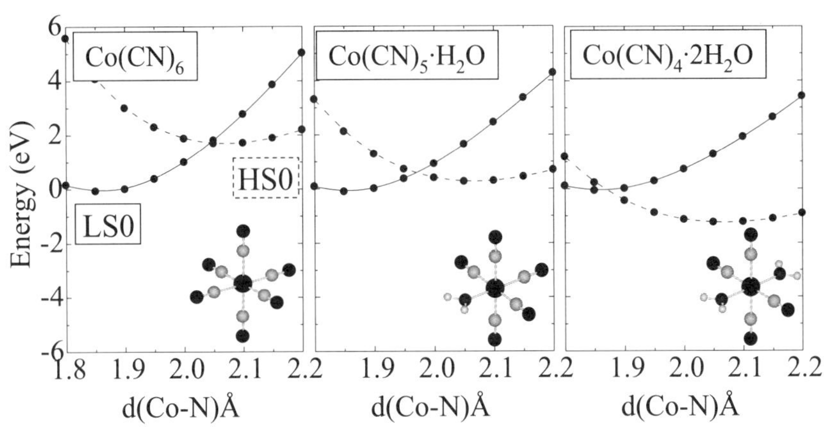

Fig. 32. Local potentials of Co-centered clusters in the Co-Fe Prussian Blue complex obtained from ab initio calculations. The used clusters are shown in each panel: (a) the Co ion is surrounded by six ligand ions CN^-, (b) one of them is substituted by H_2O, and (c) two of them. The solid and broken curves represent the potentials of the low-spin and high-spin states, respectively.

transition by IR/visible light occurs in $Pr_{0.7}Ca_{0.3}MnO_3$ under an applied electric field,[34] while the photoinduced metallic state goes back to the ground state in turning the applied field off. Recently, persistent photoconductivity without a field has been observed in $Pr_{0.75}Na_{0.25}MnO_3$.[35] These materials are regarded as an insulating matrix containing ferromagnetic metallic clusters. The size of such a metallic cluster grows under light illumination, because a surrounding region of a metallic cluster tends to be easily converted to a metallic state. This situation is similar to the easy nucleation of a photoinduced phase discussed above for a spatial structure with different relative stabilities. We expect an accelerated phase transition if an appropriate superlattice is constructed in the perovskite manganese oxide.[36]

Our proposal of combining different local units is advantageous for photoinduced phase transitions in the sense that availability of materials can be extended substantially, because even an appropriate combination of inferior materials can produce a superior material. It is also possible to search an optimized structure to reduce the threshold of the excitation strength, e.g., by adjusting the volume ratio of the competing units such that the free energies of the two phases are as close to each other as possible. This can be done, for example, by controlling the layer thickness in constructing the 1D superstructure.

For the comparison of simulations with experiments, we should comment on the excitation condition. We have assumed in our simulations that photoexcitation is exerted uniformly on both kinds of units. This condition can be fulfilled in real materials, if the photoexcitation wavelengths of the two kinds of units are close to each other, as in some spin-crossover complexes, or if we use a couple of light sources to match the two separated wavelengths. But it is also possible to excite only one of the two kinds of units resonantly by the light of a specific frequency. The forward and backward transitions by different light frequencies observed in Fe-Co Prussian blue analogues are such examples. Monte Carlo simulations for such an excitation condition have also been performed, indicating that the essential behavior is not much different, although the conversion efficiency is lowered.

We emphasize that the Ising-like model is quite useful in investigating the photoinduced phase transitions in various systems including the spin crossover complexes. Monte Carlo simulations on this model not only reproduce the typical behavior observed experimentally, but also can demonstrate a new design principle of materials for efficient photoinduced phase transitions. We note that the Ising model has been applied to wider problems such as the motion of atoms on surface, spin glass, neural networks, and even economic problems. The most essential difference between the original Ising model and the one discussed in this chapter for photoinduced phase transitions is the effect of external stimuli, which causes compulsive one-way flip of spin. It may be very interesting to search analogous effects in many other systems described by the Ising model.

References

1. T. Tayagaki and K. Tanaka, *Phys. Rev. Lett.* **86** 2886 (2001).
2. Y. Ogawa, S. Koshihara, K. Koshino, T. Ogawa, C. Urano, and H. Takagi, *Phys. Rev. Lett.* **84** 3181 (2000).
3. T. Suzuki, T. Sakamaki, K. Tanimura, S. Koshihara, and Y. Tokura, *Phys. Rev.* **B60** 6191 (1999).
4. N. Shimamoto, S. Ohkoshi, O. Sato, and K. Hashimoto, *Chem. Lett.* **2002** 486 (2002).
5. A. Bousseksou, J. Nasser, J. Linares, K. Boukheddaden, and F. Varret, *J. Phys. I* **2** 1381 (1992).
6. H. Spiering, T. Kohlhaas, H. Romstedt, A. Hauser, C. Bruns-Yilmaz, J. Kusz, and P. Gutlich, *Coord. Chem. Rev.* **190** 629 (1999).
7. H. Romstedt, A. Hauser, and H. Spiering, *J. Phys. Chem. Solids.* **59** 265 (1998).
8. K. Boukheddaden, I. Shteto, B. Hôo, and F. Varret, *Phys. Rev.* **B62** 14806 (2000).
9. P. Gütlich, *Struct. Bond.* **44** 83 (1981).
10. S. Decurtins, P. Gütlich, C.P. Köhler, H. Spiering, and A. Hauser, *Chem. Phys. Lett.* **105** 1 (1984).
11. S. Decurtins, P. Gütlich, K.M. Hasselbach, H. Spiering, and A. Hauser, *Inorg. Chem.* **24** 2174 (1985).
12. A. Bousseksou, J. J. McGarvey, F. Varret, J. A. Real, J. P. Tuchagues, A. C. Dennis, and M. L. Boillot, *Chem. Phys. Lett.* **318** 409 (2000).
13. N. Willenbacher and H. Spiering, *J. Phys.* **C21** 1423 (1988).
14. B. K. Chakrabarti and M. Acharyya, *Rev. Mod. Phys.* **71** 847 (1999).
15. Y. Ozeki and N. Ito, *Phys. Rev.* **B64** 024416 (2001).
16. H. Hinrichsen and E. Domany, *Phys. Rev.* **E56** 94 (1997).
17. O. Sakai, T. Ogawa, and K. Koshino, *J. Phys. Soc. Jpn.* **71** 978 (2002).
18. O. Sakai, M. Ishii, T. Ogawa, and K. Koshio, *J. Phys. Soc. Jpn.* **71** 2052 (2002).
19. M. Nishino, K. Yamaguchi, and S. Miyashita, *Phys. Rev.* **B58** 9303 (1998).
20. M. Nishino and S. Miyashita, *Phys. Rev.* **B63** 174404 (2001).
21. T. Kawamoto and S. Abe, *Appl. Phys. Lett.* **80** 2562 (2002).
22. T. Kawamoto and S. Abe, *Phase Transit.* **75** 753 (2002).
23. T. Kawamoto and S. Abe, *Phys. Rev.* **B68**, 235112 (2003).
24. A. Hauser, *Coord. Chem. Rev.* **111** 275 (1991).

25. A. Hauser, *Chem. Phys. Lett.* **192** 65 (1992).
26. A. Hauser, *J. Chem. Phys.* **94** 2741 (1991).
27. O. Sato, Y. Einaga, T. Iyoda, A. Fujishima, and K. Hashimoto, *J. Electrochem. Soc.* **144** L13 (1997).
28. S. Koshihara, Y. Takahashi, H. Sakai, Y. Tokura, and T. Luty, *J. Phys. Chem.* **B103** 2592 (1999).
29. X. Liu, Y. Moritomo, T. Kawmaoto, A. Nakamoto, and N. Kojima, *Phys. Rev.* **B67** 012102 (2003).
30. X. Liu, Y. Moritomo, T. Kawamoto, A. Nakamoto, and N. Kojima, *J. Phys. Soc. Jpn.* **72** 1615 (2003).
31. O. Sato, S. Hayami, Y. Einaga, and Z. Z. Gu, *Bull. Chem. Soc. Jpn.* **76** 443 (2003).
32. H. W. Liu, K. Matsuda, Z. Z. Gu, K. Takahashi, A. L. Cui, R. Nakajima, A. Fujishima, and O. Sato, *Phys. Rev. Lett.* **90** 167403 (2003).
33. T. Kawamoto, Y. Asai, and S. Abe, *Phys. Rev. Lett.* **86** 348 (2001).
34. K. Miyano, T. Tanaka, Y. Tomioka, and Y. Tokura, *Phys. Rev. Lett.* **78** 4257 (1997).
35. T. Satoh, Y. Kikuchi, K. Miyano, E. Pollert, J. Hejtmanek, and Z. Jirak, *Phys. Rev.* **B65** 125103 (2002).
36. J. Z. Sun, W. J. Gallagher, P. R. Duncombe, L. Krusin-Elbaum, R. A. Altman, A. Gupta, Y. Lu, G. Q. Gong, and G. Xiao, *Appl. Phys. Lett.* **69** 3266 (1996).

CHAPTER 6

PHOTOINDUCED PHASE TRANSITIONS IN ONE-DIMENSIONAL CORRELATED ELECTRON SYSTEMS

Hiroshi Okamoto

Department of Advanced Materials Science
Faculty of Frontier Sciences, University of Tokyo
Kashiwanoha, Kashiwa, Chiba 277-8561, Japan

and

Correlated Electron Research Center (CERC)
National Institute of Advanced Industrial Science and Technology (AIST)
Tsukuba, Ibaraki 305-8564, Japan
E-mail: okamotoh@k.u-tokyo.ac.jp

Shin-ichiro Iwai

Correlated Electron Research Center (CERC)
National Institute of Advanced Industrial Science and Technology (AIST)
Tsukuba, Ibaraki 305-8564, Japan

and

Nanostructure and Material Property Group
PRESTO, JST, Kawaguchi 332-0012, Japan

Hiroyuki Matsuzaki

Department of Advanced Materials Science
Faculty of Frontier Sciences, University of Tokyo
Kashiwanoha, Kashiwa, Chiba 277-8561, Japan

One-dimensional (1D) correlated electron systems are good targets for the exploration of photoinduced phase transitions (PIPTs). It is because charge-transfer excitations and/or carrier generations by lights can stimulate instabilities inherent to 1D nature of electronic state through strong electron (spin)-electron (spin) interactions. In this chapter, we review dynamical aspects of PIPTs observed in several typical 1D correlated electron systems; (1) a photoinduced transition from a Mott insulator to a metal in halogen-bridged Ni-chain compound, $[Ni(chxn)_2Br]Br_2$ (chxn=cyclohexanediamine); (2) a photoinduced diamagnetic to paramagnetic transition in a spin-Peierls system of an organic radical crystal, 1,3,5-trithia-2,4,6-triazapentalenyl (TTTA); (3) a photoinduced transition between neutral and ionic states in an organic charge-transfer complex, tetrathiafulvalene-*p*-chloranil (TTF-CA); (4) a photoinduced transition between a diamagnetic charge-density-wave state and a paramagnetic charge-polarization state in an iodine-bridged binuclear Pt compound, $R_4[Pt_2(pop)_4I]$ ($R=(C_2H_5)_2NH_2^+$, $pop=P_2O_5H_2^{2-}$). The dynamics and mechanism of these PIPTs are discussed on the basis of the experimental results obtained by femtosecond pump-probe spectroscopy and other several kinds of laser spectroscopy.

Contents

1. Introduction

The control of phase transitions and related macroscopic properties by photo-irradiation has recently attracted much attention.[1] This phenomenon is called photo-induced phase transition (PIPT), and is important not only as a new phenomenon in the fields of physics and chemistry, but also as a useful mechanism applicable to future optical switching devices. We can imagine several interesting PIPTs such as insulator to metal, antiferroelectric to ferroelectric, and diamagnetic to para- or ferromagnetic phase transitions. When one considers applications of these phenomena to switching devices, it is essentially important to control these physical properties by weak intensity of lights and in picosecond (ps) or sub-ps time scale.

A key strategy toward realizing such PIPTs is the exploration of one-dimensional (1D) materials.[1–10] 1D electronic (spin) states essentially include instabilities inherent to electron-electron (e-e) and/or electron-lattice (e-l) interactions and sometimes produce characteristic phase transitions at low temperatures. Under the influence of these e-e and e-l interactions, a small density of photoexcitations will be able to stimulate instability of electronic states, and then dramatic PIPTs may be observed. Especially, in the 1D systems with strong e-e interactions, which are called as 1D correlated electron systems, PIPTs are expected to be driven in ultrafast time scale, since purely electronic processes, such as charge-transfer (CT) processes of electrons, occur in femtosecond (fs) or sub-ps time domain. In this chapter, we will review PIPTs observed in several typical 1D correlated electron systems.

1.1. *Fundamental feature of 1D correlated electron systems*

First, we will briefly review the fundamental properties of 1D correlated electron systems. Here, we consider a simple 1D electronic chain with one electron in each atom schematically shown in Fig. 1.1(a). We assume that the system can be described by the half-filled Hubbard model,[11] in which only the nearest neighbor transfer energy t and the Coulomb repulsion energy U between two electrons residing on the same site are taken into account. When $U = 0$, the system is a metal having a simple half-filled band with the bandwidth of $4t$ represented in Fig. 1.1(c). In the case that U is not equal to zero, the electronic property of the system changes strongly depending on the magnitude of U.[12] When U is larger than the bandwidth, the electrons tend to localize on each site to minimize Coulomb repulsion energy. As a result, the system becomes a so-called Mott insulator.

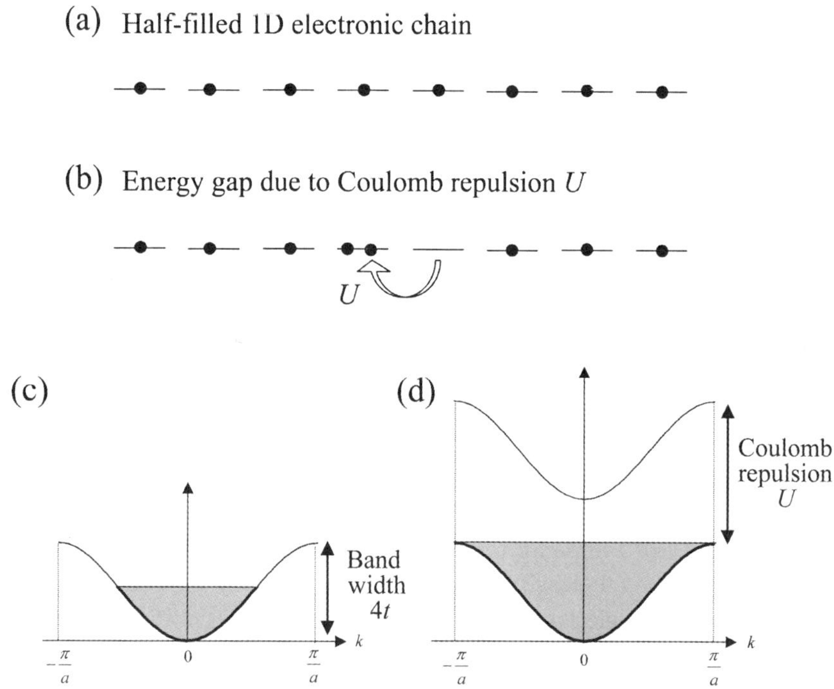

Fig. 1.1. (a,b) Schematic representations of the half-filled 1D electronic chain for $U = 0$ (a) and $U > 0$ (b). Schematic band structures corresponding to (a) and (b) are presented in (c) and (d), respectively.

Such a metal-insulator transition is called a Mott transition. In the case that $U > 4t$, the electronic excitation corresponds to the transfer of an electron to another site already occupied by one electron, as shown in Fig. 1.1(b). In the excited state, the energy increases by U. The excited electron can move to other sites occupied by one electron with no additional energy. It is, therefore, natural to consider that the excited electron occupy on the second conduction band, which is shown in Fig. 1.1(d). If $U > 4t$, a finite gap will be produced between the first and second conduction bands. This gap is called a Hubbard gap, and the first and second bands are called lower and upper Hubbard bands, respectively. More detail theoretical treatments are necessary to discuss exactly the electronic state of the system with a finite U. The picture presented in Fig. 1.1, however, can explain well qualitative feature of the 1D correlated electron systems.

With increase of U, the spin degree of freedom becomes important. When $U \gg 4t$, the spin-spin interaction can be described by the following

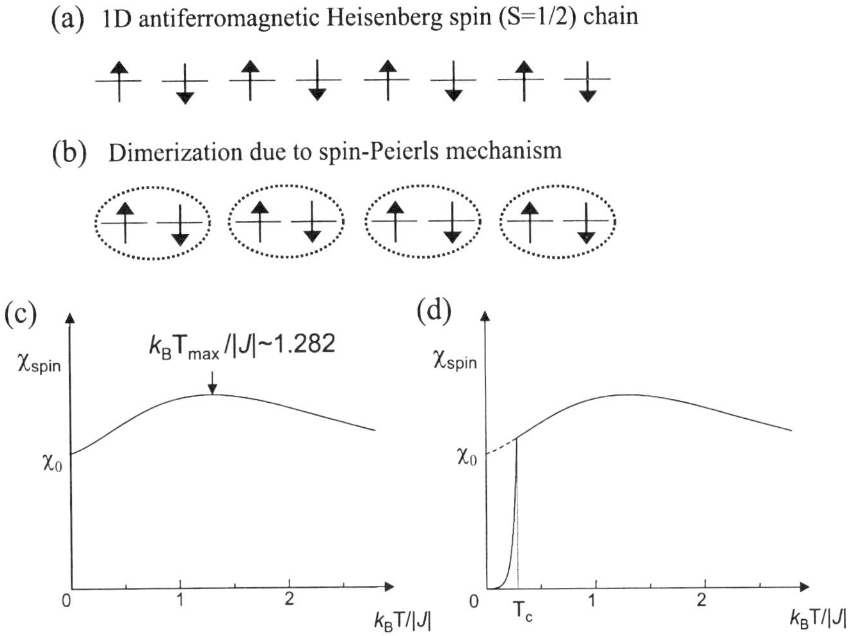

Fig. 1.2. (a,b) Schematic representations of the 1D uniform antiferromagnetic spin (S = 1/2) chain (a) and the dimerized one (b). (c) Temperature dependence of spin susceptibility χ_{spin} in the 1D uniform antiferromagnetic spin (S = 1/2) chain. J is the antiferromagnetic exchange interaction energy between the two neighboring spins. (d) Temperature dependence of χ_{spin} when a spin-Peierls transition occurs at T_c.

effective spin Hamiltonian, which corresponds to the 1D Heisenberg antiferromagnetic spin (S = 1/2) model.[13,14]

$$H = J \sum_{l} S_l \cdot S_{l+1} \tag{1}$$

J is the exchange interaction between the neighboring two spins and equal to $4t^2/U$ for $U \gg 4t$. As shown in Fig. 1.2(a), spins tend to arrange antiferromagnetically to gain the kinetic energy of electrons. The general feature of the magnetic susceptibility (χ_{spin}) of 1D Heisenberg antiferromagnetic spin (S = 1/2) system is well described by the Bonner–Fisher curve[15] shown in Fig. 1.2(c). When temperature is much higher than J/k_B (k_B: Boltzmann constant), χ_{spin} follows the Curie-Weiss law with antiferromagnetic interaction. As temperature T is decreased from the high temperature, χ_{spin} increases gradually, and then reach the maximum value at around T_{max}, which is given by $k_B T_{\text{max}}/|J| \sim 1.282$. When the temperature is decreased lower than T_{max}, χ_{spin} is suppressed through the antiferromagnetic exchange interaction J, although χ_{spin} remains to be a finite value χ_0 even at T = 0K due to fluctuations of the 1D spin system. If the lattice is dimerized at T_c as shown in Fig. 1.2(b), the two neighboring spins form a singlet state and then χ_{spin} decreases abruptly at T_c, vanishing as $T \to 0$ as shown in Fig. 1.2(d).[14,16] Such a magnetic transition is called spin-Peierls (SP) transition[14] and is actually observed in several materials with 1D spin (S = 1/2) chains. This transition is caused by the fact that the gain of magnetic energy coming from the formation of the spin-singlet states overcomes the lattice distortion energy.

1.2. *Concepts of photoinduced phase transitions in 1D correlated electron systems*

In this subsection, we introduce the materials discussed in this chapter and summarize the concepts of the PIPTs they exhibit. The first example is a photoinduced insulator-metal transition (or a Mott transition) in a halogen-bridged nickel chain compound, $[Ni(chxn)_2Br]Br_2$ (chxn=cyclohexanediamine), having nickel (Ni^{3+})-bromine (Br^-) chain as schematically shown in Fig. 1.3A.[17] This compound has an unpaired electron in a 3*d* orbital of Ni forming 1D electronic state (Fig. 1.3A(a)). It is, however, not a metal due to the large Coulomb repulsion (U) among 3*d* electrons of Ni. Each electron is localized on the Ni sites, and then this compound becomes a 1D Mott insulator.[18] When this compound is irradiated by lights, the electrons are excited to another sites as shown in Fig. 1.3A(b). This corresponds to a photo-generation of electron and hole carriers. A number of carriers may change a Mott insulator to a metal. In Sec. 2, we will show that a photoinduced Mott transition indeed occurs in an ultrafast time scale in the halogen-bridged Ni compound.[10]

The second example is an organic radical (S = 1/2) crystal, 1,3,5-trithia-2,4,6-triazapentalenyl (TTTA).[19,20] In this material, TTTA molecules form 1D chains with spin S = 1/2 as shown in Fig. 1.3B. The overlap of π orbitals between neighboring TTTA molecules (the transfer energy t) is much smaller than the on-site Coulomb repulsion (U), so that TTTA is a typical 1D Mott insulator. At around room temperature, this material exhibits a SP-like magnetic transition, which is schematically illustrated in Fig. 1.3B(a). The lowest electronic excitation of this material is the intermolecular (or CT) transition as shown in Fig. 1.3B(b). Such an excitation breaks spin-singlet states. Since the gain of the magnetic energy stabilizes the dimerized SP state, the breaking of the singlet states may make the SP state unstable and induce a photoinduced transition from a diamagnetic SP phase to a paramagnetic non-distorted phase. In TTTA, such a PIPT is indeed observed. Dynamics and mechanism of this PIPT will be elucidated in Sec. 3.[9]

The third example is a neutral to ionic transition in an organic charge-transfer (CT) compound, tetrathiafulvalene-*p*-chloranil (TTF-CA).[21,22] This compound has 1D chains composed of donor (D) molecules (TTF) and acceptor (A) ones (CA) stacking alternately as shown in Fig. 1.3C. In this type of mixed-stack CT compounds, overlap of π orbitals between adjacent donor and acceptor molecules is also essentially small as compared with both on-site and nearest-neighbor e-e Coulomb interactions. TTF-CA is, therefore, regarded as a 1D correlated electron system. Its electronic structure is reasonably expressed by a localized picture as shown in

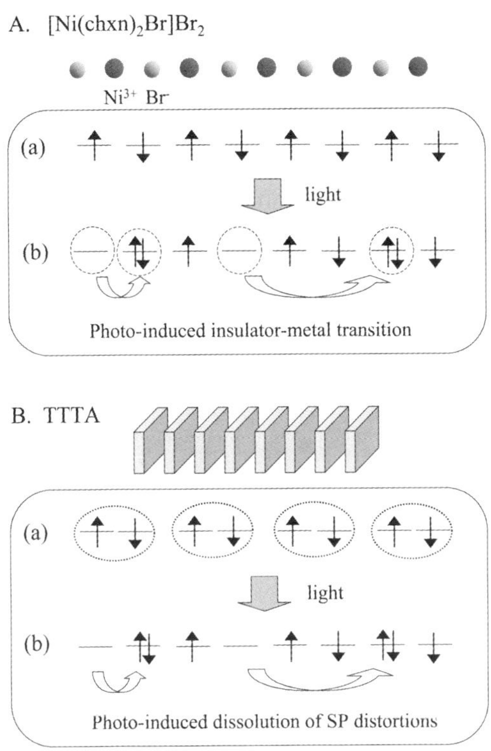

Fig. 1.3. Schematic illustrations of the chain structures and the concepts of the PIPTs in the materials discussed in this chapter; A: $[Ni(chxn)_2Br]Br_2$, B: TTTA, C: TTF-CA and D: $R_4[Pt_2(pop)_4I]$.

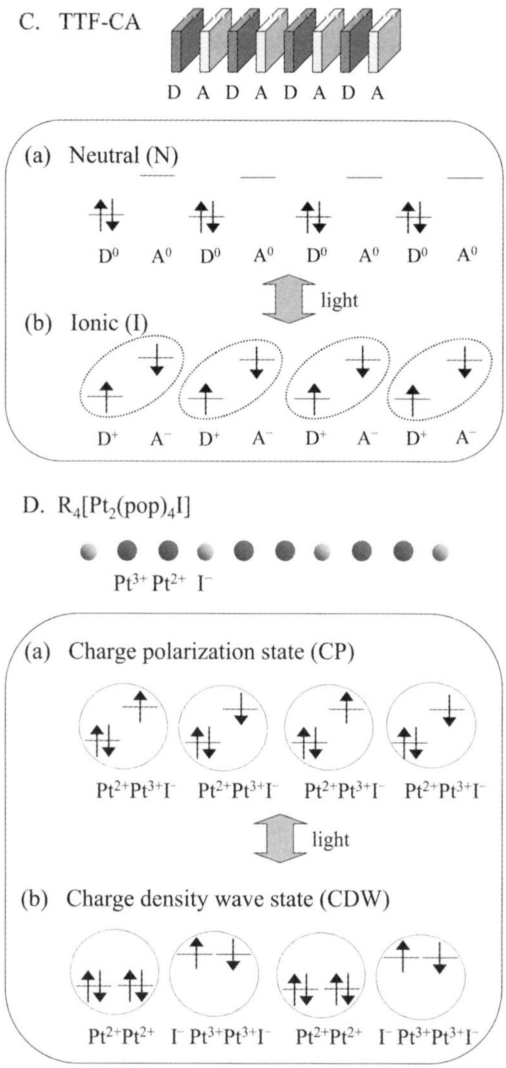

Fig. 1.3. (*Continued*)

Fig. 1.3C(a,b), in which the highest occupied molecular orbital (HOMO) of donor and the lowest unoccupied molecular orbital (LUMO) of acceptor are taken into account. In the DA-type CT compounds, the ionic state is stabilized by long-range Coulomb interaction. In TTF-CA, the neutral (N) state (Fig. 1.3C(a)) and the ionic (I) state (Fig. 1.3C(b)) are almost degenerate and the two states can be switched to each other by photo-irradiations.[2,4–7,23,24] In the I state of TTF-CA, the DA chain is dimerized due to the SP mechanism.[25–30] Therefore, the e-l interaction also plays an important role on the photoinduced transition of TTF-CA as well as the electron Coulomb interaction. Photoexcitation in the I phase generates a CT excited state, which stimulates the instability to the N state and the PIPT from the I state to the N state, occurs. The PIPT from the N state to the I state is also observed in this system. Dynamical aspects of the photo-induced I to N and N to I transition will be discussed in Sec. 4.[6,24]

The last example is iodine-bridged binuclear platinum compounds, $R_4[Pt_2(pop)_4I]nH_2O$ and $R'_2[Pt_2(pop)_4I]nH_2O$ ($(pop)=P_2O_5H_2^{2-}$), in which PtPtI chains dominate their electronic properties.[8,31–34] The Pt dimmer has three electrons per two d orbitals forming the 1D electronic state as schematically illustrated in Fig. 1.3D. In these compounds, electronic structures on the PtPtI chains can be controlled between a diamagnetic charge-density-wave (CDW) state and a paramagnetic charge-polarization (CP) state as shown in Fig. 1.3D, by modification of the counterions (R, R′) located between chains.[8] In the PtPtI chain, correlation between e-e Coulomb repulsion and e-l interaction determines the stability of the ground state, CDW or CP. In the R = $(C_2H_5)_2NH_2^+$ compound, a pressure-induced CP to CDW transition occurs. This transition is accompanied by a large hysteresis loop within which photoinduced transition between CDW and CP can be driven. Dynamics and mechanism of this photoinduced transition is discussed in Sec. 5.[8]

2. Ultrafast Photoinduced Transition from Mott Insulator to Metal in Halogen-Bridged Nickel-Chain Compound

Since the discovery of high-T_c superconductivity, doping-induced insulator (I)-metal (M) transitions or filling-control Mott transitions in $3d$ transition metal compounds have been attracting much attention. In most undoped $3d$ transition-metal oxides, electrons are localized on atomic sites due to the strong on-site Coulomb repulsion energy U, forming antiferromagnetic insulators (Mott insulators). Their electronic and magnetic properties can, however, be modified to a large extent by chemical doping.[35–37] The high-T_c superconductivity that emerges in the hole- or electron-doped layer-structured cuprates is the most dramatic example. Photoirradiation is another effective method of producing carriers.[38] In this section, we review the photoinduced I-M transition of the halogen-bridged Ni-chain compound, $[Ni(chxn)_2Br]Br_2$, which is a prototypical 1D Mott insulator. Photocarrier doping by using a 130 femtosecond laser pulse on the Ni chain induces a remarkable change of the electronic structure. When the photoexcitation density exceeds 0.1 per Ni site, a Drude-like high-reflection band appears in the infrared (IR) region, signaling the formation of a metallic state. Ultrafast dynamics of the photoinduced metallic state will be discussed on the basis of the results of temporal and doping-density dependence of the reflectivity spectra.

2.1. *Crystal and electronic structures of halogen-bridged nickel-chain compound*

Figure 2.1(a) shows the crystal structure of $[Ni(chxn)_2Br]Br_2$. The Ni^{3+} and Br^- ions line up alternately along the b-axis.[15] Four N atoms of the two ligand units (chxn=cyclohexanediamine) coordinating a Ni ion produce such a strong ligand field that the Ni^{3+} ion is in a low spin state (d^7: S = 1/2) with an unpaired electron in the d_{z^2} orbital. The 1D electronic state is

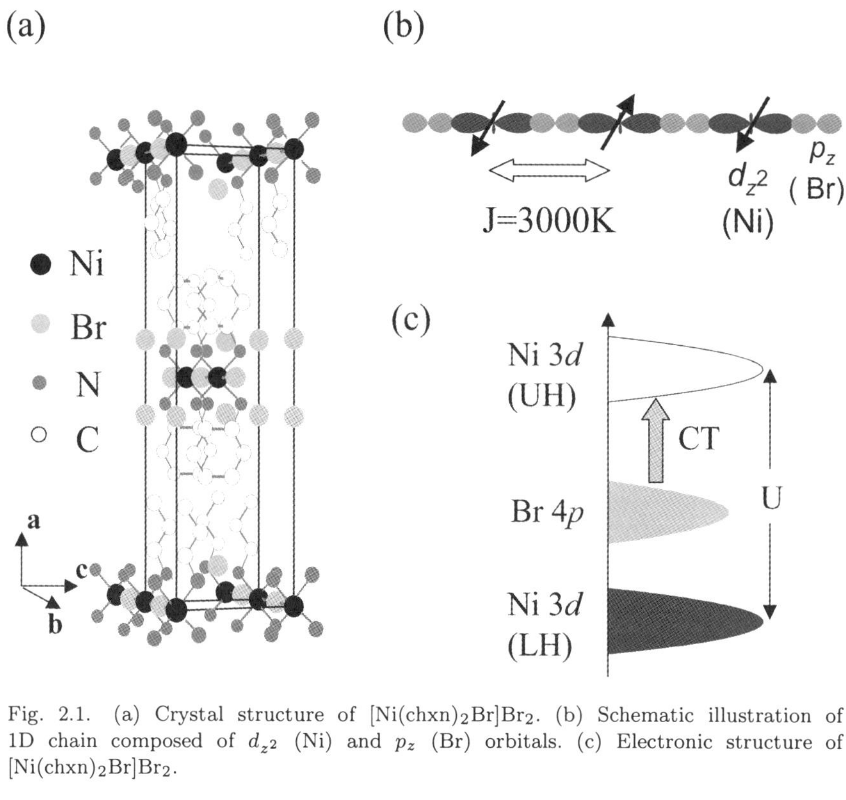

Fig. 2.1. (a) Crystal structure of $[Ni(chxn)_2Br]Br_2$. (b) Schematic illustration of 1D chain composed of d_{z^2} (Ni) and p_z (Br) orbitals. (c) Electronic structure of $[Ni(chxn)_2Br]Br_2$.

formed via the hybridization between Ni d_{z^2} and Br p_z orbitals as shown in Fig. 2.1(b).[39–42] The large U of the Ni $3d$ electrons opens a gap between the Ni $3d$ upper Hubbard (UH) band and the lower Hubbard (LH) one (Fig. 2.1(c)). The occupied Br $4p$ band positions itself within the Mott-Hubbard gap, and hence the lowest-energy electronic excitation is a CT transition from the Br $4p$ valence band to the Ni $3d$ UH band.[18] Thus, the electronic structure of this system is similar to that of the 1D cuprates such as M_2CuO_3 (M=Sr or Ca).[35,43] Strong one-dimensional antiferromagnetic interaction J $\sim$ 3000 K[44] works between the spins on the neighboring Ni site, similarly to the case of Sr_2CuO_3 (J $\sim$ 2200 K).[45]

It has recently been reported[46] that this Ni compound has a gigantic third-order nonlinear susceptibility $\chi^{(3)}$. The enhancement of $\chi^{(3)}$ is due to the large dipole moment ($\approx$20 Å) between the odd- and even-parity CT excited states, which are nearly degenerate with each other.[46,47] Such a large dipole moment and large J are consequences of strong *p-d* hybridization.[48] In addition, the ultrafast decay of the photoexcited states within a few ps has been reported in another 1D Mott insulator, Sr_2CuO_3.[49,50] The strong *p-d* hybridization and the ultrafast relaxation of the photoexcited states expected for the Ni-chain compound will provide a good arena for testing an ultrafast optical switching from I to M.

2.2. *Change of electronic structure in two-dimensional cuprate by chemical carrier doping*

Before the discussion about the photoinduced IM transition, we will review the studies of the filling control Mott transition in the 2D cuprates, as a prototypical example of the IM transition in correlated electron systems. In the 2D cuprate, La_2CuO_4, the crystal structure of which is shown in Fig. 2.2(a), hole carries can be introduced to the CuO plane (the ab plane) by substituting La^{3+} ions by Sr^{2+} ones.[35,51] With increase of the hole-carrier concentration (x) in $La_{2-x}Sr_xCuO_4$, the transition from Mott insulator to metal occurs at around $x = 0.05$. This IM transition is schematically illustrated in Fig. 2.3. Figure 2.3(a) shows schematically the electronic state of the Mott insulator, in which $3d$ electrons are localized in each Cu^{2+} ion due to the large on-site Coulomb repulsion energy U between the $3d$ electrons. The arrows indicate the antiferromagnetic spin arrangement. Figure 2.3(b) shows the holes-doping regime. When the holes (the open circles) are introduced by the carrier doping, the electrons can move to another sites as indicated by the arrows due to the presence of the electron vacancies. In this case, a finite density of state is produced near the Fermi level

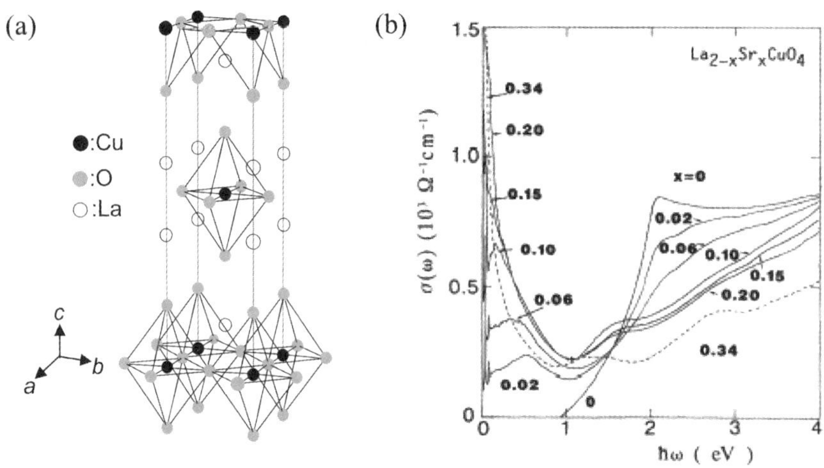

Fig. 2.2. (a) Crystal structure of the 2D cuprate, La_2CuO_4. (b) Optical conductivity $\sigma(\omega)$ spectrum of $La_{2-x}Sr_xCuO_4$ with various hole carrier concentrations (x).[51]

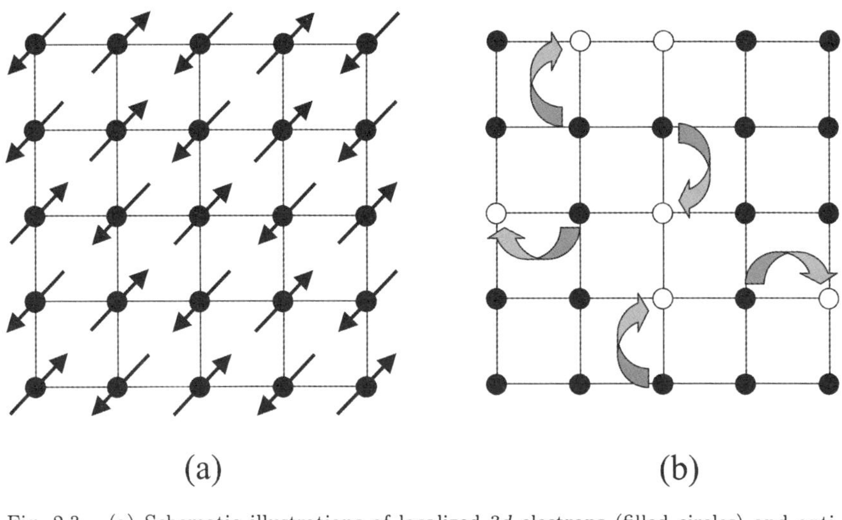

Fig. 2.3. (a) Schematic illustrations of localized $3d$ electrons (filled circles) and antiferromagnetic spin order (arrows) in 2D Mott insulator. (b) Conducting (metallic) state realized by the introduction of holes (vacant circles).

within the optical gap and then the IM transition occurs. As well-known, $La_{2-x}Sr_xCuO_4$ locating near the IM transition boundary exhibits superconductivity at low temperatures. In another 2D cuprate, Nd_2CuO_4, a similar IM transition and superconductivity at low temperature is observed by substituting Nd by Ce through the electron-carrier doping.[52]

In these materials, the IM transition and the collapse of the gap are clearly reflected by the optical conductivity spectra in the near-infrared to visible region. Figure 2.2(b) shows the doping concentration (x) dependence of the optical conductivity (σ) spectra in $La_{2-x}Sr_xCuO_4$, which are obtained from the reflectivity spectra by the Kramers–Kronig (KK) transformation.[51] The parent compound (La_2CuO_4) has a clear peak at around 2 eV corresponding to the CT gap. With increase of x, the spectral weight of the CT gap transition is transferred into the intragap region. Such a huge spectral change over a wide energy region is a most characteristic feature of the Mott transition in the 2D cuprate and also other strongly correlated electron systems of $3d$ transition metal oxides.[35] Our purpose is to drive a similar Mott transition by a photocarrier doping in the Ni-chain compound.

2.3. *Ultrafast insulator-metal transition observed in femtosecond pump-probe reflection spectroscopy*[10]

To probe the change of the electronic structure induced by a photocarrier doping, we adopted femtosecond (fs) pump-probe reflection spectroscopy, the experimental setup of which is schematically illustrated in Fig. 2.4.

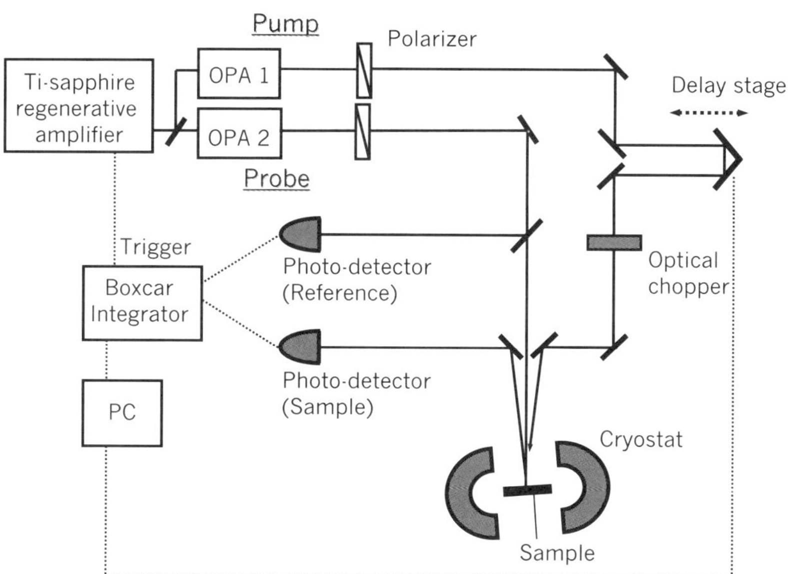

Fig. 2.4. Experimental setup for reflection detected pump-probe measurement.

In the measurement, a Ti : sapphire (Al_2O_3) regenerative amplifier system operating at 1 kHz was employed as a light source. Output from the amplifier (800 nm : 1.55 eV) with the pulse width of 130 fs was divided into two beams, which are used for the excitations of two optical parametric amplifier (OPA) systems. From the two OPA systems, the probe light pulses ranging from 0.1 eV to 2.5 eV and the pump light pulses ranging from 0.5 eV to 2.5 eV are obtained. (When the pump light is set at 1.55 eV (800 nm) as the case for the Ni-chain compound discussed below, the output from the regenerative amplifier system itself is used. For the experiments on TTF-CA discussed in Sec. 4, the double OPA system shown in Fig. 2.4 is used.) The time resolution of the apparatus is about 200 fs.

In Fig. 2.5(a), the polarized reflectivity spectrum of the Ni-chain compound is presented. A sharp peak at 1.3 eV is due to the CT gap transition. The transient reflectivity (TR) spectra observed at the delay time t_d after the photoirradiation are shown by the dots and lines in Fig. 2.5(c). The excitation energy is 1.55 eV (800 nm) just above the CT gap. The intensity of the irradiated light was 3.6 mJ/cm^2. Under this condition, the average excitation density x_{ph} of the absorbed photon is 0.5 per Ni site within the absorption depth (460 Å), as evaluated by taking account of the reflection loss (30%) and the unit cell volume (8.68×10^{-22} cm^{-3}). Immediately after the photoirradiation ($t_d = 0.1$ ps), the reflectivity in the mid-IR region remarkably increases, being reminiscent of the Drude-like response, while the reflectivity around the CT band decreases due to photoinduced bleaching. The magnitude of the reflectivity R' at $t_d = 0.1$ ps reaches about 70% at the lowest photon energy of the probe light (0.12 eV), where the change of reflectivity ($\Delta R/R = (R' - R)/R$) is as large as 260% of the original reflectivity R. The optical conductivity σ spectrum was obtained by performing the Kramers–Kronig (KK) transformation of the original reflectivity spectrum and the transient ones, which are shown in Fig. 2.5(b) and (d), respectively. As seen in Fig. 2.5(d), the σ at $t_d = 0.1$ ps monotonically increases with lowering the probe photon energy to 0.12 eV, suggesting the closing of the optical gap. Such a remarkable photoinduced feature is observed only for the probe light polarization (E) parallel to the Ni-Br chain (E//b) and not for E $\perp$ b at all, suggesting the photogeneration of a quasi-one-dimensional metallic state.

To clarify the photoinduced change of the electronic state in more detail, the excitation density x_{ph} dependence has been investigated. Spectra of the TR and σ at $t_d = 0.1$ ps for various x_{ph} are presented in Fig. 2.6(a,b). For

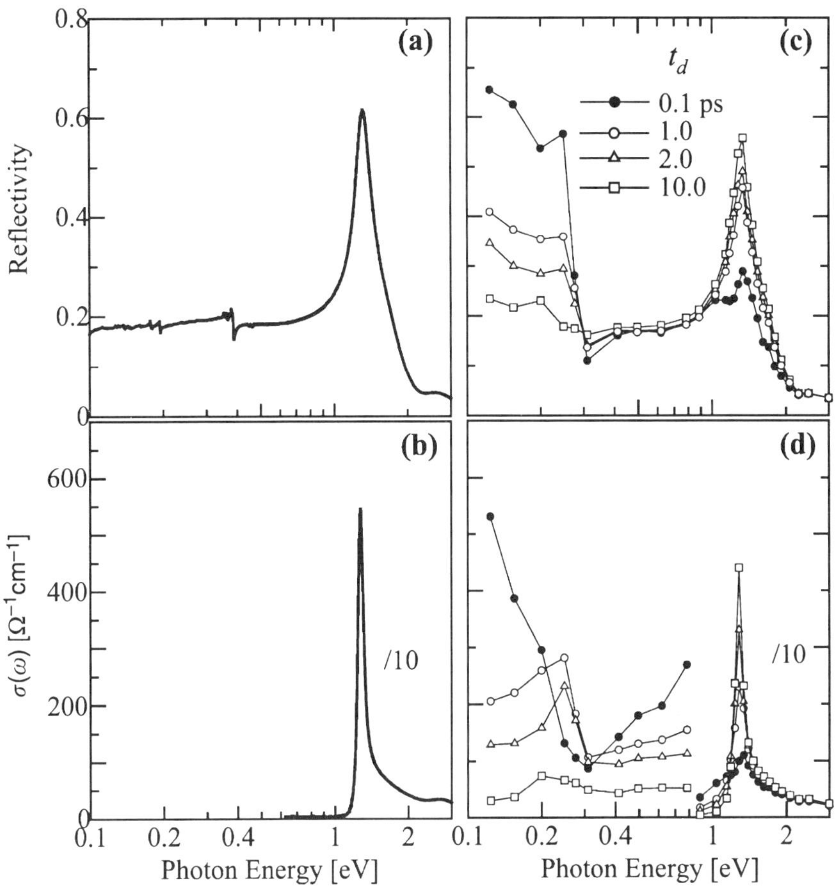

Fig. 2.5. Polarized reflectivity spectra in $[Ni(chxn)_2Br]Br_2$ before the photoexcitation (a) and at the delay time t_d after the photoexcitation (c) at room temperature. The excitation density x_{ph} is 0.5 photon/Ni site. Polarizations of both the pump and probe lights are parallel to the chain axis b. Optical conductivity ($\sigma(\omega)$) spectra obtained by the Kramers–Kronig transformation of the reflectivity data in (a) and (c) are presented in (b) and (d), respectively.

the weak excitation of $x_{ph} = 6.2 \times 10^{-4}$, a midgap absorption with a peak at 0.4–0.5 eV is observed in the σ spectrum as shown in Fig. 2.6(b). A similar photoinduced mid-gap absorption has also been reported in the 2D cuprates such as Nd_2CuO_4 and La_2CuO_4.[53] As x_{ph} increases, the low-energy part of σ below 0.2 eV significantly grows, and for $x_{ph} > 0.1$, the optical gap seems to disappear.

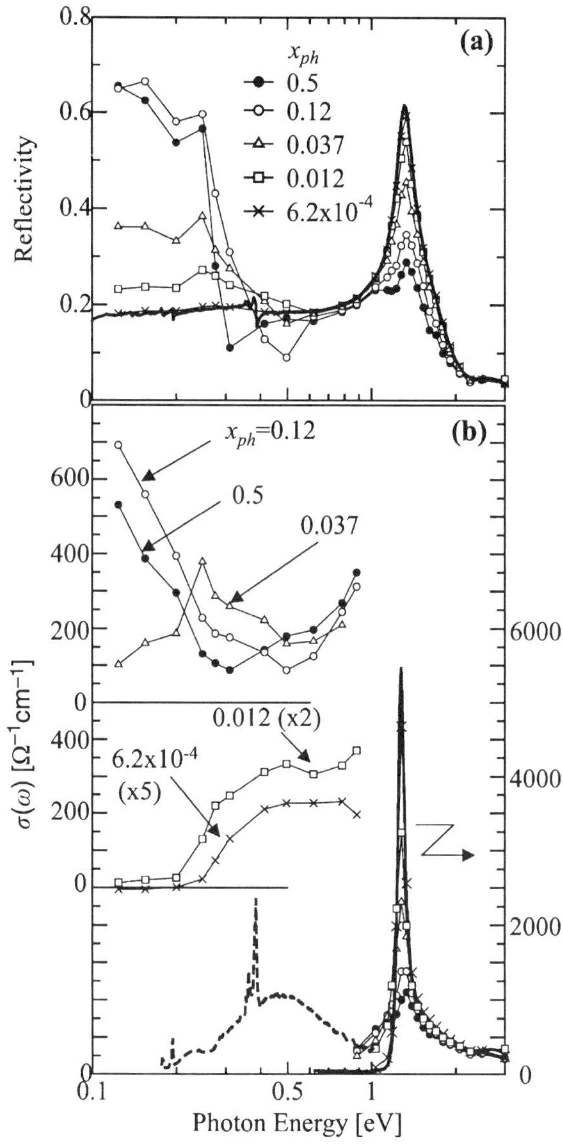

Fig. 2.6. (a) Reflectivity spectra in $[Ni(chxn)_2Br]Br_2$ before (the bold line) and immediately after the photoexcitation (t_d = 0.1 ps) for the various excitation densities x_{ph} at room temperature. (b) $\sigma(\omega)$ spectra before (the bold line) and immediately after the photoexcitation (t_d = 0.1 ps). The broken line shows the polarized absorption spectrum before the photoexcitation (given in arbitary units), which was directrly measured by the transmission configuration.

To investigate the evolution of the photoinduced Mott transition, it is useful to examine the transfer of the spectral weight from the CT gap region to the inner-gap one. The spectral weight can be quantitatively analyzed in terms of the effective number of electrons $N_{\text{eff}}(\omega)$ defined as follows.[51]

$$N_{\text{eff}}(\omega) = \frac{2m_0}{\pi e^2 N} \int_0^{\omega} \frac{\sigma(\omega')}{4\pi} d\omega'. \tag{2}$$

Here, m_0 is the free electron mass and N the number of Ni atoms per unit volume. $N_{\text{eff}}(\omega)$ is the measure for the kinetic energy of electrons on an energy scale of $\hbar\omega$. Since the lower energy bound of the measured photoinduced signals is 0.12 eV, the photoinduced change in N_{eff} (ΔN_{eff}) was calculated using σ from 0.12 eV to $\hbar\omega$. The results for various x_{ph} at $t_{\text{d}} = 0.1$ ps are presented in Fig. 2.7. $\Delta N_{\text{eff}}(\omega)$ monotonically increases with $\hbar\omega$ up to 1 eV, reflecting the accumulation of the spectral weight below 1 eV. Then, it abruptly drops almost to zero at 1.3 eV due to the bleaching around the CT band, indicating that the spectral weight of the CT gap transition is transferred to the inner-gap region by the photocarrier doping. The values of ΔN_{eff} above 1.5 eV for $x_{\text{ph}} = 0.12$ and $x_{\text{ph}} = 0.5$ are negative as seen in Fig. 2.7. With increase of x_{ph} from 0.12 to 0.5, the photoinduced bleaching observed around the CT band is enhanced, while the spectral intensity in

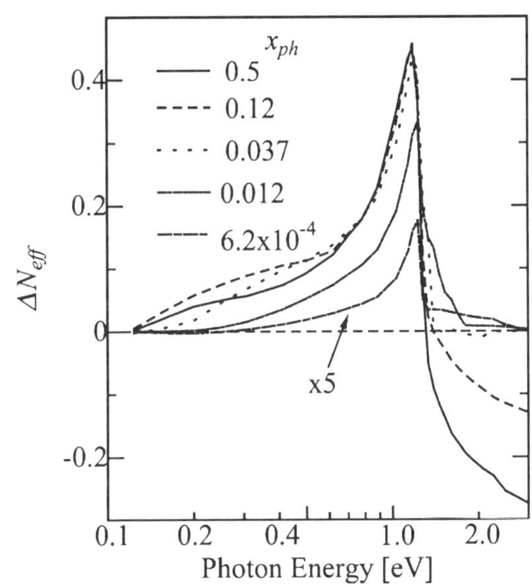

Fig. 2.7. Photoinduced changes of the effective number of electrons $N_{\text{eff}}(\omega)(\Delta N_{\text{eff}}(\omega))$ at t_{d}=0.1 ps for various x_{ph} in $[\text{Ni(chxn)}_2\text{Br}]\text{Br}_2$.

the infrared region is rather decreased as seen in Fig. 2.6. These results suggest that the negative values of ΔN_{eff} above 1.5 eV for $x_{\text{ph}} > 0.1$ (especially for $x_{\text{ph}} = 0.5$) would be attributable to the miscount of the spectral weight in the infrared region below the lower energy bound (0.12 eV) of the measured spectral range.

Here, we will comment on the validity of the KK transformation on the TR spectra. When applying the KK transformation on the TR spectra, there are two important effects to be considered;[54–56] (1) carrier concentration changes depending on the distance from the sample surface, and (2) absorption depths of the probe light l_r and the pump light l_p are different. It is, therefore, necessary to check carefully the validity of the analyses. In the experimental results of the Ni-chain compound, the spectral shape of the TR due to the mid-gap absorption observed for $x_{\text{ph}} < 0.012$ is almost unchanged and, therefore, will not be so affected by those two effects. It is, however, reasonable to consider that the absolute values of the TR and σ are somewhat underevaluated, since $l_r > l_p$ (= 460 Å). To evaluate the two effects on the Drude-like reflection band observed for $x_{\text{ph}} > 0.1$, we postulated a metallic state expressed by a simple Drude model with the thickness l_t (200–2000 Å) on the surface of the Ni-chain compound and simulated the R and σ spectra. The result of the simulation shows that for $l_t > 1000$ Å, spectral shape and absolute value of R and σ are independent of l_t. In the Ni-chain compound, the thickness of photoinduced metallic state (region with carrier concentration >0.1) exceeds 1000 Å for $x_{\text{ph}} > 0.2$. So we can consider that the Drude-like reflection band observed for $x_{\text{ph}} > 0.2$ will not be so influenced by the two effects. For the intermediate excitation density ($0.2 > x_{\text{ph}} > 0.02$), it might be necessary to take account of some errors in the analysis. Nevertheless, the observed systematic changes of $\Delta N_{\text{eff}}(\omega)$ and the approximate holding of the sum rule over the wide range of x_{ph} ensure that the analysis using the KK transformation presented here reflects well the photoinduced changes of the electronic state.

In Fig. 2.8(a), we plot ΔN_{eff} (1 eV) and ΔN_{eff} (0.2 eV) at $t_d = 0.1$ ps as a function of x_{ph}. ΔN_{eff} (1.0 eV), which represents the total spectral weight transferred from the CT band to the inner-gap region, saturates for $x_{\text{ph}} > 0.04$. On the other hand, ΔN_{eff} (0.2 eV), i.e. the spectral weight accumulated in the lower energy region between 0.12 and 0.2 eV, increases almost linearly with x_{ph} up to $x_{\text{ph}} = 0.1$. The steady increase of ΔN_{eff} (0.2 eV) for $x_{\text{ph}} > 0.04$ will be attributable to the growth of the Drude weight. Let us compare the x_{ph} dependence of ΔN_{eff} with the chemical-doping-density (x) dependence of N_{eff} in the 2D cuprates

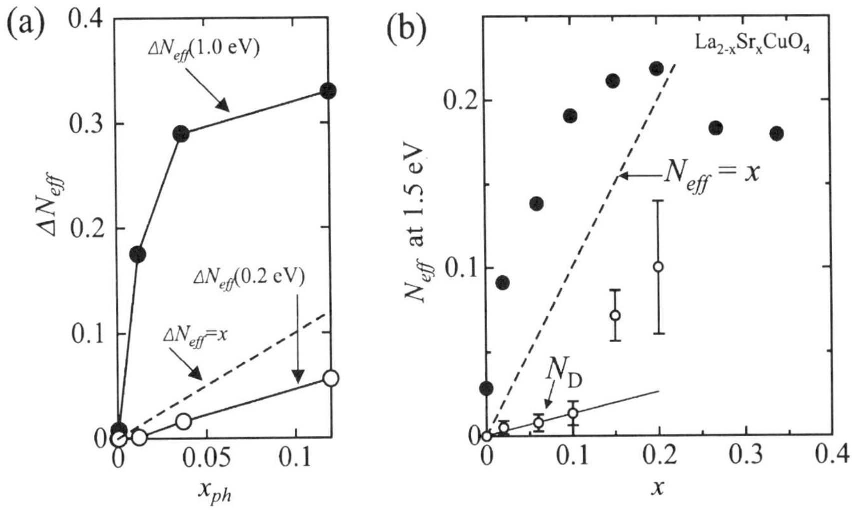

Fig. 2.8. (a) Photoinduced changes of the effective number of electrons $\Delta N_{\rm eff}$ (1.0 eV) and $\Delta N_{\rm eff}$ (0.2 eV) at $t_{\rm d} = 0.1$ ps as a function of the photo-doping concentration $x_{\rm ph}$ in $[\mathrm{Ni(chxn)_2Br}]\mathrm{Br_2}$. (b) The effective number of electrons as a function of the chemical-doping concentration x for $\mathrm{La}_{2-x}\mathrm{Sr}_x\mathrm{CuO_4}$.[51] $N_{\rm eff}$ and $N_{\rm D}$ are the total spectral weight in the infrared region and the Drude component, respectively.

previously reported, which is presented in Fig. 2.8(b).[52] In Fig. 2.8(b), the solid and open circles show the x dependence of $N_{\rm eff}$ at 1.5 eV and that for the Drude component in the infrared region, respectively. The observed clear resemblance of the $x_{\rm ph}$ dependence of $\Delta N_{\rm eff}$ in the Ni-chain compound with the x dependence of $N_{\rm eff}$ in the 2D cuprate demonstrates that the Mott transition is driven by the photocarrier doping in the Ni-chain compound.

A noteworthy aspect of the photoinduced Mott transition is the influence of the electron-hole asymmetry on the transient optical spectra. Being distinct from the chemical doping case, the photoirradiation creates both electrons and holes. In the Ni-chain compound, it has been clarified from the measurement of the DC conductivity[57] and the Seebeck coefficient[58] that a small number of electron carrier exists in an as-grown sample. A weak midgap absorption due to such carriers or small polarons is indeed observed in the σ spectrum (E//b) as shown by the broken curve in Fig. 2.6(b). The photoinduced absorption observed for small $x_{\rm ph}$ ($= 6.2 \times 10^{-4}$) resembles this midgap absorption in spectral shape and energy position. This resemblance suggests that electron-type carriers are responsible for the photoinduced midgap absorption. The hole-type carriers with the Br

p-character and the resultant strong electron-lattice interaction may be strongly localized and hence make the least contribution to the spectral weight of the mig-gap absorption. When the metallic state is formed for $x_{\text{ph}} > 0.1$, however, there should be no distinction between electrons and holes for the strongly d-p hybridized state near the Fermi level.

2.4. *Relaxation dynamics of the photoinduced metallic state*

The ultrashort lifetime of the metallic state is another important aspect of the present photoinduced Mott transition. Figure 2.9 shows the temporal characteristics of ΔR at 0.12 eV and 1.39 eV, respectively. Two results are the same with each other. In Fig. 2.9(a), we also present the time evolutions of ΔN_{eff} (0.5 eV) for $x_{\text{ph}} = 0.012$ and $x_{\text{ph}} = 0.5$ by the open circles. ΔR (0.12 eV) and ΔN_{eff} (0.5 eV) are also in good agreement with each other. Therefore, the time characteristics of ΔR can be considered to reflect those of ΔN_{eff} (0.5 eV). The time characteristics of ΔR (0.12 eV) can be reproduced by the sum of the three exponential functions whose time constants τ_{d} (weights) are 3 ps (40%), 8 ps (50%), and $\gg$500 ps (10%) for $x_{\text{ph}} = 0.012$, and 0.4 ps (60%), 3 ps (15%), and 8 ps (25%) for $x_{\text{ph}} = 0.5$.

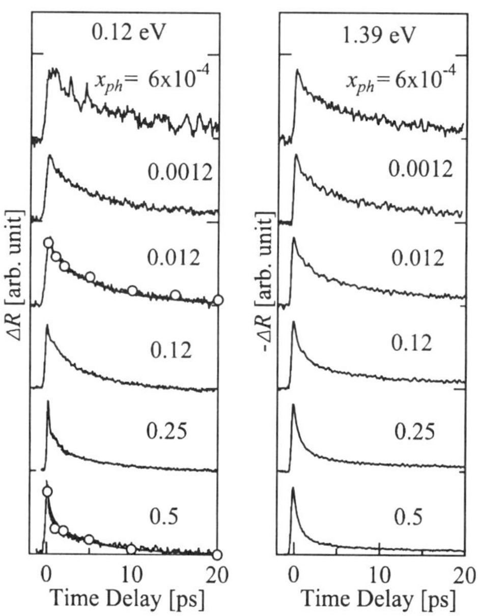

Fig. 2.9. Time evolutions of ΔR measured at 0.12 eV and at 1.39 eV for various excitation density x_{ph} in $[\text{Ni(chxn)}_2\text{Br}]\text{Br}_2$. Time evolutions of ΔN_{eff} (0.5 eV) for $x_{\text{ph}} = 0.5$ and 0.012 are also shown by the open circles.

The component with $\tau_d \gg 500$ ps for $x_{ph} = 0.012$ is attributable to some long-lived trapped carriers. The ultrafast decay component with $\tau_d = 0.5$ ps is dominant for $x_{ph} = 0.5$, while such a fast decay was not discerned for $x_{ph} = 0.012$. Therefore, it is likely that the ultrafast decay is characteristic of the photoinduced metallic states.

The decay dynamics of the photoexcited state in Mott insulators has also been studied on the 2D cuprates (Nd_2CuO_4 and $Sr_2CuO_2Cl_2$)[59,60] and 1D cuprate (Sr_2CuO_3).[49] In these cuprates, the lifetime τ of the photoproducts has been revealed to be 1 ps to several tenth ps which is significantly faster than that in conventional semiconductors. The electron-electron scattering with emission of spin-excitations (or spinons) is considered as a possible mechanism for the ultrafast relaxation in the Mott insulators.[49,59] In the present halogen-bridged Ni-chain compound, enhancement of similar electron-electron scattering in the quasi-1D metallic state may play a major role in the increase of the recombination rate of photocarriers.

3. Photoinduced Phase Transition in a Spin-Peierls System of Organic Radical Crystal, TTTA

Photocontrol of magnetic properties is an important subject not only in terms of fundamental science, but also with regard to potential applications in novel magneto-optical devices. Photoinduced changes in magnetization have been reported for several kinds of materials, including molecule-based magnets,[61] magnetic semiconductor heterostructures,[62] manganese oxides,[63] and spin-crossover complexes.[64,65]

A key strategy toward realizing photoswitching of magnetism is the exploration of materials near magnetic phase transition boundaries. The optical control of magnetic transitions requires the magnetic instability of the material to be enhanced or suppressed by light. Since light fields do not couple directly with spin states, it is necessary to find an effective mechanism that combines electronic excitations induced by light irradiations with magnetic instability.

A promising candidate for the optical control of magnetic transitions is the family of spin-Peierls (SP) system, in which 1D paramagnetic spin ($S = 1/2$) states are converted to dimerized diamagnetic states with lowering temperature as mentioned in Subsec. 1.1. Chemical doping of carriers or nonmagnetic impurities can cause SP states to become unstable, as observed in $(DMe\text{-}DCNQI)_2Li$(DMe-DCNQI=2,5-dimethyl-N,N′-dicyanoquinonediimine)[66] and $CuGeO_3$.[67] Therefore, it can be expected that photocarrier doping and the resultant modified spin states in SP systems would lead to a macroscopic change in magnetic properties. A number of organic compounds with 1D spin (S = 1/2) chains such as M-TCNQ (M = Na, K, Rb, and NH_4, TCNQ = tetracyanoquinodimethane),[68] $MEM(TCNQ)_2$ (MEM = N-methyl-N-ethylmorpholinium),[69] TTF-CuBDT (BDT = bis-dithiolene),[14] DAP-TCNQ (DAP = 1,6-pryrendiamine),[70] and $(BCPTTF)_2X$ (X = PF_6 and AsF_6, BCPTTF = benzocyclopentyl-tetrathiafulvalene),[71] form spin singlet states through lattice dimerizations at low temperatures. Most of the transitions in these organic materials except for $MEM(TCNQ)_2$ and TTF-CuBDT have not been established as typical SP transitions because of the lack of the experiments about magnetic field effects on their transitions. The SP instability, however, will play a significant role in the transitions of these materials, so that we call them SP systems in this chapter. An attempt to control the SP instability by lights was in fact performed in K-TCNQ.[72] According to the study, it was revealed from the molecular vibrational spectroscopy that the molecular dimerizations were weakened by light irradiations. In K-TCNQ, however, the realization of the photoinduced phase transition from the SP state to the paramagnetic state has not been fully clarified yet. In this section, we report the photoinduced phase transition in the SP system of an organic radical (S = 1/2) crystal, 1,3,5-trithia-2,4,6-triazapentalenyl (TTTA), which exhibits optical and magnetic bistability at around room temperature.[19,20] A nanosecond laser pulse is found to induce a phase transition from diamagnetic low-temperature (SP) phase to a paramagnetic high-temperature phase both inside (296 K) and outside (11 K) the hysterisis loop. In addition, it has been demonstrated that the photoinduced transition is driven by suppression of the spin-Peierls instability by the accumulation of photocarriers from the comparison of the excitation energy dependence between transition efficiency and photoconductivity.

3.1. *Room-temperature optical and magnetic bistability in TTTA*

In this subsection, we review the crystal structure, and the optical and magnetic properties of TTTA. Figure 3.1(c) shows the temperature dependence of the paramagnetic susceptibility χ_{spin} for a polycrystalline sample of TTTA previously reported by Fujita and Awaga.[19] In the cooling run, χ_{spin} begins to decrease at around 230 K, and approaches zero at 170 K. The gradual increase in χ_{spin} below 100 K is due to Curie spins on lattice defects. When heated from the lowest temperature, χ_{spin} remains at almost zero over the range of 80–300 K, increasing abruptly at 305 K. Thus, TTTA

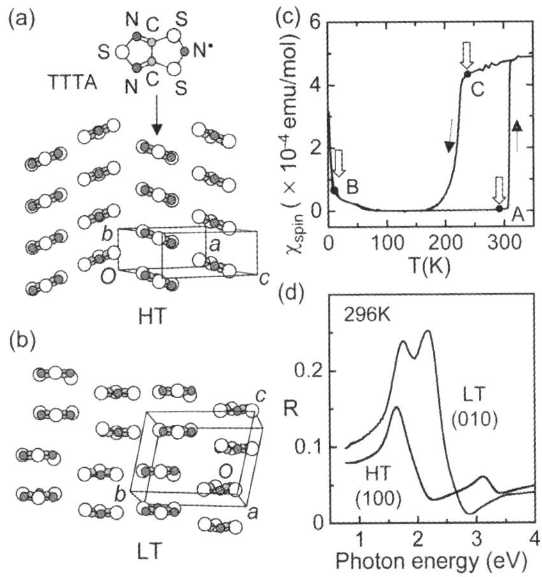

Fig. 3.1. Crystal structures of the (a) HT phase and (b) LT phase for TTTA with its molecular structure. (c) Temperature dependence of χ_{spin} upon cooling and heating processes. The photoinduced effects were investigated at the points A (296 K), B (11 K), and C (240 K). (d) Polarized reflectivity spectra for the light polarization E parallel to (//) the stacking axis in the HT and LT phases at 296 K.

exhibits a first-order magnetic phase transition near room temperature with an anomalously large hysteresis loop ($T_{c\downarrow} = 230$ K and $T_{c\uparrow} = 305$ K).

The crystal structure of TTTA at room temperature has also been reported by Fujita and Awaga, which is presented in Fig. 3.1(a) and (b), for the high-temperature (HT) and low-temperature (LT) phase, respectively.[19] In the HT phase, TTTA molecules stand face-to-face to form 1D regularly stacking columns along the b axis. Calculation of the intermolecular overlap integrals using the single-occupied molecular orbital (SOMO) obtained by the extended Huckel method reveals that the intercolumn overlap cannot be neglected, as compared with the intracolumn one. In fact, the temperature dependence of χ_{spin} in the HT phase (230–365 K) can be well reproduced by the modified Bonner–Fisher model with intercolumn interactions as the mean-field component.[19,20] The evaluated intra- and intercolumn antiferromagnetic exchange energies in the HT phase are $J/k_B = 320$ K and $J'/k_B = 60$ K, respectively.

In the LT phase, TTTA molecules are strongly dimerized along the c-axis, giving a phase transition that is analogous to an SP transition discussed in the Subsec. 1.1. The steep drop of χ_{spin} at $T_{c\downarrow}$ is attributable to the formation of the magnetic gap due to the molecular dimerizations. The molecular packing in the LT phase differs considerably from that in the HT phase; the intracolumn molecular overlaps in the LT phase are considerably larger than those in the HT phase, while the intercolumn ones are almost identical. This results in an enhancement of one-dimensionality in the LT phase. Actually, intradimer antiferromagnetic exchange energy is evaluated to be $J/k_B = 650\,K$ by analyzing the slight increase in χ_{spin} above 200 K in the LT phase with the Bleaney–Bowers equation.[20,73] This value is much larger than $J/k_B = 320\,K$ in the HT phase. Thus, the transition of TTTA is considered to include a degree of structural character as well as the SP mechanism. This is the reason why the transition observed for TTTA is not a second-order one expected in typical SP systems.

Another remarkable feature in TTTA is that its magnetic transition accompanies a drastic color change in the crystal. Figure 3.1(d) shows the polarized reflectivity spectra (E//stacking axis) at 296 K. The strong bands observed below 2.5 eV are assigned to the charge-transfer (CT) transition between the neighboring TTTA molecules in both phases. The two peaks around 1.8 and 2.1 eV observed in the LT phase can be assigned to the intradimer and interdimer CT transitions, respectively. The corresponding microscope images for E//stacking axis are shown in Figs. 3.2(a) and (b). The presence of clear chromism from yellow-green in the LT phase to red-purple in the HT phase makes it possible to observe the PIPT visually. We

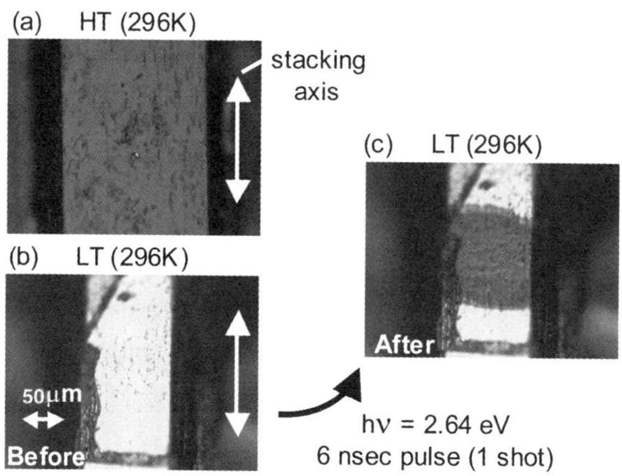

Fig. 3.2. Microscope images (E//stacking axis) for (a) HT phase, (b) LT phase, and (c) after a single-shot irradiation at 2.64 eV (E_{exc}//stacking axis) at 296 K.

have investigated the photoinduced effect in the LT phase at 296 K (the point A) and 11 K (the point B), and in the HT phase at 240 K (C) [see Fig. 3.1(c)].

3.2. *Photoinduced phase transition in TTTA*[9]

Figure 3.2(c) shows an image of the crystal after a single-shot irradiation at 2.64 eV (E_{exc}//stacking axis) at the point A in Fig. 3.1(c). A tunable optical parametric oscillator (OPO) pumped with the third harmonics of a *Q*-switched Nd:YAG laser was employed to irradiate the sample with 6-ns pulses. The crystal exhibits a permanent color change from yellow-green to red-purple in the irradiated area. In addition to the color change, PIPT was measured based on Raman spectra, as shown in Fig. 3.3(a) for the polarization configuration of $z(xx)\bar{z}$ (x $\perp$ stacking axis) at 296 K. The peaks observed around 1350 cm^{-1} in both phases can be assigned to the C=N stretching mode. After the thermal transition from the LT phase to

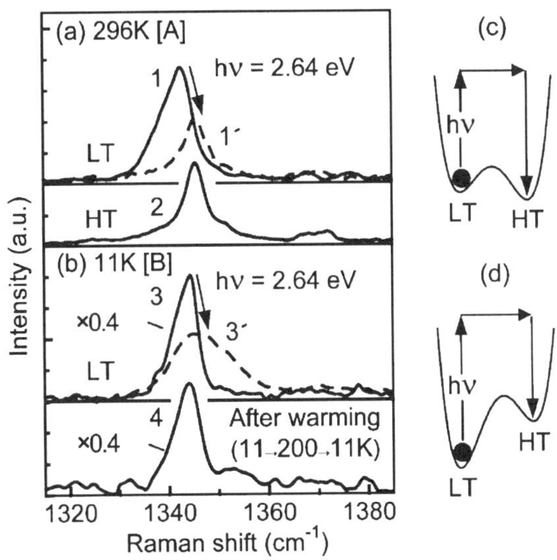

Fig. 3.3. (a) Raman spectra with polarization configuration of $z(xx)\bar{z}$ (x $\perp$ stacking axis) in TTTA at 296 K (the point A). Solid curves 1 and 2 represent the spectra in the LT phase and HT phase, respectively. Broken curve 1 represents the spectrum after a single-shot irradiation at 2.64 eV (E_{exc}//stacking axis) in the LT phase. (b) Raman spectra at 11 K (the point B). Solid curve 3 and broken curve 3 represent the spectra before and after the single-shot irradiation at 2.64 eV (E_{exc}//stacking axis). Solid curve 4 represents the spectrum after heating to about 200 K and then cooling to 11 K. (c) and (d) show the schematic energy potential curves at the points A and B, respectively.

the HT phase, the frequency of this mode increases slightly with a substantial reduction in intensity. The broken curve 1 shows the Raman spectrum after irradiation at 2.64 eV (E_{exc}//stacking axis). The penetration depths of the pump laser (2.64 eV) and the probe laser (1.96 eV) are evaluated to be almost equal (~800 Å) from the values of the absorption coefficient (α) obtained by the KK analysis of the polarized reflectivity spectra. Thus, the probe laser is able to detect photoinduced changes in almost the entire volume irradiated by the pump laser pulse. As can be seen in Fig. 3.3(a), the shape of the Raman signal for the irradiated region coincides well with that of the HT phase, showing clearly that a conversion from the LT phase (metastable state) to the HT phase (ground state) can be induced by irradiation with a laser pulse, as illustrated in Fig. 3.3(c).

The photoinduced LT-to-HT conversion with a similar color change was also observed at 11 K (the point B). The solid curve 3 and broken curve 3′ in Fig. 3.3(b) are the Raman spectra at 11K before and after a 2.64 eV laser pulse irradiation. The spectrum 3′ resembles the spectrum 2 in the HT phase, except for a spectral broadening. When heated to about 200 K and then returned to 11 K, the spectrum 3′ in the irradiated region changes to the spectrum 4, which is in good agreement with the spectrum 3 before photoirradiation. Such a recovery of the initial LT phase excludes the possibility of sample damage as the origin for the broadening of the spectrum 3. The spectrum 3′ cannot be fully reproduced by superposition of the HT spectrum 2 and the LT spectrum 3, even if the ratio of the two spectra is varied. It suggests that the photoinduced phase differs from the HT phase and that the coexistence of the LT phase and the photoinduced phase will be responsible for the observed spectral broadening. Judging from the clear color change to red-purple in the irradiated area, however, the photoinduced phase is essentially the same as the HT phase. Namely, the difference between the photoinduced phase and the HT phase will be small. Such a difference between the photoinduced phase and thermally induced HT phase has been reported in the PIPT of a spin-crossover complex, $[Fe(2-pic)_3]Cl_2EtOH$.[65] The photoinduced HT phase of TTTA persists at least for several hours at 11 K, suggesting that the system preserves the double potential minima in the free energy as illustrated in Fig. 3.3(d).

The conversion efficiency (ϕ) of the PIPT is shown in Figs. 3.4(a)-(i) and (ii) for the points A, B, and C as a function of the absorbed photon density per unit volume (I). The energy of the pump laser is 2.64 eV. Here, I was evaluated by considering the absorption coefficient α and the reflection loss of the excitation light, and ϕ was estimated from the photoinduced

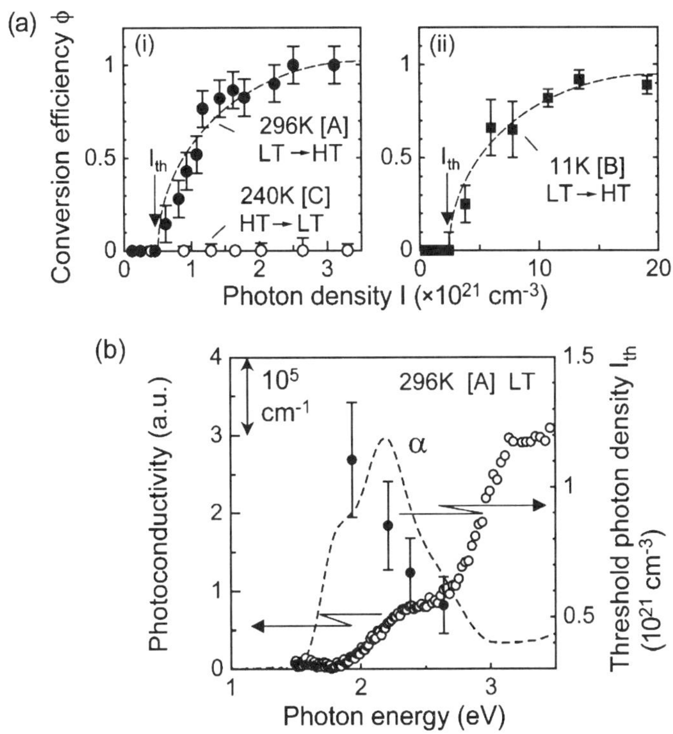

Fig. 3.4. (a) Excitation density dependence of photoinduced LT-to-HT and HT-to-LT conversion efficiency at (i) the points A (296 K) and the point C (240 K), and (ii) the point B (11 K). Excitation energy was 2.64 eV (E_{exc}//stacking axis). (b) Excitation energy dependence of the threshold photon density I_{th} for LT-to-HT conversion (right ordinate) at 296 K. Open circles show the excitation profile of photoconductivity along the stacking axis, and the broken curves shows the absorption coefficient α for E//stacking axis in the LT phase at 296 K (left ordinate).

change in the C=N Raman stretching bands. At the point A, ϕ increases nonlinearly with I, with a clear threshold at $I_{th}(A) \sim 5.5 \times 10^{20}\,cm^{-3}$. At the point B, ϕ also exhibits a threshold at $I_{th}(B) \sim 2.4 \times 10^{21}\,cm^{-3}$. $I_{th}(B)$ is about 4 times as large as $I_{th}(A)$, reflecting that the LT phase is much more stable than the HT phase at the point B (11 K), as schematically shown in Fig. 3.3(d). The observed threshold behavior at the points A and B suggests that a cooperative effect between photoexcited species plays an important role in the growth of the macroscopic phase change, as discussed in relation to the PIPT of the CT complex, TTF-CA (see Sec. 4). In contrast, at the point C (240 K), where the material is in the metastable HT phase, the HT-to-LT transition could not be driven by the irradiation at 2.64 eV, even when the excitation densities are increased up to $5 \times 10^{21}\,\mathrm{cm}^{-3}$ [see Fig. 3.4 (a)-(i)].

To deduce the mechanism of the LT-to-HT transition observed at the points A and B, ϕ was also measured as a function of the absorbed photon density per unit volume I for various excitation photon energies (1.95–2.64 eV) at the point A. The results reveal that ϕ is strongly dependent on the excitation energy. The most notable feature is that I_{th} increases gradually as the excitation energy decreases, as shown in Fig. 3.4(b). The observation of this phenomenon excludes the possibility that laser-heating effects due to light irradiations are responsible for the photoinduced LT-to-HT phase transition. It is reasonable to consider that the observed excitation energy dependence of I_{th} is associated with changes in the nature of the initial photoexcited states. As for the nature of the initial photoexcited states, the excitation profile of photoconductivity (PC) gives important information.

In the PC measurements, light from a Xenon lamp was monochromized through a grating monochromator. The monochromized light was polarized by a Glan–Taylor prism and focused on the surface of the sample. The applied dc electric field F and the polarization of the exciting light E were set parallel (//) to the stacking axis. The photo-current, which is modulated by the incident light chopped with the frequency f ($\sim$270 Hz), is detected by a lock-in amplifier. The spectral intensity of the irradiated lights was corrected by using a Si photodiode. The thickness of the sample is much larger than the absorption depths of the incident lights. Therefore, we can consider that all the incident photons were absorbed in the sample. The excitation profile of the photoconductivity was normalized to the incident photon number by taking account of the reflection loss of the incident light. In the measurements, we have confirmed that the chopping frequency f ($\sim$270 Hz) of the incident light is high enough that the contribution of thermally excited carriers is negligible compared to the photo-generated ones. Linearity of the photocurrent against the applied voltage was also carefully checked.

The excitation profile of PC along the stacking axis was measured at the point A (296 K). The result is plotted by the open circles in Fig. 3.4(b). The PC is almost negligible around the intradimer CT band ($\sim$1.8 eV), suggesting excitonic character of the photoexcited state. The PC then gradually increases with excitation photon energy. As the PC around the interdimer CT band ($\sim$2.2 eV) is much smaller than the saturation value around 3 eV, the excitonic effect is considered to be still important in the interdimer

CT excited state. I_{th} decreases as the PC increases, suggesting that photogenerated charge carriers can promote the PIPT more effectively than excitons.

Considering these findings, a possible scenario for the LT-to-HT transition is considered as follows. Under the influence of electron (spin)-lattice interaction, dissociated charge carriers generated in the LT phase will be relaxed to polarons. Previous theoretical studies in the SP systems have expected that doped carriers can exist as charged (spinless) solitons (CS's).[74] Stabilization of CS's has in fact been observed in a segregated stacked CT complex, DAP-TCNQ, which is known as a doping regime for an SP system.[70] Taking account of these studies, CS's are considered to be another possible charged species in TTTA, as well as polarons. Both polarons and CS's are spinless species, so that they will break the spin singlet states (or equivalently the dimerizations) and then cause the dimerized states to become unstable. In Fig. 3.5(a), the photogeneration of a polaron pair is schematically illustrated. When the excitation power is weak and the amount of such spinless charge carriers is small, molecular dimerizations will be suppressed only around carriers. In this case, the PIPT hardly occurs. When the excitation density is increased and spinless charge carriers concentrate above a critical value, the SP state will become unstable, leading to the photoinduced LT to HT transition. Such a degradation of the SP state is

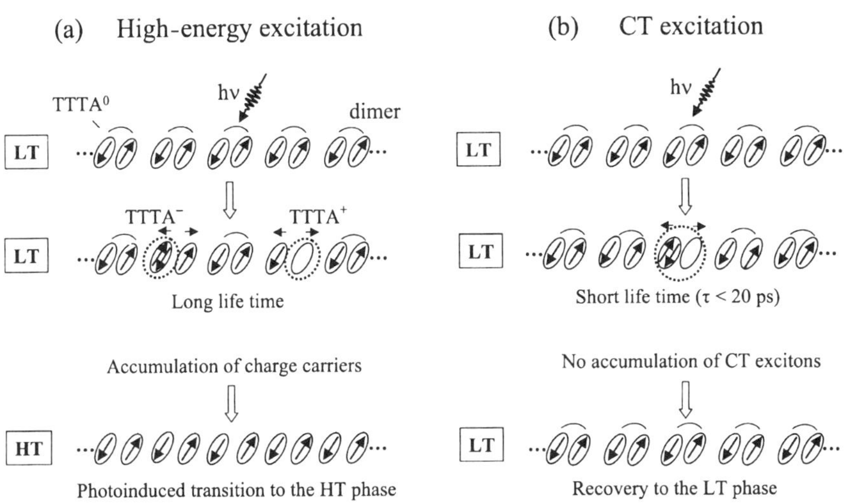

Fig. 3.5. Schematic dynamics of the photoexcited states in TTTA for the high-energy excitation (a) and the CT excitation (b).

analogous to the suppression of the SP state in $(DMe\text{-}DCNQI)_2Li_{1-x}Cu_x$[66] by doping of charge carriers, as well as in $Cu_{1-x}Zn_xGeO_3$[67] by doping of nonmagnetic impurities.

The CT excitons are also spinless species as illustrated in Fig. 3.5(b) and may contribute the photoinduced LT-to-HT transition. Important information related to the CT exciton has been obtained from the luminescence measurements. In Fig. 3.6(a), the steady-state luminescence spectrum for the 1.96 eV excitation measured at 296K in the LT phase is given by the solid line. The broken line shows the absorption spectrum. The peak observed at around 1.4 eV is attributable to the luminescence from the lowest intradimer CT excited state. To evaluate the life time of the CT excited state, we also measured the time-resolved luminescence spectrum. In the measurement, the output from the optical parametric amplifier (OPA) system excited by the light from the $Ti:Al_2O_3$ regenerative amplifier system was used as a light source. The excitation energy is set to be 2.06 eV and its pulse width is about 130 fs. For the detection of the luminescence, a streak camera is used. The solid line in the lower panel of Fig. 3.6(b) is the response of the laser pulse used for the excitation, which shows that the time-resolution of the detection system is about 20 ps. The dot and line in the upper panel is the time-characteristics of the luminescence in the LT phase (296 K) at the peak energy of the luminescence spectrum. The

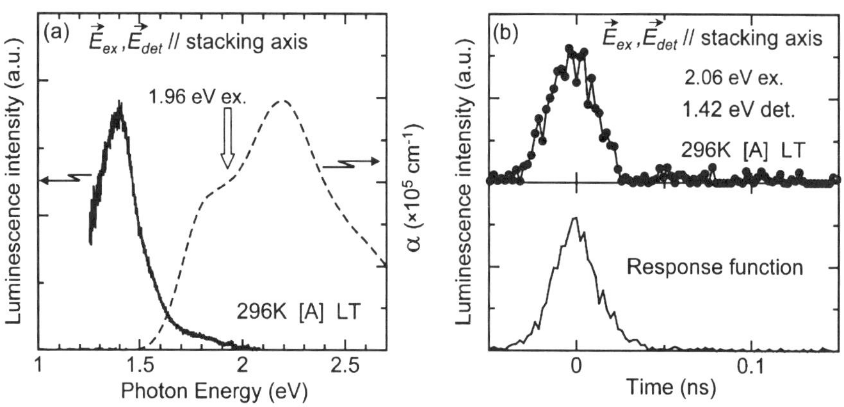

Fig. 3.6. (a) The luminescence spectra (the solid curve, left ordinate) for the excitation photon energy of 1.96 eV and the absorption coefficient α (the broken curve, right ordinate) for E//stacking axis in TTTA at 296 K (the point A). The excitation light and observed luminescence is polarized parallel to the stacking axis. The open arrow indicates the energy position of the excitation light. (b) Time evolution of luminescence detected at 1.42 eV for the excitation photon energy of 2.06 eV at 296 K (the point A) (the upper figure). The response function of the system detected at 2.06 eV (the lower figure).

decay time of the exciton luminescence is evaluated to be less than 20 ps. It is much smaller than the radiative life time (~60 ns) estimated from the oscillator strength of the CT band.[75] These results suggest that the CT exciton should decay nonradiatively. Because of the fast decay, excitons hardly accumulate in the time domain of a laser pulse (~6 ns) used as the excitation lights for the photoinduced transition. Dissociated charge carriers (CS's or polarons) will be rather stabilized as self-trapped carriers due to the strong electron (spin)-lattice interaction and have relatively longer life time. This might be a reason why photocarriers induce the PIPT more effectively than CT excitons. The scenario of the PIPT in TTTA presented here is summarized in Fig. 3.5. To detect the dynamics of the photoexcited states more directly, the femtosecond pump-probe spectroscopy is a most effective method. The pump-probe experiments under weak excitation condition, in which the permanent PIPT does not occur, are essentially important to clarify the nature and dynamics of the photoexcited states initially produced.

Optical excitation in the HT phase will also produce photocarriers as well as CT excitons. However, it is quite natural that such spinless excitations never enhance the SP instability in the HT phase. Hence, the HT-to-LT transition cannot be driven by photoirradiations, as observed in the experiments.

4. Dynamical Aspects of Photoinduced IN and NI Transitions in TTF-CA

The material discussed in this section is a mixed-stack charge-transfer (CT) complex, tetrathiafulvalene-*p*-chloranil (TTF-CA), which is one of the typical examples showing PIPT.[2,4–7] The molecular structure and the electronic structure of TTF-CA are shown in Fig. 4.1(a) and (b), respectively. By lowering temperature, it undergoes a neutral (N) to ionic (I) phase transition at $T_c = 81$ K.[22,25,28,76] The effective ionization energy of TTF donor (D) and CA accepter (A) pair is equal to I_D-E_A. Here, I_D and E_A are the ionization potential of D and the electron affinity of A, respectively. The NI transition is caused by the energy gain of the long-range Coulomb attractive interaction overcoming the effective ionization energy of DA pairs.[21] The degree of CT (ρ) between TTF and CA changes from 0.3 to 0.7 at T_c.[22,25,28,77] In the I phase, each molecule has spin $S = 1/2$ constituting the 1D spin chains, which are dimerized due to the SP mechanism.[78] The recent x-ray and neutron studies revealed that the directions of dimeric molecular displacements are three-dimensionally ordered and then ferroelectric ground state is stabilized.[30,79]

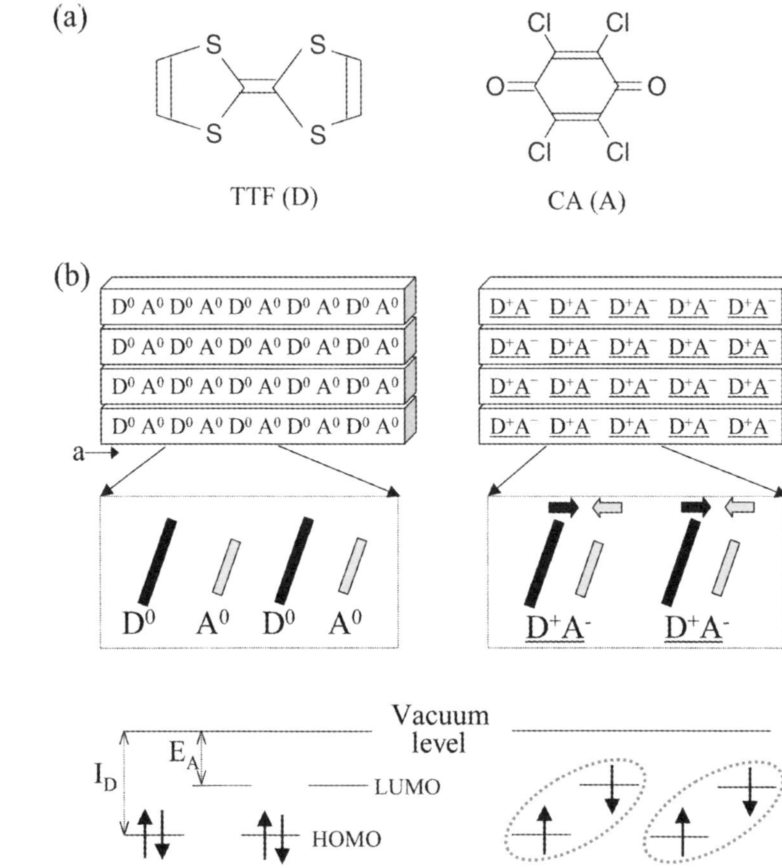

Fig. 4.1. (a) Structures of TTF (donor) and CA (acceptor) molecules. (b) Schematic illustrations of D^0A^0 (neutral) and D^+A^- (ionic) chains along the a axis and the energy level structures of neutral and ionic states. I_D and E_A are the ionization potential of the donor and the electron affinity of the acceptor, respectively.

The PIPT of TTF-CA was reported first about 10 years ago by Koshihara *et al.*[2] The most striking feature of the PIPT they reported is that more than 100 DA pairs are changed from ionic to neutral by one photon. From the subsequent studies, they reported that the time scale of the photo-induced IN transition is the order of 100 ps.[4] They also suggested that the photoinduced NI transition occurs within the similar time scale as the photoinduced IN transition. In these studies, the excitation lights were set at 1.5–2.5 eV and polarized perpendicular to the DA stacking axis a ($E \perp a$), which produce the local intramolecular transition of

TTF. Their energies are much higher than the peak energy of the lowest CT transition (0.65 eV). The observed dynamics of PIPT, therefore, are quite complicated and its mechanism is unclear up to the present. More recently, Suzuki *et al.* reported the behaviors of the IN transition induced by the irradiation of lights with the energy of 1.2 eV for E//a.[5] They suggested that the relative efficiency of the photoinduced IN transition with the CT excitation is smaller than that with the intramolecular excitation. The used excitation energy (1.2 eV), however, corresponds to the off-resonant CT excitation with a large excess energy. To clarify clearly the mechanism of the PIPT in TTF-CA, it is essentially important to detect the dynamical behaviors of the transition under the resonant excitation of the CT band.

Here, we review the recent studies of the PIPT in TTF-CA using femtosecond pump-probe reflection spectroscopy.[6,24] It has been found that the photo-induced IN transition is driven in a ps time scale by the resonant excitation to the lowest CT transition, which corresponds to an excitation of neutral donor (D)-acceptor (A) (D^0A^0) pair in the ionic DA stacks such as [—$D^+A^-D^+A^-D^+A^-$—] → [—$D^+A^-D^0A^0D^+A^-$—]. In this condition of photoexcitations, we can investigate directly the multiplication process of the D^0A^0 pairs leading to the semi-macroscopic photo-induced IN transition. To unravel the mechanism of the multiplication of the D^0A^0 pairs, we have studied the excitation-density, temperature, and excitation-energy dependence of the time characteristics of the photogenerated N states. The results clearly demonstrate that a microscopic 1D N domain, that is, a sequence of D^0A^0 pairs, is initially produced from a CT excitation, and then macroscopic IN transition occurs within 20 ps through the cooperative multiplication of the 1D N domains. When the excitation energy is increased, the dynamical behaviors of the photoinduced IN transition are considerably modified. The systematic studies of the pump-probe reflection spectroscopy reveal that the natures of the photoexcited states initially produced by the resonant CT excitation and by the higher energy excitation (or off-resonant CT excitation) are different from each other. As a result, both the dynamics and the efficiency of the IN transition depend on the excitation energy. In addition, we will report that the three kinds of coherent oscillations with the periods of ~0.6 ps, ~50 ps and ~85 ps are observed in the photoinduced reflectivity change during the photoinduced IN transition. These oscillations are attributable to the dynamical dimeric displacements of molecules associated with the dissolution of the SP distortions, the shock wave driven by the sudden volume change due to the photoinduced IN transition, and the oscillation of the NI domain boundary. At the final part of this section, dynamical behaviors of the photoinduced NI transition (the reverse process of the IN transition) will be also reported.

4.1. *Dynamical aspect of photoinduced IN transition*

In Fig. 4.2(a), we present polarized reflectivity spectra for E//a and ⊥ a of TTF-CA crystal on (100) surface measured at 4 K (I phase) and at 90 K (N phase). The prominent peak structures observed around 0.65 eV in both I and N phases are due to the CT transition. The structure around 2–2.5 eV in Fig. 4.2(a) is attributable to the intra-molecular transition of TTF, that shifts sensitively depending on ρ; its peak energy is 2.25 eV in the I phase with $\rho = 0.7$ and 2.4 eV in the N phase with $\rho = 0.3$.[76] By measuring photo-induced reflectivity changes of this band, photo-conversion of I state (D^+A^-) to N state (D^0A^0) can be detected.[2,5] For this purpose, we used femtosecond pump-probe reflection spectroscopy, the detail of which is described in the previous section (see Fig. 2.4).

The lower part of Fig. 4.2(b) shows the transient differential reflectivity ($\Delta R/R$) spectra at 4 K with the resonant excitation of the CT band (0.65 eV) for E//a. Excitation density (photon number per unit area N_{ex}) is $1.2 \times 10^{16}/cm^2$. The observed spectral changes $\Delta R/R$ at any delay time t_d are similar to the differential spectrum $((R_N(90K) - R_I(4K))/R_I(4K))$ shown in the upper panel of Fig. 4.2(b), which is calculated from the reflectivity spectra in the I phase at 4 K and the N phase at 90 K. It indicates that the N states are photo-generated in the I stacks. The spectral shape of $\Delta R/R$ is almost independent of the excitation density in the range of $0.01–1.2 \times 10^{16}/cm^2$. The magnitudes of the spectral changes increase with t_d up to 40 ps and then saturate as seen in Fig. 4.2(b).

To discuss the dynamics of the photo-generated N states, we present the time evolutions of $-\Delta R/R$ at 4 K and 77 K in Figs. 4.3(a) and (b), respectively. The detection energy is set to the peak energy (2.25 eV) of the intramolecular transition of TTF in the I phase, so that the decrease of R (or the increase of $-\Delta R/R$) indicates the generation of the N states. The $-\Delta R/R$ signals are accompanied by the prominent oscillations. The origin of these oscillations will be discussed in the Subsec. 4.3. The initial responses are presented in the right hand figures of Figs. 4.3(a,b). At 4 K, the time characteristics strongly depend on the excitation density (see the left panel of Fig. 4.3(a)). For the lower excitation density ($N_{ex} = 0.02 \times 10^{16}/cm^2$),

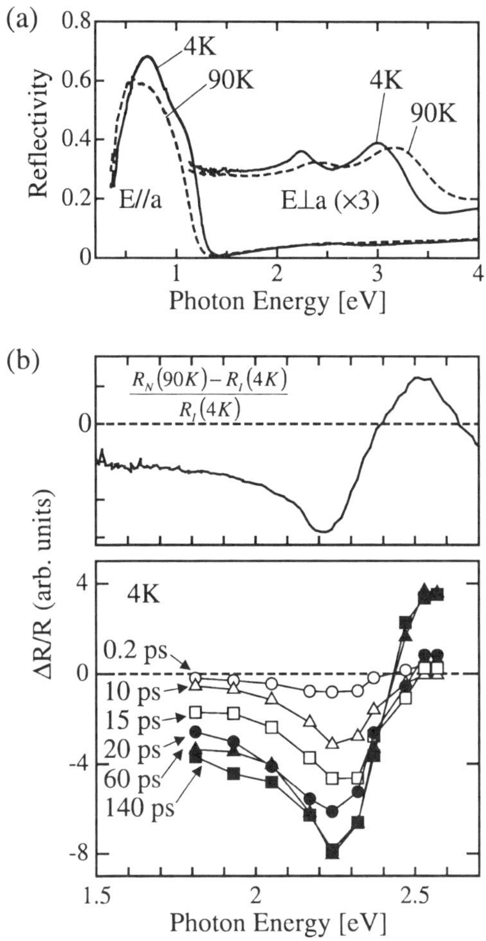

Fig. 4.2. (a) Polarized reflectivity spectra for E//a and E ⊥ a of TTF-CA on (100) surface in the I (ionic) phase (4 K, the solid line) and in the N (neutral) phase (90 K, the broken line). (b) Transient differential reflection spectra (ΔR/R) measured at 4 K for various time delays after CT excitation. Excitation density is 1.2×10^{16} photons/cm^2. The upper panel shows the differential spectrum $(R_N - R_I)/R_I$. R_I and R_N are the reflectivity spectra of the I and N phases measured at 4 K and 90 K, respectively.

the initial rise of the signal is very fast as seen in the right lower panel of Fig. 4.3(a). The rise time τ_r is less than the time resolution of our system (~200 fs). The initial rise is accompanied by a sharp drop up to 2 ps and then the signal decays with the decay time of about 300 ps. For the higher excitation density ($N_{ex} = 1.2 \times 10^{16}/\text{cm}^2$), we also observe the fast initial rise that is, however, followed by the drastic increase of the signal. Such an

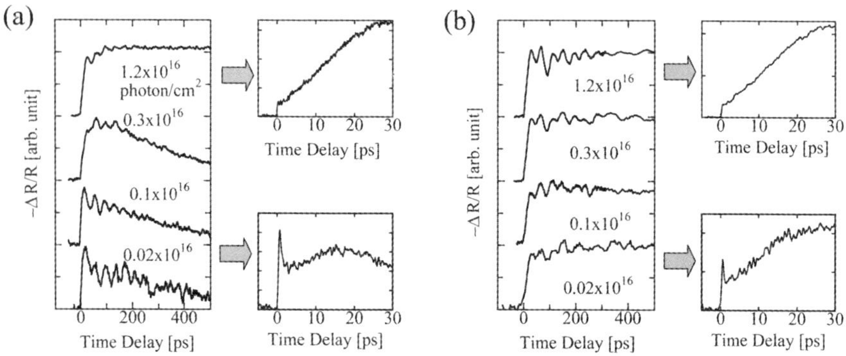

Fig. 4.3. Time evolutions of the transient reflectivity changes ($-\Delta R/R$) measured at 2.25 eV for $E \perp a$; (a) 4 K and (b) 77 K. The pump energy is 0.65 eV. The right side panels show the dynamics in the short time region.

increase saturates at $t_d \sim 30$ ps, but the signal does not decay at least to 500 ps. It suggests the formation of a stable N state, which is different from the initial N state produced just after the photo-irradiation.

At 77 K just below T_c, on the other hand, the time evolutions of the signals are rather independent of the excitation density as seen in Fig. 4.3(b) that are characterized by the fast rise within 200 fs and the following increase to $t_d \sim 20$ ps. Even for the lowest excitation density ($N_{ex} = 0.01 \times 10^{16}/cm^2$), the signal does not decay, that is in contrast with the fast decay with the life time $\tau \sim 300$ ps at 4 K. The photogenerated stable N states completely decay within at least $t_d = 500\,\mu s$ at both 4 K and 77 K, although their decay time is dependent on the excitation density.

To clarify the excitation-density dependence in the dynamics of the photo-induced N states, we plotted $-\Delta R/R$ at $t_d = 2$ ps and 500 ps as a function of N_{ex} in Fig. 4.4. The $-\Delta R/R$ values at $t_d = 2$ ps indicate the magnitudes of the initial rise, which show the amount of the D^0A^0 pairs just after the photoirradiation. When N_{ex} is less than $0.15 \times 10^{16}/cm^2$, the initial rise is proportional to N_{ex} at 4 K as indicated by the broken line. In the same region of N_{ex}, the signals almost disappear at $t_d = 500$ ps which corresponds to the fact that the photo-induced N states decay with the life time of ~300 ps. When N_{ex} exceeds $0.2 \times 10^{16}/cm^2$, the initial rise saturates and the signal at $t_d = 500$ ps increases sharply indicating the multiplication of the N states. At 77 K, the initial rise also shows the linear dependence on N_{ex} (the dashed-dotted line) for small N_{ex} less than $0.05 \times 10^{16}/cm^2$ and the saturation for the larger N_{ex}. The N states produced just after the photoirradiation never decay but always multiply.

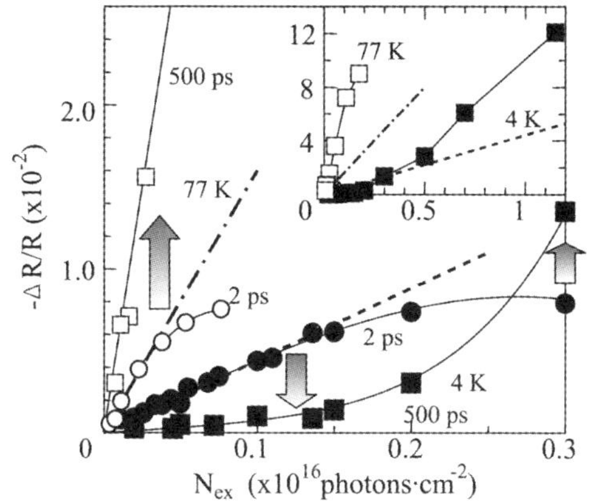

Fig. 4.4. The photoinduced changes in reflectivity ($-\Delta R/R$) at 2.25 eV for $E \perp a$ as a function of the excitation density N_{ex}; the closed circles: $T = 4$ K and $t_d = 2$ ps; the closed squares: $T = 4$ K and $t_d = 500$ ps; open circles: $T = 77$ K and $t_d = 2$ ps; open squares: $T = 77$ K and $t_d = 500$ ps. In the inset, only the squares are plotted.

The saturated value of the signals $-\Delta R/R$ at $t_d = 2$ ps is equal to 0.008 which is in common at 4 K and 77 K, therefore, it is reasonable to consider that the saturation of the initial rise is related with the space-filling of the N states generated in the I stacks. To estimate the amount of the initial D^0A^0 pairs, we assume that the saturation occurs when half of the molecules within the absorption depth L for the pump lights are neutralized. The amount of the D^0A^0 pairs will not exceed that of the I state, since the two CT processes, $D^+A^- \to D^0A^0$ and $D^0A^0 \to D^+A^-$, balance with each other, both of which are resonantly excited by the 0.65 eV pump. Using the saturation densities ($N_s = 0.3 \times 10^{16}/\text{cm}^2$ at 4 K and $0.1 \times 10^{16}/\text{cm}^2$ at 77 K) and $L = 400$ Å evaluated from the analysis of the polarized reflectivity spectrum, we can estimate the amount of the initial D^0A^0 pairs produced by one photon (a single CT excitation) to be $8D^0A^0$ at 4 K and $24\,D^0A^0$ at 77 K. These initial N states are considered to be 1D confined N domains relaxed from the Franck-Condon type CT excited state. This relaxation process is schematically shown below.

$$[\text{—}D^+A^-D^+A^-D^+A^-D^0A^0D^+A^-D^+A^-D^+A^-\text{—}]$$
$$\longrightarrow [\text{—}D^+A^-D^0\,A^0\,D^0\,A^0\,D^0\,A^0\,D^0\,A^0\,D^0\,A^0\,D^+A^-\text{—}]$$

It has been theoretically demonstrated that the energy of a 1D N domain is lower than that of a single CT state, when the N and I phases are almost degenerate.[80–83] The size of the initial 1D N domain at 77 K is larger than that at 4 K. This is quite natural, since the valence instability is enhanced at

77 K near T_c. A large size of a 1D N domain can be regarded as an excitation of an NI domain wall (DW) pair. In the time evolution of $-\Delta R/R$ shown in Fig. 4.3, the spike-like rise and decay is observed within 2 ps for the small excitation density, that might be related with the relaxation dynamics of a 1D N domain. Characteristic dynamics within several ps time region observed in the transient reflectivity change will be discussed later in the Subsec. 4.3.

At 4 K, the initial N states decay rapidly for the low excitation density, while they are multiplied for the high excitation density. The N states produced after the multiplication have a life time much longer than 500 ps, suggesting the formation of the stable N states, which are different from the 1D N domains initially produced. It is reasonable to consider that the 1D N domains become the neighboring I states unstable and change them to N states, resulting in the production of semi-macroscopic N states. The I phase in TTF-CA is stabilized by the interchain interactions as well as the intrachain one, as demonstrated by the detail structural analysis using the x-ray and neutron scattering.[30] As for the interchain interaction, there are two kinds of the energy gain stabilizing the I phase; one is due to the interchain Coulomb interaction, another due to the 3D ordering of the dimeric displacements giving rise to the ferroelectric nature of this material. Both of these two energy gains will be lost by the photogeneration of the 1D N domains cooperatively. That would be the reason why the multiplication of the 1D N domains to the semi-macroscopic N states occurs. The photoinduced behaviors for the low and high excitation densities observed at 4 K are schematically illustrated in the middle and lower panels of Fig. 4.5, respectively.

At 77 K, the multiplication is observed for all the excitation densities we used. The lowest one is $0.005 \times 10^{16}/\text{cm}^2$, which corresponds to 1 photon/1000 DA pairs, and neutralizes only 0.1% of molecules. Such a promotion of the multiplication process of the N states at 77 K is again attributable to the enhancement of the valence instability near T_c. The photoinduced behaviors for the low excitation density observed at 77 K is schematically illustrated in the upper panel of Fig. 4.5.

4.2. *Excitation energy dependence of photoinduced IN transition*

In the previous subsection, we report the photoinduced IN transition under the resonant CT excitation. The study about the excitation-energy dependence of the dynamics is indispensable for the overall understanding of the mechanism of the photoinduced IN transition.[84]

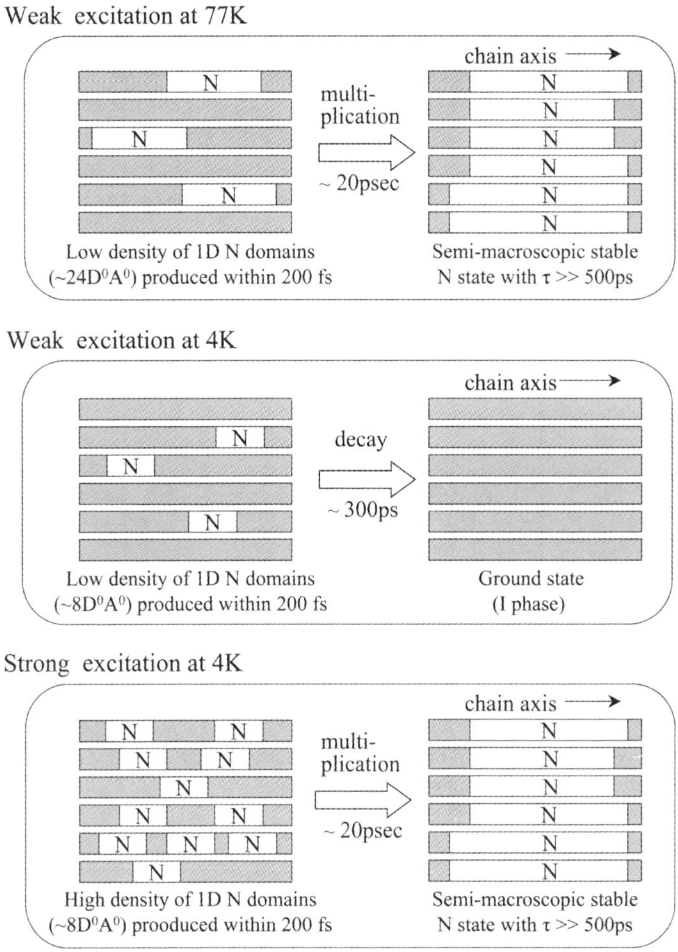

Fig. 4.5. Schematic illustrations of the photoinduced behaviors observed in TTF-CA; the upper panel: the low excitation density, T = 77 K; the middle panel: the low excitation density, T = 4 K; the lower panel: the high excitation density, T = 4 K.

Figure 4.6 shows the time evolutions of $-\Delta R/R$ at 2.25 eV, which reflect the generation of the N states, for various excitation energies ranging from 0.65 eV to 1.55 eV at 4K. For each excitation, the excitation density is set to be very low, in which the observed dynamics is independent of the excitation density. As seen in Fig. 4.6, the decay of the photoinduced N states becomes slower with increase of the excitation energy. For the 1.55 eV excitation, the signal does not decay up to 500 ps. In this excitation condition, the N states are never multiplied. Namely, the dynamics observed for the 1.55 eV excitation with low excitation density is completely different from

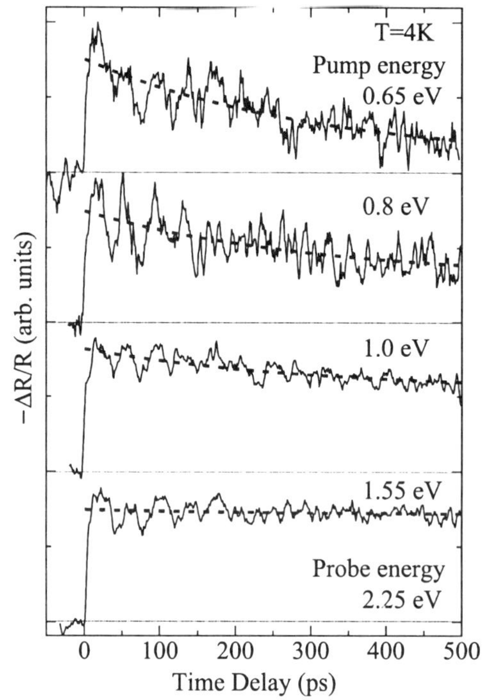

Fig. 4.6. Time evolutions of $-\Delta R/R$ at 2.25 eV for various excitation energies ranging from 0.65 eV to 1.55 eV at 4K. The excitation density is set to be very low, in which the dynamics is independent of the excitation density.

those observed for the resonant (0.65 eV) excitation at 4 K and 77 K shown in Fig. 4.3. The time characteristics of the N states in Fig. 4.6 can be reproduced by the sum of the component with the decay time $\tau \sim 300$ ps and the constant term; $\mathrm{A}\exp(-t/\tau) + \mathrm{B}$. The latter term corresponds to the component with the decay time much longer than 500 ps. Its decay time was not exactly evaluated. However, it has been ascertained that the signal completely decays within at least $t_{\mathrm{d}} = 500\,\mu$s. The amplitude of the slow decay component relative to the total amplitude of the signal $(\mathrm{B}/(\mathrm{A}+\mathrm{B}))$ is plotted by the solid circles as a function of the excitation energy in Fig. 4.7, together with the ε_2 spectrum along the DA stack at 77 K (the solid line) obtained from the polarized reflectivity spectrum. With increase of the excitation energy, the ratio of the slow decay component gradually increases, indicating that the nature of the initial excited states produced by lights changes depending on the excitation energy. For the 1.55 eV excitation, the fast decay component (A) is negligible and $\mathrm{B}/(\mathrm{A}+\mathrm{B})$ is almost equal to 1.

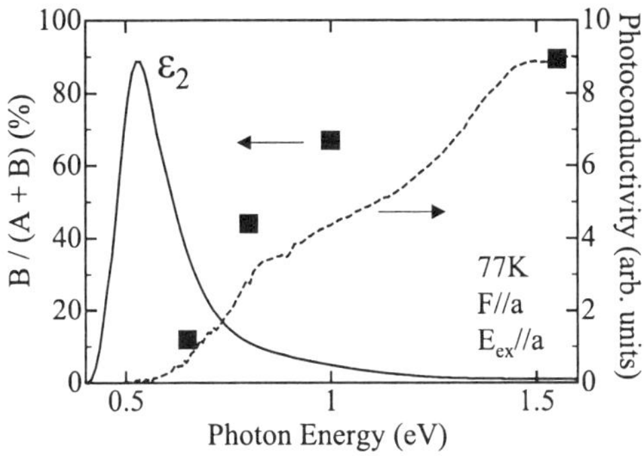

Fig. 4.7. The amplitude of the slow decay component relative to the total amplitude of the signal (B/(A + B)) as a function of the excitation energy (the solid circles). The solid line is the ε_2 spectrum along the DA stack at 77 K obtained from the polarized reflectivity spectrum and the broken line is the excitation profile of the photoconductivity along the DA stack.

As discussed above, the initial excited state for the 0.65 eV excitation is a microscopic 1D N domain generated from a CT excited state, which decays with the decay time $\tau \sim 300$ ps and almost disappears at $t_d = 500$ ps. When the excitation energy increases, a photoexcitation may not produce a CT excited state ($[\text{—}D^+A^-D^+A^-D^+A^-\text{—}] \rightarrow [\text{—}D^+A^-D^0A^0D^+A^-\text{—}]$), but generate possibly positively and negatively charged species. To investigate the contribution of the charged species, we measured the excitation profile of the photoconductivity (PC) along the DA stacks.

The experimental details of the PC measurements are presented in the previous section. The applied dc electric field F and the polarization of the exciting light E were set parallel(//) to the DA stacking axis (the a axis). Linearity of the photocurrent against the intensity of the incident light and the applied voltage was carefully checked. In our experimental condition, the intensity of the incident light is extremely low, so that the photoinduced transition to semi-macroscopic N state never occurs.

The excitation profile of the photoconductivity at 77 K is presented in Fig. 4.7 by the broken line. The photoconductivity is very small at the peak energy of the CT band, suggesting that the charged species are not generated from the CT excited state. The photoconductivity then gradually increases with increase of the excitation energy. The excitation energy dependence of the slow decay component (the solid circles) is in good agreement with that of the photoconductivity. It suggests that the charged N states are generated by the higher energy excitation and they have long life time.

Previous theoretical studies about the TTF-CA reported that a possible charged excited state in the ionic phase is a charged soliton,[81] which is schematically shown as follows:

Negatively charged soliton

$$[-\underline{D^+A^-}\ \underline{D^+A^-}\ \underline{D^+A^-}\ \underline{D^+A^-}\ D^0\ \underline{A^-D^+}\ \underline{A^-D^+}\ \underline{A^-D^+}\ \underline{A^-D^+}-]$$

Positively charged soliton

$$[-\underline{A^-D^+}\ \underline{A^-D^+}\ \underline{A^-D^+}\ \underline{A^-D^+}\ A^0\ \underline{D^+A^-}\ \underline{D^+A^-}\ \underline{D^+A^-}\ \underline{D^+A^-}-].$$

When the energy of the N phase is close to that of the I phase, the energy gain of the Coulomb interaction due to the distribution of the charge will overcome the energy cost for the generation of the N states (D^0A^0 pairs). As a result, the width of the charged soliton is expected to increase. In the case that the I and N phase are almost degenerate, we should consider such charged excited states are not charged solitons, but charged domains composed of the odd number of N molecules shown below.

Negatively charged N domain

$$[-\underline{D^+A^-}\quad \underline{D^+A^-}\ D^0\ A^0\ D^0\ A^0\ D^0-A^0\ D^0\ \underline{A^-D^+}\ \underline{A^-D^+}-]$$

Positively charged N domain

$$[-\underline{A^-D^+}\quad \underline{A^-D^+}\ A^0\ D^0\ A^0\ D^0\ A^0-D^0\ A^0\ \underline{D^+A^-}\ \underline{D^+A^-}-].$$

From the magnitude of the photoinduced reflectivity change for the 1.55 eV excitation, the amount of the N states generated by one photon is evaluated to be about 40 D^0A^0 pairs at 4 K, which is larger than that ($\sim 8D^0A^0$ pairs) for the 0.65 eV excitation (the resonant CT excitation) at the same temperature. As mentioned above, for the 1.55 eV excitation, the time characteristic of the reflectivity change is dominated only by the long decay component. Taking account of this result and the nature of soliton-like charged species, we propose a possible scenario as follows. A photoexcited state is separated to negatively and positively charged excited states immediately after the photoirradiation. The charged species will be converted to large size of charged N domains, probably during the carrier separation process. In other words, the charge carrier injection (or equivalently the injection of D^0 and A^0) in the I phase makes the neighboring I molecules unstable and then a relatively large size of 1D N domain which is evaluated to be $\sim$20 D^0A^0 pairs per each (positively or negatively) charged state, is produced. Such an excited state will be strongly bound to the lattice due to its net charge, as compared with an N domain with no charge. This may be the reason why the life time of the N states for the higher energy excitation is much longer than that for the resonant CT excitation.

When the excitation density is increased for the 1.55 eV excitation at 4 K, the multiplication of the N states is found to occur within about 20 ps following the 1D N domain formation just after the photoirradiation and the semi-macroscopic stable N state is also produced. From the excitation energy dependence of the magnitudes of the photoinduced reflectivity changes, it has also been revealed that the efficiency of the IN transition (or the formation of the semi-macroscopic N state) for the 1.55 eV excitation is larger than that for the resonant CT (0.65 eV) excitation. It might reflect the differences in the size and the life time of the 1D N domains initially produced between the two excitation energies.

4.3. *Coherent oscillations in photoinduced IN transition*

In this sub-section, let us discuss the coherent oscillation of the photoinduced signals. Figures 4.8(a), (b) and (c) show the time characteristics of $-\Delta R/R$ for the 0.65 eV excitation at 77 K in three typical time domains, 0–6 ps, 0–18 ps and 0–500 ps, respectively. In the fastest time domain of

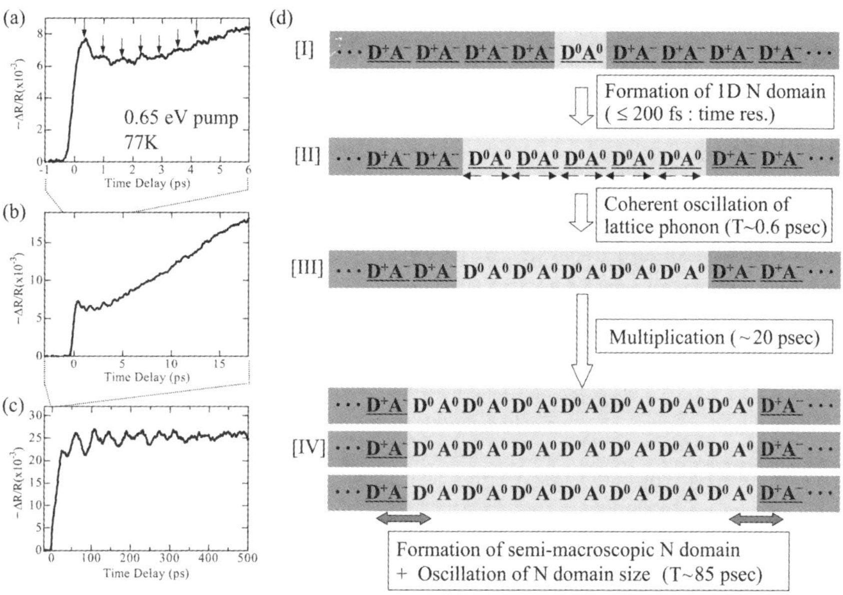

Fig. 4.8. Time evolutions of $-\Delta R/R$ at 2.25 eV in the time regions of 0–6 ps (a), 0–18 ps (b) and 0–500 ps (c) at 77 K. The pump energy is 0.65 eV. Schematic illustration of the observed dynamics in the photoinduced IN transition is presented in (d).

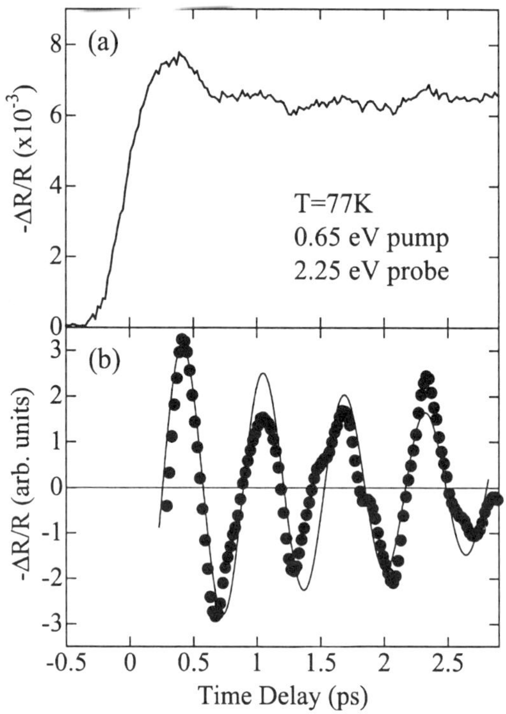

Fig. 4.9. (a) Initial response of the reflectivity change ($-\Delta$ R/R) at 2.25 eV at 77 K. The pump energy is 0.65 eV. (b) The oscillating component obtained by subtracting background rise and decay from the time profile in (a) (the solid circles).

0–6 ps, a rapid oscillation is observed as indicated by the solid arrows in Fig. 4.8(a). The magnified figure is presented in Fig. 4.9(a). By subtracting background rise and decay from the time profile, the oscillating component is obtained as shown in Fig. 4.9(b) by the dots. This oscillating component was analyzed by a damped oscillator represented as a following simple formula.

$$-\frac{\Delta \mathrm{R}}{\mathrm{R}} = A_0 \cos(\omega_0 t - \phi_0) \exp(-t/\tau_0). \qquad (3)$$

In Fig. 4.9(b), the calculated time profile is shown by the solid line, which reproduces well the experimental result. The obtained frequency ω_1 (the period T) of this oscillation is $\sim$56 cm^{-1}($\sim$0.6 ps). The decay time τ_0 of the oscillation is about 3.5 ps. Taking into account the frequency value of the oscillation (56 cm^{-1}), it is natural to consider that this oscillation is related with an optical mode of lattice phonon.

The observed rapid oscillation can be qualitatively interpreted as shown in Fig. 4.8(d). A CT excited state is initially photogenerated (Fig. 4.8(d)-(I)). Just after the photoirradiation, a 1D N domain with the size of about 24 D^0A^0 pairs is produced from the CT excited state (Fig. 4.8(d)-(II)). This process corresponds to a sequence of the electron transfer processes, so that it will occur within the time scale of the transfer energy t. If t is assumed to be 0.1 eV, which is a typical value for the mixed-stack CT compounds, its time scale is ~40 fs. In the formation process of the 1D N domain, the dimeric molecular displacements will persist as shown in Fig. 4.8(d)-(II), since the time scale of the lattice phonon, that is 0.6 ps, is much longer than that of the transfer energy (~40 fs). In the 1D N domain, each molecule has no spin, so that the dimeric displacements should be dissolved. Remember that the molecular dimerization is caused by the SP instability. In the dissolution process of the dimeric molecular displacements, the coherent oscillation does occur as shown by the arrows in Fig. 4.8(d)-(II). Such an oscillation modulates the transfer energy between the neighboring D and A molecules, and, therefore, modulates the degree of charge-transfer ρ. As a result, the coherent oscillation is detected as the reflectivity change of the intramolecular transition, which is sensitive to ρ. As mentioned in the Subsec. 4.1, the spike-like rise and decay is observed within 2 ps in the time evolution of $-\Delta R/R$ for the small excitation density (see the right lower panels of Fig. 4.3(a) and (b)). It is natural to consider that these structures will be also related with the dynamical behavior of molecular displacements. In fact, the similar analyses have revealed that several periods of the coherent oscillation can be distinguished in the results for the low excitation density.

As detailed in the Subsec. 4.1, the microscopic 1D N domains are multiplied and then the semi-macroscopic stable N state is produced within 30 ps. After such semi-macroscopic N state is stabilized, other oscillations with the periods of several tenth ps are observed in the time evolution of the photoinduced reflectivity change $-\Delta R/R$, as seen in Fig. 4.3(a,b). Here, we focus on the oscillations observed for the excitation density of 1.2×10^{16} photon/cm^2 at 77 K. The oscillation for the probe energy of 2.25 eV can be seen in the left panel (top) of Fig. 4.3(b). Such oscillations are also observed, when the probe energies are changed from 1.8 eV to 2.5 eV. To analyze these oscillations, we subtracted the background rise and decay from the observed time profiles. Figure 4.10(a) shows the residual oscillating components at several typical energies of the probe lights. The obtained oscillating components are composed of two oscillations having different periods. We have tried to

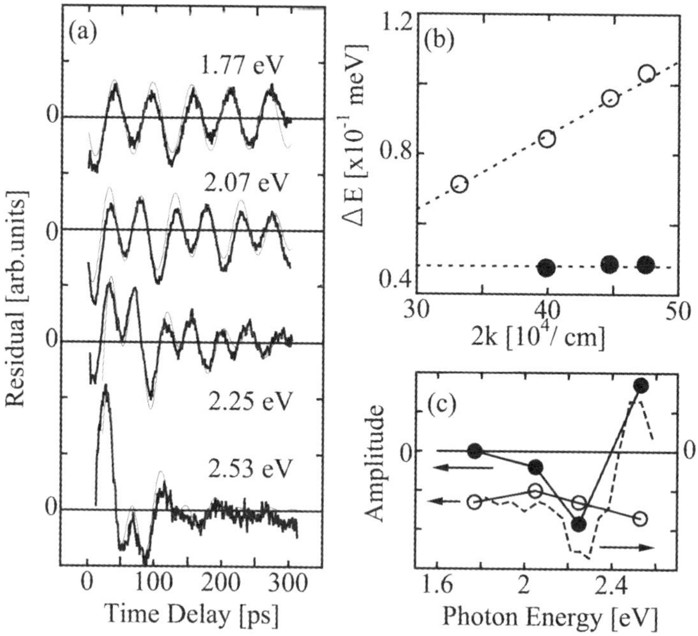

Fig. 4.10. (a) Oscillatory components of the reflectivity change ($-\Delta$R/R) for various probe energies at 77 K. The pump energy is 0.65 eV and the excitation density is $1.2 \times 10^{16}/\text{cm}^2$. (b) Quantum energy of the oscillation ($\Delta E = \hbar\omega$) as a function of 2 k; closed circles: the low-frequency oscillation; open circles: the high-frequency oscillation. k is the wave number of the probe light. (c) Amplitudes of the high-frequency oscillation (the open circles) and the low-frequency oscillation (the closed circles) as a function of the probe energy. The broken line is the Δ R/R spectrum at $t_d = 500$ ps.

fit the data by a superposition of two exponentially damped oscillations represented as follows.

$$-\frac{\Delta R}{R} = \sum_i A_i \cos(\omega_i t - \phi_i) \exp(-t/\tau_i)\text{: } (i = 1, 2). \tag{4}$$

The thin solid lines are the calculated results, which reproduce well the experimental ones in Fig. 4.10(a).

Figure 4.10(b) shows the quantum energy of the oscillation ($\Delta E = \hbar\omega$) as a function of the twice (2 k) of the wave number of the probe lights k (k $= 2\pi$n$/\lambda$) in the sample. In Fig. 4.10(c), the amplitudes of the low- and high-frequency oscillations are presented by the solid and open circles, respectively. In the low-frequency oscillations, the phase (ϕ) is almost constant for the probe energies of 1.77–2.25 eV, but shifts relatively by π at 2.53 eV. Such a shift of ϕ is taken into account as a change of the sign of the amplitudes in Fig. 4.10(c). In the high frequency oscillations, on the other hand, ϕ is independent of the probe energy. As seen in Fig. 4.10(b,c), the observed two oscillations show apparently different behaviors.

The amplitudes of the high frequency oscillation (the open circles in Fig. 4.10(c)) are almost constant. ΔE rather depends on the probe energy, being proportional to k. From these features, this oscillation is attributable to the shock wave, which is generated from the photoinduced impulsive mechanical stress. In the temperature induced NI transition, the unit cell volume decreases by about 0.6% across T_c.[30] Therefore, photo-induced IN conversion will also produce a sudden volume change, which acts as a broad band source of acoustic phonons.[85] In this case, the monochromatic probe light should be modulated by the phonon having the wave number 2 k. A relation between the modulation frequency and 2 k corresponds to a dispersion of the acoustic phonon, which gives a linear relation for small k. The linear relation between ΔE and 2 k shown in Fig. 4.10(b) demonstrates that the observed high frequency oscillation is attributed to a shock wave. The sound velocity is evaluated $v = 0.65 \times 10^5$ cm/s. Tanimura and Akimoto also detected the same oscillations due to the shock wave in TTF-CA.[23] The generation of the shock wave and the dispersion of the acoustic phonon modulating the probe lights are schematically illustrated in Fig. 4.11. This kind

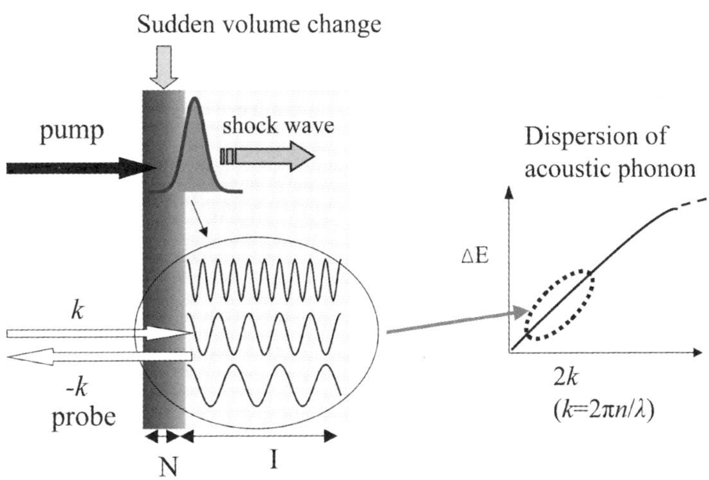

Fig. 4.11. Schematic illustrations of the shock wave and the dispersion of the acoustic phonon. The shock wave acts as a broad band source of acoustic phonons. The probe light having the wave number k is modulated by the coherent acoustic phonon with 2 k. ΔE as a function of 2 k gives the dispersion of the acoustic phonon.

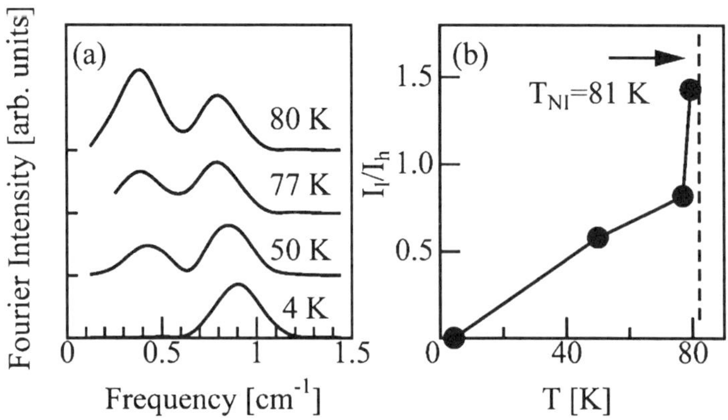

Fig. 4.12. (a) Fourier power spectra of the oscillatory components observed at various temperatures. (b) The ratio between the amplitudes of low- (I_l) and high-frequency peak (I_h).

of oscillation in the photo-induced reflectivity change has been previously reported in the study on the photo-induced melting of a charge-ordered state of perovskite manganite.[86]

As for the low-frequency oscillation, ΔE does not depend on the probe energies as seen in Fig. 4.10(b) and the profile of the amplitudes is in good agreement with that of $-\Delta R/R$ shown by the broken line in Fig. 4.10(c). Therefore, this oscillation should be related with the change of the molecular ionicity. A possible origin is a modulation in the amount of the N states. To investigate this possibility, we have measured temperature dependence of the oscillations. Fig. 4.12(a) shows the power spectra at various temperatures obtained from the time profiles of $-\Delta R/R$ at 2.25 eV by using the Fourier transformation procedures. Intensity of the low-frequency band (I_l) relative to the high-frequency band (I_h) is considerably enhanced with increase of temperature up to T_c as shown in Fig. 4.12(b). From the result, we can consider that the low-frequency oscillation is related with the valence instability, and, therefore, is attributed to the modulation in the amount of the N states. It should be noted that this oscillation starts at $t_d \approx 20$ ps just after the multiplication of the 1D N domains to the semi-macroscopic N state is completed. This indicates that the observed oscillation is not due to the microscopic 1D N domains. The modulation in the size of the semi-macroscopic N domain, or equivalently, the coherent motion of the NI domain boundary over the macroscopic scale is responsible for the low-frequency oscillation.

The dynamical behaviors of the photo-induced IN transition clarified in our study are summarized in Fig. 4.8(d). Within the time resolution of 200 fs just after the photoirradiation, 1D N domain is produced (Fig. 4.8(d)-(II)). Then the coherent oscillation of an optical lattice phonon with the period of 0.6 ps is driven by the dissolution of the dimeric displacements. After that, the multiplication of the N states occurs within about 20 ps through the cooperative interaction among a number of 1D N domains. It results in the formation of semi-macroscopic N state with a life time much longer than 500 ps (Fig. 4.8(d)-(IV)). This stable N domain shows an oscillation of its size with a period T of 85 ps.

4.4. *Dynamical aspect of photoinduced NI transition*

Next, we will discuss the photo-induced N to I transition. Figure 4.13 shows the spectra of the photoinduced reflectivity change ΔR/R at 90 K (the N phase) with the 0.65 eV excitation. The solid circles and triangles are the results at the delay time t_d = 0.3 ps and 500 ps, respectively. In Fig. 4.13,

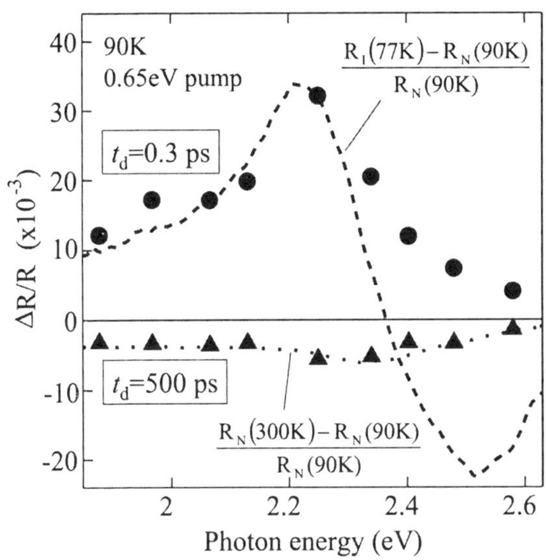

Fig. 4.13. The photoinduced reflectivity change ΔR/R at 90 K (the N phase) with the 0.65 eV excitation. The filled circles and triangles are the results at the delay time t_d = 0.3 ps and 500 ps, respectively. The excitation density is 0.77 × 10^{16} photon/cm^2. The broken line and the dotted line are the differential spectra, (R_I(77 K) − R_N(90 K))/R_N(90 K) and (R_N(300 K) − R_N(90 K))/R_N(90 K), respectively.

the broken line is the differential spectrum calculated from the reflectivity spectra $((R_I(77\,K) - R_N(90\,K))/R_N(90\,K))$ in the I phase at 77 K and the N phase at 90 K. The $\Delta R/R$ spectrum at $t_d = 0.3$ ps is almost equal to the calculated differential spectrum except for in the high-energy region (2.3–2.6 eV). The recent detail analysis of the $\Delta R/R$ spectrum has revealed that such a spectral difference is attributable to the difference of the absorption depths between the pump and probe lights and that the observed $\Delta R/R$ spectrum can be interpreted as the photogeneration of the I state in the N phase.[87] At $t_d = 500$ ps, however, the $\Delta R/R$ spectrum becomes to be negative and completely different from the calculated differential spectrum $((R_I(77\,K) - R_N(90\,K))/R_N(90\,K))$.

Figures 4.14(a) and (b) show the time characteristics of $\Delta R/R$ at 2.25 eV for several typical excitation densities, which are considered to reflect the amounts of the I states in the N phase. In each time characteristic, there is a fast rise within the time resolution, which decays very rapidly, and then the signal becomes to be negative, being accompanied by

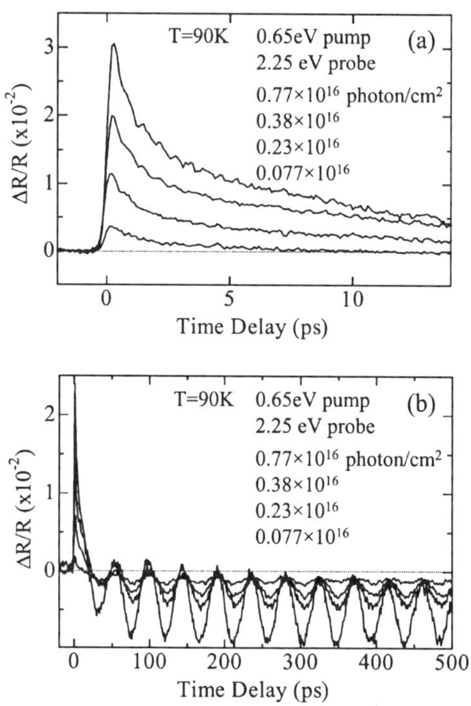

Fig. 4.14. Time characteristics of the photo-induced reflectivity changes $\Delta R/R$ at 2.25 eV (90 K) for several typical excitation densities; (a) short time region; (b) long time region. The pump energy is 0.65 eV.

a prominent oscillation. In Fig. 4.15, the excitation density dependence of the ΔR/R signals at $t_d = 0.3$ ps and 500 ps is presented. The ΔR/R signals at $t_d = 0.3$ ps are almost proportional to the excitation density up to 0.4×10^{16} photon/cm^2 and then saturate, similarly to the case of the photoinduced IN transition. Therefore, the initial rise of ΔR/R is attributable to the formation of a 1D I domain[88] as shown schematically in Fig. 4.16(b). Figure 4.17(a) is the magnified time profile in 0–5 ps for the excitation of 0.9×10^{16} photon/cm^2. A rapid oscillation is clearly observed similarly to the case of the photoinduced IN transition. By subtracting the background rise

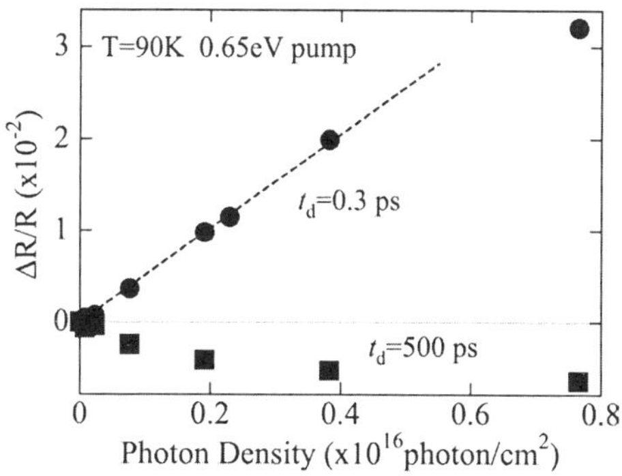

Fig. 4.15. The excitation density dependence of the ΔR/R signals at $t_d = 0.3$ ps and 500 ps (90 K, 0.65 eV excitation). The ΔR/R signals at $t_d = 0.3$ ps are proportional to the excitation density up to 0.4×10^{16} photon/cm^2 as shown by the broken line.

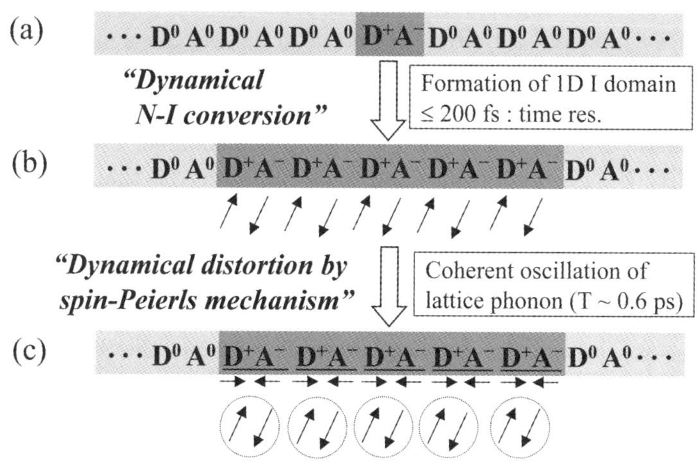

Fig. 4.16. The schematic illustration of the observed dynamics in the photoinduced NI transition.

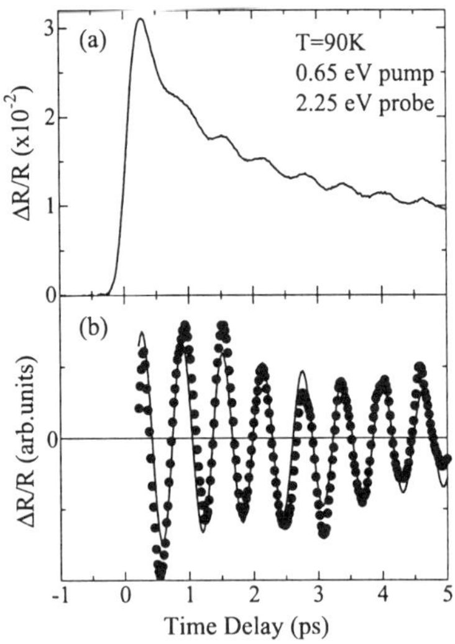

Fig. 4.17. (a) The time evolution of $\Delta R/R$ at 2.25 eV (90 K). The pump energy is 0.65 eV and the excitation density is 0.9×10^{16} photon/cm^2. (b) The oscillating component obtained by subtracting the background rise and decay from the time evolution in (a) (the solid circles).

and decay from the time profile, we have obtained the oscillating component as shown in Fig. 4.17(b). The frequency of this oscillation is 54 cm^{-1}, which is nearly equal to that observed in the photoinduced IN transition. Therefore, it can also be attributed to the lattice phonon corresponding to the dimeric molecular displacement. As shown in Fig. 4.16(b), the 1D I domain is produced immediately after the photoirradiation. Since each molecule in the 1D I domain has spin (S = 1/2), the 1D I domain has the SP instability, which would lead to the dimeric displacements. It can be considered that such dimeric displacements are detected as the coherent oscillation shown in Fig. 4.17(b). The important point to be emphasized is that the dynamical NI conversion (Fig. 4.16(b)) and the dynamical SP distortion (Fig. 4.16(c)) can be discriminated in the time domain.

In the time profile of $\Delta R/R$ up to $t_d = 500$ ps (Fig. 4.14(b)), the prominent oscillating structures with the period of 45 ps are clearly observed. They are attributable to the shock wave discussed in the Subsec. 4.3. The $\Delta R/R$ signal decreases to zero at $t_d \sim 25$ ps and are negative for $t_d > 30$ ps. The magnitudes of the negative signals ($\Delta R/R$ at $t_d = 500$ ps) saturate at very low excitation density as seen in Fig. 4.15. A possible origin of the negative signals is the thermal effect. To evaluate the thermal effect, we calculate the differential spectrum, $(R_N(300\,K) - R_N(90\,K))/R_N(90\,K)$, which is shown by the dotted line in Fig. 4.13. $(R_N(300\,K) - R_N(90\,K))/R_N(90\,K)$ is found to be negative over the wide energy range. Its spectral shape is almost equal to the $\Delta R/R$ spectrum experimentally obtained. These results support the following scenario; the photoinduced I states decay within $\sim$30 ps through the nonradiative processes and their energies are transferred to the lattice. Then the lattice temperature is increased and the spectral change corresponding to the sample heating is observed for $t_d > 30$ ps. As seen in Figs. 4.14(a,b), the life time of the photoinduced I states is very small. In addition, in the excitation density dependence of Fig. 4.15, there is no indication of multiplication of the I domain, when the excitation density is increased. It has been ascertained that these features are almost independent of the excitation energy from 0.65 eV to 1.55 eV.

It is clear that there is a significant difference of the dynamical behaviors between the photoinduced IN and NI transitions. Such a difference may be simply explained as follows. The ionic state of TTF-CA is the ferroelectric ordered state with the 3D ordering of dimeric displacements. Namely, the photo-induced formation of the N domain in the I phase corresponds to a dissolution process of the ordering, so that this process is easily driven by photoirradiations. On the other hand, the N to I transition is a formation process of a 3D ordered I state. Considering this point, it is natural that the stable semi-macroscopic I state is difficult to be produced by photoirradiations. Very recently, however, Collet *et al.* has reported from the time-resolved x-ray study that the photoirradiation can stabilize the ferroelectrically ordered I state in the N phase.[7] The reason for the discrepancy between our results and theirs is not clear at present. It may be attributable to the difference of the excitation conditions, e.g. the pulse width of the pump lights, the spot size of the pump lights relative to the sample size. Further studies should be necessary to clarify this problem.

5. Phase Control and Photoinduced Phase Transition in 1D Halogen-Bridged Binuclear Platinum Compound

The material discussed in this section is an iodine-bridged binuclear platinum compound (the PtPtI chain compound). This compound has a purely 1D chain composed of repeating metal-metal-halogen (M-M-X) units with M = Pt and X = I, called the MMX chain [Fig. 5.1(a)]. In the MMX chain, a 1D electronic state is formed by d_{z^2} orbitals of M and p_z orbitals of X. The average valence of M is 2.5, and 3 electrons exist per two d_{z^2}

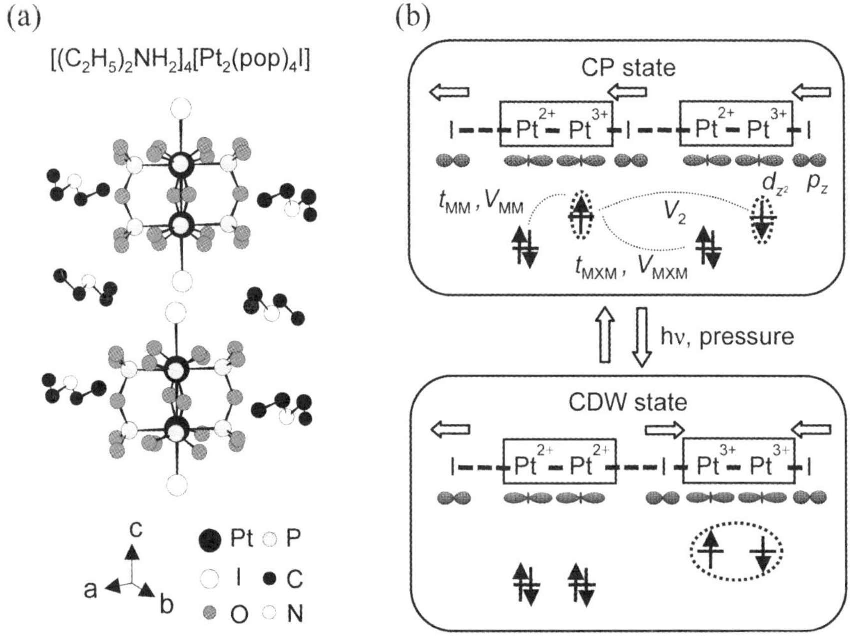

Fig. 5.1. (a) Crystal structure of $[(C_2H_5)_2NH_2]_4[Pt_2(pop)_4I]$. H atoms have been omitted for clarity. (b) Schematic electronic structures of CDW and CP. t_{MM} and t_{MXM} denote the electron transfer energies. V_{MM}, V_{MXM}, and V_2 denote the Coulomb interactions.

orbitals. Four charge-ordering states shown below are theoretically expected in this system.[31,89–92]

Average valence state (AV):
$$—X^-—M^{2.5+}—M^{2.5+}—X^-—M^{2.5+}—M^{2.5+}—X^-—$$

Charge-density-wave state (CDW):
$$—X^-—M^{2+}—M^{2+}—X^-—M^{3+}—M^{3+}—X^-—$$

Charge-polarization state (CP):
$$—X^-—M^{2+}—M^{3+}—X^-—M^{2+}—M^{3+}—X^-—$$

Alternating charge-polarization state (ACP):
$$—X^-—M^{2+}—M^{3+}—X^-—M^{3+}—M^{2+}—X^-—$$

CDW and CP are stabilized by the displacements of X, while ACP is stabilized by the displacements of M dimers. Two classes of MMX chains with different ligand molecules, $R_4[Pt_2(pop)_4X]nH_2O((pop) = P_2O_5H_2^{2-})$,

R = K, NH_4 and X = Cl, Br)[32,93] and $Pt_2(dta)_4I$(dta = $CH_3CS_2^-$)[94,95] have been studied thus far. The ground state of $R_4[Pt_2(pop)_4X]nH_2O$ (R = K, NH_4 and X = Cl, Br) has been revealed to be CDW.[96] For $Pt_2(dta)_4I$, it has been suggested that ACP is stabilized at low temperature,[95] but their electronic states have not been fully clarified yet. As a general trend, the hybridization between the *p* orbital of halogens and the *d* orbital of metals in the I-bridged compounds is larger than that in the Cl- or Br-bridged compounds. The large *pd* hybridization will stabilize the various oxidation states (or charge-ordering states) of the metals.

In this chapter, we will show that charge and spin states on the PtPtI chains ($R_4[Pt_2(pop)_4X]nH_2O$ and $R'_2[Pt_2(pop)_4I]nH_2O$) can be controlled between a diamagnetic charge-density-wave (CDW) state and a paramagnetic charge-polarization (CP) state by modification of the counterions (R, R′) located between chains. Moreover, in $[(C_2H_5)_2NH_2]_4[Pt_2(pop)_4I]$, the pressure-induced transition from CP to CDW accompanied by a large hysteresis loop has been observed. It has been found that the transition between the two charge-ordering states can be driven within this hysteresis loop by photo-irradiation.

5.1. *Phase control in PtPtI chains*

To control the electronic structures of the MMX chain compounds, Yamashita and their collaborators have synthesized about 20 compounds of $R_4[Pt_2(pop)_4I]nH_2O$ and $R'_2[Pt_2(pop)_4I]nH_2O$, with counterions (R^+ or R'^{2+}) of the alkyl-ammonium (R = $C_nH_{2n+1}NH_3$, $(C_nH_{2n+1})_2NH_2$ and R′ = $H_3N(C_nH_{2n})NH_3$) and the alkali-metal (R=Na, K, NH_4, Rb and Cs).[8,32,97] The substitution of the counterions greatly alters the distance d(Pt-I-Pt) between two Pt ions bridged by the I ion. Such control of d(Pt-I-Pt) makes it possible to control the various electronic parameters such as inter-site e-e interaction and e-l interaction, and to realize the paramagnetic CP state as well as the diamagnetic CDW state.

Figure 5.1(a) shows the structure of a typical PtPtI chain, $[(C_2H_5)_2NH_2]_4[Pt_2(pop)_4I]$. Two Pt ions are linked by four pop molecules, forming a binuclear $Pt_2(pop)_4$ unit. The two neighboring $Pt_2(pop)_4$ units are bridged by I and the PtPtI chain structure is formed along the c axis. The iodine ion deviates from the midpoints between the two neighboring Pt ions, indicating that this compound is in CDW or CP, as illustrated in Fig. 5.1(b). Figure 5.1(a) shows the two possible positions of I, as the displacements of I are not three-dimensionally ordered. X-ray structural analysis was unable to resolve whether the ground state is CDW or CP. CP is composed of Pt^{2+}-Pt^{3+} units, and the Pt^{3+} ions have spin (S = 1/2), forming a 1D antiferromagnetic spin chain. CDW on the other hand is composed of Pt^{2+}-Pt^{2+} and Pt^{3+}-Pt^{3+} units and is dimagnetic because the two neighboring Pt^{3+} ions form a singlet state. The ground states of the PtPtI chains can then be discussed keeping these characteristics of the two charge ordering states in mind.

Figure 5.2(a) shows the optical conductivity spectra of $[H_3N(C_6H_{12})NH_3]_2[Pt_2(pop)_4I]$ and $[(C_2H_5)_2NH_2]_4[Pt_2(pop)_4I]$ as typical examples. The optical gap energies E_{CT} (1.0 eV and 2.4 eV) of the two compounds differ considerably. The Raman spectra are shown in the insets

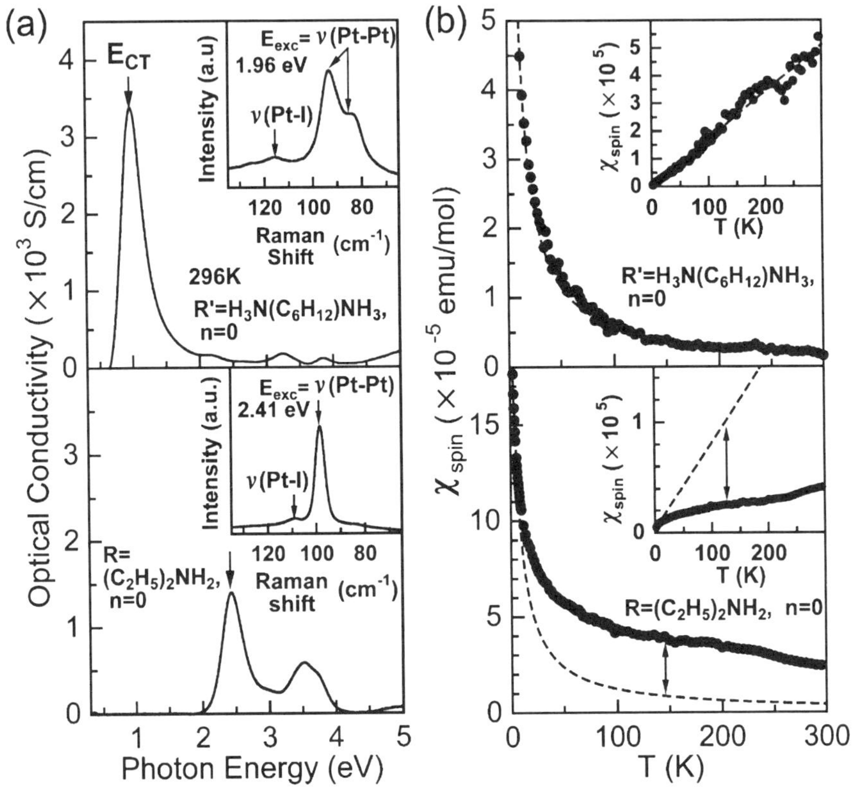

Fig. 5.2. (a) Optical conductivity spectra with light polarization parallel to the chain axis in $[(C_2H_5)_2NH_2]_4[Pt_2(pop)_4I]$ and $[H_3N(C_6H_{12})NH_3]_2[Pt_2(pop)_4I]$, and Raman spectra (insets) for the polarization of $z(xx)\bar{z}$ (x//chain axis). (b) Temperature dependence of χ_{spin} and $1/\chi_{spin}$ (insets). Broken lines represent Curie components.

to provide information concerning the valence of M.[32,96] Strong bands at 80–100 cm^{-1} are attributed to the Pt-Pt stretching mode of the Pt-Pt unit. The splitting of this mode indicates the formation of two kinds of Pt-Pt units, in this case Pt^{2+}-Pt^{2+} and Pt^{3+}-Pt^{3+}. The ground state of $[H_3N(C_6H_{12})NH_3]_2[Pt_2(pop)_4I]$ can therefore be considered to be CDW. In $[(C_2H_5)_2NH_2]_4[Pt_2(pop)_4I]$, this Pt-Pt stretching mode is a single band, suggesting that the ground state is CP. The weak bands around 110 cm^{-1} can be attributed to the Pt-I stretching mode, which is activated by the displacements of I in both compounds.

The magnetic properties of the materials were then investigated to confirm these assignments. The temperature dependence of χ_{spin} ($1/\chi_{spin}$) is presented in Fig. 5.2(b). In $[H_3N(C_6H_{12})NH_3]_2[Pt_2(pop)_4I]$, χ_{spin} follows the Curie law as shown by the broken line in Fig. 5.2(b) (the upper figure). Curie impurities make up 0.16% per Pt site. No paramagnetic components originate from the 1D spin chains. In $[(C_2H_5)_2NH_2]_4[Pt_2(pop)_4I]$, on the other hand, the temperature dependence of χ_{spin} cannot be explained by the Curie component alone. There is a finite component independent of temperature, attributed to a contribution from the 1D spin chain. As discussed in the Subsec. 1.1, the Bonner–Fisher (BF) curve[15] is applicable to a temperature dependence of χ_{spin} in a 1D antiferromagnetic $S = 1/2$ chain. The BF curve shows that χ_{spin} is only slightly dependent on temperature when the exchange interaction J is much larger than 300 K. The result in Fig. 5.2(b) (the lower figure) can be largely reproduced by the sum of the BF curve with $J \sim 3000$ K and Curie component with the concentration of 0.34% per Pt site. From these comparative studies, we can conclude that the ground states of $[(C_2H_5)_2NH_2]_4[Pt_2(pop)_4I]$ and $[H_3N(C_6H_{12})NH_3]_2[Pt_2(pop)_4I]$ are CP and CDW, respectively.

Systematic optical, magnetic and x-ray studies have been performed on various PtPtI chain compounds with different counterions. From the results, the phase diagram shown in Fig. 5.3 was obtained. The upper figure represents the magnitude of the splitting in the Pt-Pt stretching Raman band $\Delta\nu$(Pt-Pt), which is the evidence of CDW. The lower figure shows the optical gap energy E_{CT}. The ground state changes from CP to CDW with decreasing d(Pt-I-Pt), and there is a clear boundary at around 6.1 Å.

Here, let us discuss the stability of CP and CDW as a function of d(Pt-I-Pt), taking account of the theoretical studies based on the one-band extended Peierls-Hubbard model presented by Kuwabara and Yonemitsu.[89,98] When d(Pt-I-Pt) is large, the intradimer Coulomb interaction V_{MM} and the intradimer transfer energy t_{MM} are important parameters

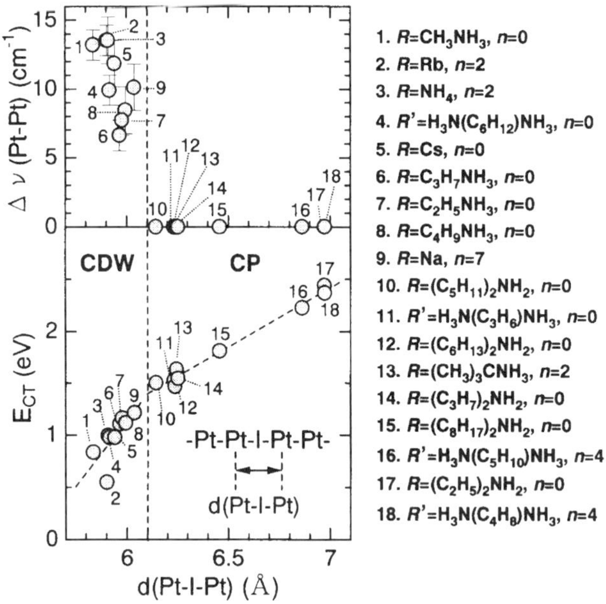

Fig. 5.3. The magnitude of the splitting in the Pt-Pt stretching Raman band $\Delta\nu$(Pt-Pt) (the upper panel) and the optical gap energy E_{CT} as a function of d(Pt-I-Pt) in various PtPtI chain compounds (the lower panel).

dominating the ground state in addition to the on-site Coulomb repulsion U on M (U_M) and the site-diagonal-type electron-lattice interaction β. The interdimer interactions such as the transfer energy t_{MXM}, the Coulomb interaction V_{MXM} and the second nearest neighbor Coulomb interaction V_2 (see Fig. 5.1(b)) can be neglected. A simple picture is given by considering the localized limit $t_{MM} = 0$. In this case, the relative energy of CP and CDW is determined by V_{MM}. The sum of the Coulomb energy for two neighboring M dimers in CDW ($5V_{MM}$) is larger than that in CP ($4V_{MM}$). Therefore, V_{MM} stabilizes CP. Under the presence of t_{MM}, the bonding orbital in the M dimer is formed in CP, but not formed in CDW because of large U_M. As a result, t_{MM} also stabilizes CP. As d(Pt-I-Pt) decreases, V_{MXM}, V_2, t_{MXM}, and β will increase. The increase of V_2 makes CP rather unstable, while V_{MXM} does not affect the relative energy of CP and CDW. The effects of t_{MXM} and β are not so straightforward. The theoretical calculations have revealed that the increase of β and t_{MXM} also stabilize CDW relatively to CP.[98] Thus, the theoretical studies can explain well the experimental result that the ground state changes from CP to CDW with a decrease of d(Pt-I-Pt).

5.2. *Pressure-induced phase transition in PtPtI chains*

From the phase diagram in Fig. 5.3, it is expected that a phase transition can be driven by applying pressure to the materials in CP. Such a pressure-induced phase transition from CP to CDW has been indeed observed for $[(C_2H_5)_2NH_2]_4[Pt_2(pop)_4I]$ (material 17 in Fig. 5.3), $[(C_5H_{11})_2NH_2]_4[Pt_2(pop)_4I]$ (material 10 in Fig. 5.3), and $[(C_3H_7)_2NH_2]_4[Pt_2(pop)_4I]$ (material 14 in Fig. 5.3) in the optical measurements using a diamond anvil pressure cell. Photographs of the pressure-induced transition in $[(C_2H_5)_2NH_2]_4[Pt_2(pop)_4I]$ are presented in Fig. 5.4(a). At ambient pressure, this material is green (Fig. 5.4(a)-(i)). When the pressure is increased to 0.6 GPa, a brown region appears (Fig. 5.4(a)-(ii)) and grows gradually (Fig. 5.4(a)-(iii), (iv)). Figure 5.4(b) shows the change in the Raman spectra accompanying this pressure-induced transition. The Pt-Pt stretching band clearly splits at above 0.6 GPa, indicating that the high-pressure brown phase is CDW. Figure 5.5(a) shows the pressure dependence of E_{CT},

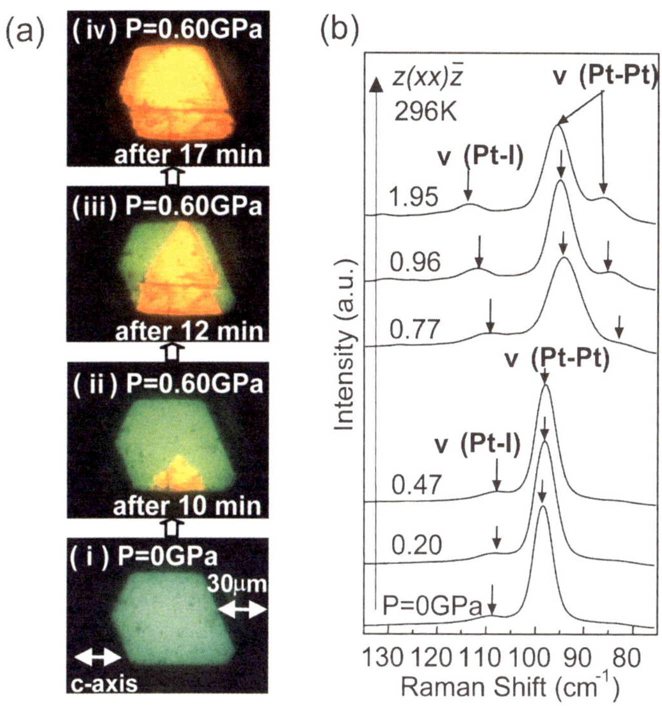

Fig. 5.4. (a) Images of $[(C_2H_5)_2NH_2]_4[Pt_2(pop)_4I]$ in the pressure-applying run. (b) Pressure dependence of Raman spectra.

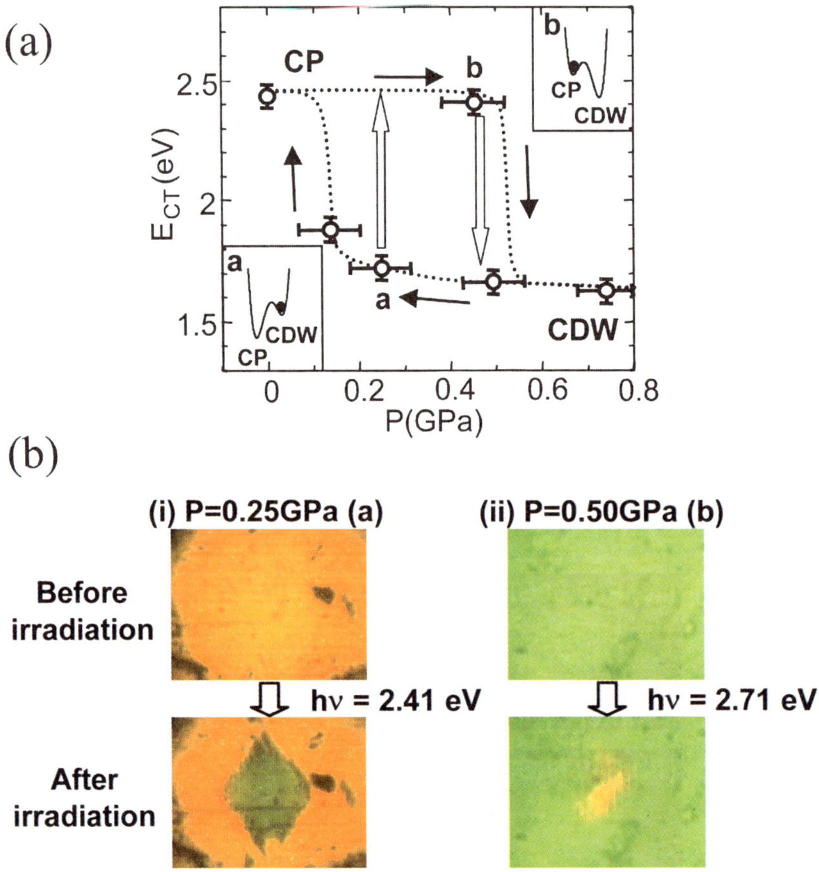

Fig. 5.5. (a) Pressure dependence of E_{CT}. The insets show the schematic energy potential curves. (b) Images at the points **a** (i) and **b** (ii) before and after photo-irradiation. Polarization of the excitation light is parallel to the chain axis.

obtained from an analysis of the polarized reflectivity spectra under pressure. The transition is a first-order reversible one with a large hysteresis loop (ca. 0.4 GPa).

Images of the crystal reveal that the sample size along the chain axis is reduced by about 8% at 0.85 GPa compared to ambient pressure. If the Pt-Pt distance d(Pt-Pt) of the Pt-Pt unit is assumed to remain constant under pressure, d(Pt-I-Pt) at 0.85 GPa is estimated to be 6.2 Å. This assumption is reasonable because d(Pt-Pt) is almost constant in various MMX chain compounds. The obtained values, d(Pt-I-Pt) = 6.2 Å and $E_{CT} = 1.6$ eV at

0.85 GPa, are very close to those in CDW (see Fig. 5.3). That is, pressure-induced transition from CP to CDW is driven by a decrease in d(Pt-I-Pt) in $[(C_2H_5)_2NH_2]_4[Pt_2(pop)_4I]$.

5.3. *Photo-induced phase transition in PtPtI chains*

At the points **a** and **b** in Fig. 5.5(a), the material is in a meta-stable state as shown in the energy potential curves (insets). The PIPT from the meta-stable state to the ground state at the points **a** and **b** were investigated. Figure 5.5(b)-(i) shows the images of the sample at **a** before and after the photo-irradiation with 2.41-eV light for 8 ms, which is obtained from a CW Ar-ion laser. As can be seen, the crystal exhibits a permanent color change from brown to green in the irradiated region. The Raman signal in the irradiated region is almost identical to that of the low-pressure phase, indicating that the PIPT from CDW to CP occurs.

To clarify whether this PIPT is induced by an optical process or a laser heating, the dependence of the PIPT efficiency on the excitation photon density N was investigated for two excitation energies E_{ex} of 1.96 eV and 2.41 eV. A He-Ne laser is used for the 1.96 eV light. The result is presented in Fig. 5.6(a). In the measurements, the excitation photon density N (photon density per pulse) was calculated considering the absorption coefficient and the reflection loss of the excitation light, both of which are obtained from the polarized reflection spectra. The converted fractions were estimated from the photoinduced changes in the integrated intensity of the Pt-Pt stretching Raman bands. The PIPT efficiency exhibits a clear threshold N_{th}, the values of which are strongly dependent on E_{ex}; $N_{th} \sim 1.4 \times 10^{25}\,cm^{-3}$/pulse at 1.96 eV and $\sim 3 \times 10^{24}\,cm^{-3}$/pulse at 2.41 eV. The absolute values of N_{th} seem to be large. However, the irradiated power of the lights is very small. Note that N_{th} is calculated for one pulse with a long duration of 8 ms. This characteristic excitation energy dependence of N_{th} demonstrates that the observed PIPT is driven not by a laser heating but by an optical process. At the point **b**, PIPT could not be driven by the irradiation of 1.96-eV or 2.41-eV light, even if the intensity and duration of the light was changed. However, irradiation with 2.71-eV light for 30 s did result in a PIPT, as shown in Fig. 5.5(b)-(ii). As this process appears to be strongly dependent on the excitation energy, the PIPT from CP to CDW is also attributed to an optical process, although the efficiency of the PIPT from CP to CDW is much lower than that from CDW to CP.

Finally, let us briefly discuss the difference of the PIPT efficiency between the CDW to CP and the CP to CDW transitions.[99,100] In Fig. 5.7,

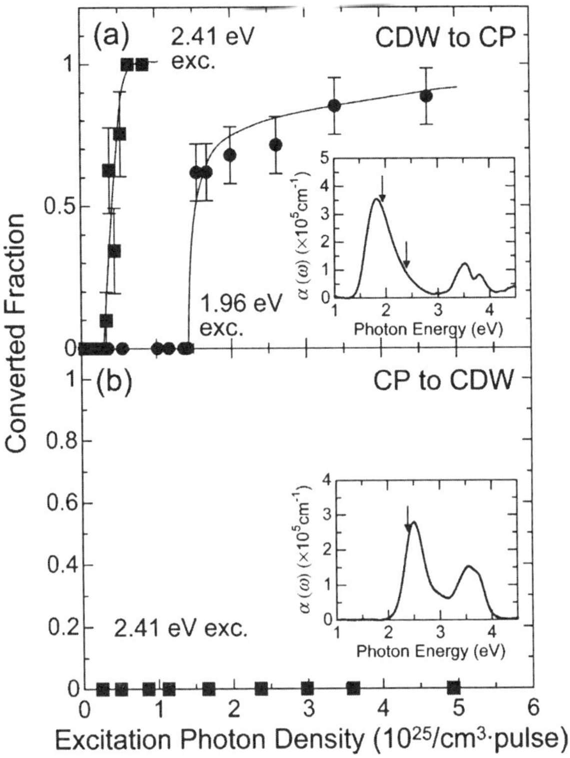

Fig. 5.6. Excitation photon density dependence of the conversion efficiency from CDW to CP at the point **a** (a) and from CP to CDW at the point **b** (b). The insets show the absorption spectra obtained from the polarized reflectivity spectra. The arrows indicate the energy positions of the excitation lights.

the optical transitions of the MMX chain in the CDW and CP ground states are schematically illustrated. The lowest transition in CDW is the inter-dimer CT excitation from [-I—Pt^{2+}-Pt^{2+}—I-Pt^{3+}-Pt^{3+}-I-] to [-I—Pt^{2+}-Pt^{3+}-I—Pt^{2+}-Pt^{3+}-I—]. The intra-dimer excitation is a higher-energy transition due to the large onsite Coulomb repulsion U_M on the Pt site. By taking into account the lattice relaxations due to the halogen displacements, the $2+3+2+3+$ valence state will be stabilized in a photoexcited state as shown in Fig. 5.7(a). Namely, in the CDW phase, an optical excitation will produce locally a CP state. This is the reason why the transition from CDW to CP is easily induced by lights. In the CP state, it has been established that the intra-dimer CT transition from [-I—Pt^{2+}-Pt^{3+}-I—Pt^{2+}-Pt^{3+}-I—] to [-I-Pt^{3+}-Pt^{2+}—I—Pt^{2+}-Pt^{3+}-I—]

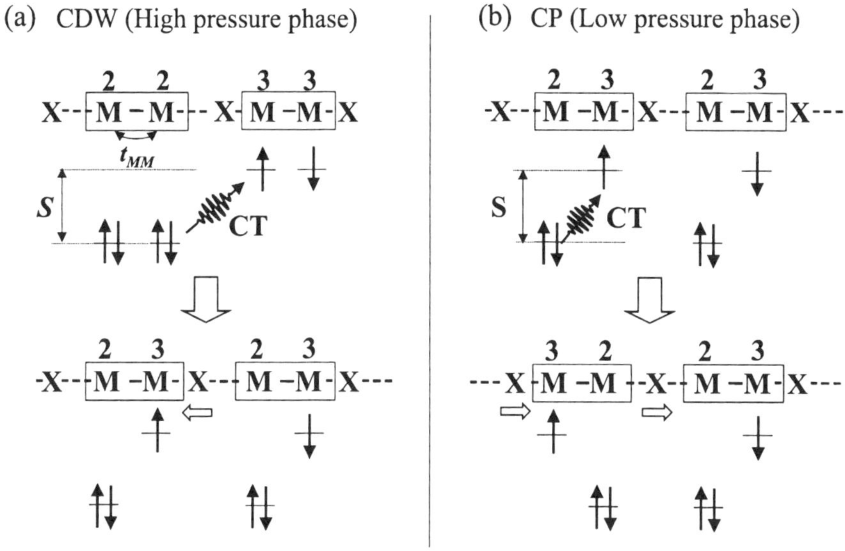

Fig. 5.7. Schematic illustration of the lowest energy optical transitions in the CDW (the high-pressure phase) (a) and in the CP (the low-pressure phase) (b).

shown in Fig. 5.7(b) is the dominant optical excitation corresponding to the gap transition. The intensity of the inter-dimer transition is smaller than that of the intra-dimer one. In this intra-dimer transition process, the CDW state is never produced, even as a local excited state. As a result, the CP to CDW transition is difficult to be achieved by the photo-irradiation with the energy of 2.41 eV nearly equal to E_{CT}. As observed in the experiments, the 2.71-eV excitation drives the PIPT from CP to CDW, although the transition efficiency is very low. Such a higher-energy excitation will induce the inter-dimer CT transition from [-I—Pt^{2+}-Pt^{3+}-I—Pt^{2+}-Pt^{3+}-I—] to [-I—Pt^{2+}-Pt^{2+}—I-Pt^{3+}-Pt^{3+}-I-] which may be relevant to the PIPT to CDW.

6. Summary

In this chapter, we have reviewed the investigations of PIPTs in the four kinds of 1D correlated electron systems. Under the influence of the strong electron-electron interaction, materials exhibit characteristic PIPTs, some of which could be driven in ultrafast (ps or sub-ps) time scale.

In Sec. 2, a photoswitching from an insulator to a metal in a 1D Mott insulator of the halogen-bridged Ni-chain compound has been detailed. The increase of the excitation density changes the photoproduct from the midgap state to the metallic state. A most important feature of this PIPT is that the response time is very fast. The photoinduced metallic state is generated within less than the time resolution of 0.2 ps and decays with the decay time of 0.4 ps. This is the first observation of the photoinduced Mott transition in the strongly correlated electron systems.

In Sec. 3, a PIPT in a spin-Peierls (SP) system of an organic radical crystal, TTTA, has been discussed. It has been demonstrated that the PIPT from a diamagnetic low-temperature phase to a paramagnetic high-temperature phase with a drastic color change is driven by a nanosecond laser pulse at both room temperature and 11 K. Suppression of the SP instability by the accumulation of photocarriers above a critical density was proposed as the mechanism driving this LT-to-HT transition.

In Sec. 4, dynamical aspects of the PIPTs from ionic (I) to neutral (N) and neutral (N) to ionic (I) phases in organic charge-transfer (CT) compound, TTF-CA, have been discussed. In the IN transition induced by the resonant excitation of the CT band, it has been clearly demonstrated that the microscopic 1D N domains are initially produced within 0.2 ps and then converted to semi-macroscopic N state within 20 ps through their cooperative multiplications. The dynamics and efficiency of the photoinduced IN transition are found to depend strongly on the excitation energy. The photoexcitations at the energy higher than the peak energy of the CT band, generate the charged excited states through the carrier separation processes, which are responsible for the dynamics and efficiency of the IN transition different from those observed in the resonant CT excitations. The dynamics of the photoinduced NI transition is found to be considerably different from that of the photoinduced IN transition. In addition, several kinds of prominent coherent oscillations with sub-ps or ps periods are observed for the first time in the photo-induced reflectivity changes. They are reasonably assigned to the dynamical dimeric displacements of molecules associated with the dissolution of the SP distortions, the shock wave driven by the sudden volume change due to the photoinduced IN transition, and the oscillation of the NI domain boundary.

In Sec. 5, the PIPT of the iodine-bridged binuclear Pt compound between the charge-density-wave (CDW) state and the charge-polarization (CP) state has been discussed. This transition behavior is an exceptional example of PIPT, demonstrating that it is possible to achieve fundamental changes in optical and magnetic properties in both directions from CDW to CP and CP to CDW through photoirradiation alone. In TTTA and the iodine-bridged binuclear Pt compound, the studies of the dynamics using the pump-probe spectroscopy have not been performed yet, which will clarify more detail pictures of their PIPTs.

Finally, we would like to propose two issues as future subjects. One is the coherent control of the material phases by ultrafast laser pulses. In Sec. 4, we have discussed about the ultrafast coherent oscillation related with the spin-Peierls instability observed in TTF-CA. If such a coherent oscillation is amplified by using double or multi-pulses excitations, not only a photoinduced phase transition may be effectively driven, but also new phases of materials, which do not appear under usual external perturbations such as temperature, pressure, and electric field, may be found. Another important issue is to control magnetism or spin states. The PIPTs of TTTA and the iodine-bridged binuclear Pt compound include the change of the magnetism. In these compounds, however, there is an antiferromagnetic interaction between the neighboring spins and the observed photoinduced changes of the magnetism are essentially small. When we consider applications to the future magneto-optical memories or switching devices, optical control of magnetism in ferromagnetic materials will be the most important. Especially, ultrafast control of the spin states in a ps time scale should be desirable. Studies from such a point of view have been started in conventional ferromagnetic metals[101] and other correlated electron systems[102] and will become more significant.

To develop the studies of the PIPTs in correlated electron systems, the collaborations of the researchers in the different fields such as synthesis and chemistry of organic and inorganic materials, physics of correlated electron systems, laser spectroscopy, and quantum optics are indispensable. Such collaborations are expected to activate this field and to lead new findings of PIPTs useful for future optical switching devices.

Acknowledgments

We thank our co-workers H. Kishida, S. Tanaka, Y. Ishige, M. Ono, and A. Maeda (Univ. of Tokyo), and Y. Tokura (CERC-AIST, Univ. of Tokyo), and M. Yamashita and K. Takizawa (Tokyo Metropolitan Univ.), and K. Awaga and W. Fujita (Nagoya Univ.). We also thank K. Yonemitsu and M. Kuwabara (IMS), S. Yamamoto (Hokkaido Univ.), N. Nagaosa (CERC-AIST, Univ. of Tokyo), A. Mishenko (CERC-AIST), Y. Okimoto (CERC-AIST), and K. Iwano and K. Nasu (Inst. Mat. Struct. Sci.) for valuable discussions.

References

1. K. Nasu (ed.), *Relaxations of Excited States and Photoinduced Structual Phase Transitions*, (Springer-Verleg, Berlin, 1997).
2. S. Koshihara, Y. Tokura, T. Mitani, G. Saito, and T. Koda, *Phys. Rev.* **B42** 6853 (1990).
3. S. Koshihara, Y. Tokura, K. Takeda, and T. Koda, *Phys. Rev. Lett.* **68** 1148 (1992); S. Koshihara, Y. Tokura, K. Takeda, and T. Koda, *Phys. Rev.* **B52** 6265 (1995).
4. S. Koshihara, Y. Takahashi, H. Sakai, Y. Tokura, and T. Luty, *J. Phys. Chem.* **B103** 2592 (1999).
5. T. Suzuki, T. Sakamaki, K. Tanimura, S. Koshihara, Y. Tokura, *Phys. Rev.* **B60** 6191 (1999).
6. S. Iwai, S. Tanaka, K. Fujinuma, H. Kishida, H. Okamoto, and Y. Tokura, *Phys. Rev. Lett.* **88** 057402 (2002).
7. E. Collet, M.H. Cailleau, M.B. Cointe, H. Cailleau, M. Wulff, T. Luty, S. Koshihara, M. Meyer, L. Toupet, P. Rabiller, and S. Techert, *Science* **300** 612 (2003).
8. H. Matsuzaki, T. Matsuoka, H. Kishida, K. Takizawa, H. Miyasaka, K. Sugiura, M. Yamashita, and H. Okamoto, *Phys. Rev. Lett.* **90** 046401 (2003).
9. H. Matsuzaki, W. Fujita, K. Awaga, and H. Okamoto, *Phys. Rev. Lett.* **91** 017403 (2003).
10. S. Iwai, M. Ono, A. Maeda, H. Matsuzaki, H. Kishida, H. Okamoto, and Y. Tokura, *Phys. Rev. Lett.* **91** 057401 (2003).
11. J. Hubbard, *Proc. Roy. Soc.* (London) **A276** 238 (1963); *ibid* **A281** 401 (1964).
12. For a review, S. Kagoshima, T. Sanbongi, H. Nagasawa, and T. Takahashi, in *Low-Dimensional Conductors. Physics of Organic Conductors and Density Waves*, ed. S. Kagoshima, SHOKABO, Tokyo, 2000, p. 143 (in Japanese).
13. I.S. Jacobs, J.W. Bray, H.R. Hart, Jr., L.V. Interrante, J.S. Kasper, G.D. Watkins, D.E. Prober, and J.C. Bonner, *Phys. Rev.* **B14** 3036 (1976).
14. For a review, J.W. Bray, L.V. Interrante, I.S. Jacobs, and J.S. Miller, in *Extended Linear Chain Compounds*, Vol. 3, ed. J.S. Miller, (Plenum Press, New York, 1983), p. 353.
15. J.C. Bonner and M.E. Fisher, *Phys. Rev.* **135** A640 (1964).
16. M.C. Cross and D.S. Fisher, *Phys. Rev.* **B19** 402 (1979).
17. K. Toriumi, Y. Wada, T. Mitani, S. Bandow, M. Yamashita, and Y. Fujii, *J. Am. Chem. Soc.* **111** 2341 (1989).
18. H. Okamoto, Y. Shimada, Y. Oka, A. Chainani, T. Takahashi, H. Kitagawa, T. Mitani, K. Toriumi, K. Inoue, T. Manabe, and M. Yamashita, *Phys. Rev.* **B54** 8438 (1996).
19. W. Fujita and K. Awaga, *Science* **286** 261 (1999).
20. W. Fujita, K. Awaga, H. Matsuzaki, and H. Okamoto, *Phys. Rev.* **B65** 064434 (2002).

21. J.B. Torrance, J.E. Vazquez, J.J. Mayerle, and V.Y. Lee, *Phys. Rev. Lett.* **46** 253 (1981).
22. J.B. Torrance, A. Girlando, J.J. mayerle, J.I. Crowley, V.Y. Lee, P. Betail, and S.J. LaPlaca, *Phys. Rev. Lett.* **47** 1747 (1981).
23. K. Tanimura and I. Akimoto, *J. Lumi.* **94–95** 483 (2001).
24. S. Iwai, S. Tanaka, H. Kishida, H. Okamoto, and Y. Tokura, *Ultrafast Phenomena XIII*, Springer Series in Chemical Physics Vol. **71**, (Springer, Berlin, 2003) p. 546.
25. A. Girlando, F. Marzola, C. Pecile, and J.B. Torrance, *J. Chem. Phys.* **79** 1075 (1983).
26. Y. Kanai, M. Tani, S. Kagoshima, Y. Tokura, and T. Koda, *Synth. Met.* **10** 157 (1984).
27. S. Kagoshima, Y. Kanai, M. Tani, Y. Tokura, and T. Koda, *Mol. Cryst. Liq. Cryst.* **120** 9 (1985).
28. Y. Tokura, Y. Kaneko, H. Okamoto, S. Tanuma, T. Koda, T. Mitani, and G. Saito, *Mol. Cryst. Liq. Cryst.* **125** 71 (1985).
29. H. Okamoto, T. Koda, Y. Tokura, T. Mitani, and G. Saito, *Phys. Rev.* **B39** 10693 (1989).
30. M. Le Cointe, M.H. Lemée–Cailleau, H. Cailleau, B. Toudic, L. Toupet, G. Heger, F. Moussa, P. Schweiss, K.H. Kraft, and N. Karl, *Phys. Rev.* **B51** 3374 (1995).
31. M.B. Robin and P. Day, *Adv. Inorg. Radiochem.* **10** 247 (1967).
32. M. Kurmoo and R.J.H. Clark, *Inorg. Chem.* **24** 4420 (1985).
33. L.G. Butler, M.H. Zietlow, C.M. Che, W.P. Schaefer, S. Sridhar, P.J. Grunthaner, B.I. Swanson, R.J.H. Clark, and H.B. Gray, *J. Am. Chem. Soc.* **110** 1155 (1988).
34. M. Yamashita, S. Miya, T. Kawashima, T. Manabe, T. Sonoyama, H. Kitagawa, T. Mitani, H. Okamoto, and R. Ikeda, *J. Am. Chem. Soc.* **121** 2321 (1999).
35. M. Imada, A. Fujimori, and Y. Tokura, *Rev. Mod. Phys.* **70** 1039 (1998).
36. Y. Tokura and N. Nagaosa, *Science* **288** 462 (2000).
37. A.J. Millis, *Nature* **392** 147 (1998).
38. G. Yu, C.H. Lee, A.J. Heeger, N. Herron, and E.M. McCarron, *Phys. Rev. Lett.* **67** 2581 (1991).
39. H. Okamoto, K. Okaniwa, T. Mitani, K. Toriumi, and M. Yamashita, *Solid State Commun.* **77** 465 (1991).
40. H. Okamoto, T. Mitani, K. Toriumi, and M. Yamashita, *Mat. Sci. Eng.* **B13** L9 (1992).
41. H. Okamoto and M. Yamashita, *Bull. Chem. Soc. Jpn.* **71** 2023 (1998).
42. H. Okamoto, H. Kishida, M. Ono, H. Matsuzaki, M. Yamashita, and T. Manabe, *J. Lumi.* **87–89** 204 (2000).
43. H. Kishida, M. Ono, K. Miura, H. Okamoto, M. Izumi, T. Manako, M. Kawasaki, Y. Taguchi, Y. Tokura, T. Tohyama, K. Tsutsui, and S. Maekawa, *Phys. Rev. Lett.* **87** 177401 (2001).
44. H. Okamoto, K. Toriumi, T. Mitani, and M. Yamashita, *Phys. Rev.* **B42** 10381 (1990).

45. N. Motoyama, H. Eisaki, and S. Uchida, *Phys. Rev. Lett.* **76** 3212 (1996).
46. H. Kishida, H. Matsuzaki, H. Okamoto, T. Manabe, M. Yamashita, Y. Taguchi, and Y. Tokura, *Nature* **405** 929 (2000).
47. H. Okamoto, H. Kishida, M. Ono, H. Matsuzaki, Y. Taguchi, Y. Tokura, and M. Yamashita, *Nonlinear Optics* **29** 687 (2002); H. Kishida, M. Ono, H. Matsuzaki, M. Yamashita, Y. Taguchi, Y. Tokura, and H. Okamoto, *Synth. Met.* **135–136** 315 (2003).
48. S. Fujimori, A. Ino, T. Okane, A. Fujimori, K. Okada, T. Manabe, M. Yamashita, H. Kishida, and H. Okamoto, *Phys. Rev. Lett.* **88** 247601 (2002).
49. T. Ogasawara, M. Ashida, N. Motoyama, H. Eisaki, S. Uchida, Y. Tokura, H. Ghosh, A. Shukla, S. Mazumdar, and M. Kuwata-Gonokami, *Phys. Rev. Lett.* **85** 2204 (2000).
50. M. Ashida, T. Ogasawara, Y. Tokura, S. Uchida, S. Mazumdar, and M. Kuwata-Gonokami, *Appl. Phys. Lett.* **78** 2831 (2001).
51. S. Uchida, T. Ido, H. Takagi, T. Arima, Y. Tokura, and S. Tajima, *Phys. Rev.* **B43** 7942 (1991).
52. Y. Tokura, H. Takagi, and S. Uchida, *Nature* **337** 345 (1989).
53. J.D. Perkins, R.J. Birgeneau, J.M. Graybeal, M.A. Kastner and D.S. Kleinberg, *Phys. Rev.* **B58** 9390 (1998).
54. J.Y. Vinet, M. Combescot, and C. Tanguy, *Solid State Commun.* **51** 171 (1984).
55. A. Lietoila and J.F. Gibbons, *Appl. Phys. Lett.* **40** 624 (1982).
56. M. Nagai, R. Shimano, and M. Kuwata-Gonokami, *Phys. Rev. Lett.* **86** 5795 (2001); M. Nagai and M. Kuwata-Gonokami, *J. Lumin.* **100** 233 (2002).
57. H. Okamoto, K. Toriumi, K. Okaniwa, T. Mitani, and M. Yamashita, *Synth. Met.* **42** 2791 (1991).
58. H. Okamoto, unpublished result.
59. K. Matsuda, I. Hirabayashi, K. Kawamoto, T. Nabatame, T. Tokizaki, and A. Nakamura, *Phys. Rev.* **B50** 4097 (1994).
60. M. Ashida, Y. Taguchi, Y. Tokura, R.T. Clay, S. Mazumdar, Yu. P. Svirko, and M. Kuwata-Gonokami, *Europhys. Lett.* **58** 455 (2002).
61. O. Sato, T. Iyoda, A. Fujishima, and K. Hashimoto, *Science* **272** 704 (1996); S. Ohkoshi, S. Yorozu, O. Sato, T. Iyoda, A. Fujishima, and K. Hashimoto, *Appl. Phys. Lett.* **70** 1040 (1997); Y. Arimoto, S. Ohkoshi, Z.J. Zhong, H. Seino, Y. Mizobe, and K. Hashimoto, *J. Am. Chem. Soc.* **125** 9240 (2003).
62. D.D. Awschalom, M.R. Freeman, N. Samarth, H. Luo, and J.K. Furdyna, *Phys. Rev. Lett.* **66** 1212 (1991); S. Koshihara, A. Oiwa, M. Hirasawa, S. Katsumoto, Y. Iye, C. Urano, H. Takagi, and H. Munekata, *Phys. Rev. Lett.* **78** 4617 (1997).
63. Y. Okimoto, Y. Ogimoto, M. Matsubara, Y. Tomioka, T. Kageyama, T. Hasegawa, H. Koinuma, M. Kawasaki, and Y. Tokura, *Appl. Phys. Lett.* **80** 1031 (2002).
64. Y. Ogawa, S. Koshihara, K. Koshino, T. Ogawa, C. Urano, and H. Takagi, *Phys. Rev. Lett.* **84** 3181 (2000).

65. T. Tayagaki and K. Tanaka, *Phys. Rev. Lett.* **86** 2886 (2001).
66. K. Hiraki and K. Kanoda, *Synth. Met.* **86** 2111 (1997).
67. M. Hase, I. Terasaki, Y. Sasago, K. Uchinokura, and H. Obara, *Phys. Rev. Lett.* **71** 4059 (1993).
68. For a review, H. Endres, in Extended Linear Chain Compounds, Vol. **3**, ed. J.S. Miller, (Plenum Press, New York, 1983), p. 263.
69. S. Huizinga, J. Kommandeur, G.A. Sawatzky, B.T. Thole, K. Kopinga, W.J.M. de Jonge, and J. Roos, *Phys. Rev.* **B19** 4723 (1979).
70. T. Sekikawa, H. Okamoto, T. Mitani, T. Inabe, Y. Maruyama, and T. Kobayashi, *Phys. Rev.* **B55** 4182 (1997).
71. Q. Liu, S. Ravy, J.P. Pouget, C. Coulon, and C. Bourbonnais, *Synth. Met.* **56** 1840 (1993).
72. S. Koshihara, Y. Tokura, Y. Iwasa, and T. Koda, *Phys. Rev.* **B44** 431 (1991).
73. B. Bleaney and K. D. Bowers, *Proc. R. Soc. London Ser.* **A214** 451 (1952).
74. M. Imada, *Prog. Theor. Phys. Suppl.* **113** 203 (1993).
75. D.L. Dexter, in Solid State Physics, Vol. **6**, eds. F. Seitz and D. Turnbull, (Academic Press, New York, 1958), p. 361.
76. Y. Tokura, T. Koda, T. Mitani, and G. Saito, *Solid State Commun.* **43** 757 (1982).
77. S. Horiuchi, Y. Okimoto, R. Kumai, and Y. Tokura, *J. Phys. Soc. Jpn.* **69** 1302 (2000).
78. J.B. Torrance, *Low-Dimentional Conductors and Superconductors*, eds. D. Jerome, L.G. Caron, **B155** of NATO ASI Series, (Plenum Press, New York, 1987).
79. M.H. Lemée-Cailleau, M. Le Cointe, H. Cailleau, T. Luty, F. Moussa, J. Roos, D. Brinkmann, B. Toudic, C. Ayache, and N. Karl, *Phys. Rev. Lett.* **79** 1690 (1997).
80. N. Nagaosa and J. Takimoto, *J. Phys. Soc. Jpn.* **55** 2745 (1986).
81. N. Nagaosa, *J. Phys. Soc. Jpn.* **55** 2754 (1986).
82. N. Nagaosa, *J. Phys. Soc. Jpn.* **55** 3488 (1986).
83. Y. Toyozawa, *Solid State Commun.* **84** 255 (1992).
84. P. Huai, H. Zheng, and K. Nasu, *J. Phys. Soc. Jpn.* **69** 1788 (2000).
85. A.M. Lindenberg, I. Kang, S.L. Johnson, T. Missalla, P.A. Heimann, Z. Chang, J. Larsson, P.H. Bucksbaum, H.C. Kapteyn, H.A. Padmore, R.W. Lee, J.S. Wark, and R.W. Falcone, *Phys. Rev. Lett.* **84** 111 (2000).
86. M. Fiebig, K. Miyano, Y. Tomioka, and Y. Tokura, *Science* **280** 1925 (1998).
87. Y. Ishige, S. Tanaka, S. Iwai, and H. Okamoto, unpublished results.
88. H. Okamoto, T. Mitani, Y. Tokura, S. Koshihara, T. Komatsu, Y. Iwasa, T. Koda, and G. Saito, *Phys. Rev.* **B43** 8224 (1991).
89. M. Kuwabara and K. Yonemitsu, *J. Phys. Chem. Solids.* **62** 435 (2001).
90. S. Yamamoto, *Phys. Rev.* **B63** 125124 (2001).
91. S. Yamamoto, *J. Phys. Soc. Jpn.* **70** 1198 (2001).
92. S. Yamamoto, *Phys. Rev.* **B64** 140402 (2001).
93. N. Kimura, H. Ohki, R. Ikeda, and M. Yamashita, *Chem. Phys. Lett.* **220** 40 (1994).

94. C. Bellitto, A. Flamini, L. Gastaldi, and L. Scaramuzza, *Inorg. Chem.* **22** 444 (1983).
95. H. Kitagawa, N. Onodera, T. Sonoyama, M. Yamamoto, T. Fukawa, T. Mitani, M. Seto, and Y. Maeda, *J. Am. Chem. Soc.* **121** 10068 (1999).
96. D. Baeriswyl and A.R. Bishop, *J. Phys.* **C21** 339 (1988).
97. K. Takizawa, T. Ishii, H. Miyasaka, H. Matsuzaka, M. Yamashita, T. Kawashima, H. Matsuzaki, H. Kishida, and H. Okamoto, *Mol. Cryst. Liq. Cryst.* **376** 159 (2002).
98. M. Kuwabara and K. Yonemitsu, *J. Mater. Chem.* **11** 2163 (2001).
99. K. Yonemitsu, N. Miyashita, and M. Kuwabara, *Synth. Met.* **135** 521 (2003).
100. K. Yonemitsu and N. Miyashita, *Phys. Rev.* **B68** 075113 (2003).
101. For example, E. Beaurepaire, J.-C. Merle, A. Daunois, and J.-Y. Bigot, *Phys. Rev. Lett.* **76** 4250 (1996); M. van Kampen, C. Jozsa, J.T. Kohlhepp, P. LeClair, L. Lagae, W.J.M. de Jonge, and B. Koopmans, *Phys. Rev. Lett.* **88** 227201 (2002).
102. For example, K. Matsuda, A. Machida, Y. Moritomo, and A. Nakamura, *Phys. Rev.* **B58** R4203 (1998); A.I. Lobad, A.J. Taylor, C. Kwon, S.A. Trugman, and T.R. Gosnell, *Chem. Phys.* **251** 227 (2000); T. Kise, T. Ogasawara, M. Ashida, Y. Tomioka, Y. Tokura, and M. Kuwata-Gonokami, *Phys. Rev. Lett.* **85** 1986 (2000); T. Ogasawara, M. Matsubara, Y. Tomioka, M. Kuwata-Gonokami, H. Okamoto, and Y. Tokura, to be published.

CHAPTER 7

PROBING PHOTOINDUCED STRUCTURAL PHASE TRANSITIONS BY FAST OR ULTRA-FAST TIME-RESOLVED X-RAY DIFFRACTION

Hervé Cailleau*, Eric Collet, Marylise Buron-Le Cointe and Marie-Hélène Lemée-Cailleau

Groupe Matière Condensée et Matériaux, UMR CNRS 6626, Université Rennes1, 35042 Rennes Cedex, France
**E-mail: herve.cailleau@univ-rennes1.fr*

Shin-ya Koshihara

Department of Materials Science, Tokyo Institute of Technology, 2-12-1 Oh-okayama, Megur-ku, Tokyo 152, Japan

A new frontier in the field of structural science is the emergence of the fast and ultra-fast X-ray science. Recent developments in time-resolved X-ray diffraction promise direct access to the dynamics of electronic, atomic and molecular motions in condensed matter triggered by a pulsed laser irradiation, i.e. to record "molecular movies" during the transformation of matter initiated by light pulse. These laser pump and X-ray probe techniques now provide an outstanding opportunity for the direct observation of a photoinduced structural phase transition as it takes place. The use of X-ray short-pulse of about 100 ps around third-generation synchrotron sources allows structural investigations of fast photoinduced processes. Other new X-ray sources, such as laser-produced plasma ones, generate ultra-short pulses down to 100 fs. This opens the way to femtosecond X-ray crystallography, but with rather low X-ray intensities and more limited experimental possibilities at present. However this new ultra-fast science rapidly progresses around these sources and new large-scale projects exist. It is the aim of this contribution to overview the state of art and the perspectives of fast and ultra-fast X-ray scattering techniques to study photoinduced phase transitions (here, the word ultra-fast is used for sub-picosecond time resolution). In particular we would like to largely present the contribution of crystallographic methods in comparison with optical methods, such as pump-probe reflectivity measurements, the reader being not necessary familiar with X-ray scattering. Thus we want to present which type of physical information can be obtained from the positions of the Bragg peaks, their intensity and their shape, as well as from the diffuse scattering beyond Bragg peaks. An important physical feature is to take into consideration the difference in nature between a photoinduced phase transition and conventional homogeneous photoinduced chemical or biochemical processes where molecules transform in an independent way each other. Actually the photoinduced phase transition with the establishment of the new electronic and structural macroscopic order is preceded by precursor co-operative phenomena due to the formation of nano-scale correlated objects. These are the counterpart of pre-transitional fluctuations at thermal equilibrium which take place above the transition temperature (short range order preceding long range one). Moreover ultra-fast X-ray scattering will play a central role within the fascinating field of manipulating coherence, for instance to directly observe coherent atomic motions induced by a light pulse, such as optical phonons. In the first part of this contribution we present what experimental features are accessible by X-ray scattering to describe the physical picture for a photoinduced structural phase transition. The second part shows how a time-resolved X-ray scattering experiment can be performed with regards to the different pulsed X-ray sources. The first time-resolved X-ray diffraction experiments on photoinduced phase transitions are described and discussed in the third part. Finally some challenges for future are briefly indicated in the conclusion.

Contents

1. Introduction

X-ray scattering experiments have largely contributed to the understanding of the physics of structural phase transitions at thermal equilibrium since they have the potential to give a global picture of structural changes.[1,2] Diffraction pattern with Bragg peaks provides direct information on the change of symmetry, allowing in particular to determine the nature of the order parameter in case of spontaneous symmetry breaking and so it has greatly contributed to the success of the theoretical analysis based on Landau theory. The temperature evolution of the order parameter as well as hysteresis and phase coexistence phenomena can be investigated by X-ray diffraction. In the same way, the determination of the crystal structure by diffraction methods is well established and has largely been utilized to know the nature of the structural change through the phase transition. Furthermore, high resolution X-ray diffraction measurements, i.e. up to large Bragg angles, now allow to accurately determine the electron density. However, only the average structure can be probed by Bragg diffraction itself whereas structural fluctuations give rise to diffuse scattering in addition to Bragg peaks. This is particularly important in the case of pretransitional phenomena which are at the origin of large interatomic or intermolecular correlations. With respect to inelastic and quasi-elastic neutron scattering available to directly study the related critical dynamics, such as soft mode behavior and critical relaxation, new frontiers are recently appearing around synchrotron source with the emergence of inelastic X-ray scattering and dynamical analysis of coherent X-ray scattering.

Another fascinating new frontier in the field of structural science is the growing up of the fast and ultra-fast X-ray science.[3–5] Recent developments in time-resolved X-ray diffraction promise direct access to the dynamics of electronic, atomic and molecular motions in condensed matter triggered by a pulsed laser irradiation, i.e. to record "molecular movies" during the transformation of matter initiated by light pulse. The "laser pump and X-ray probe" techniques now provide an outstanding opportunity for the direct observation of a photoinduced structural phase transition as it takes place. The use of X-ray short-pulse of about 100 ps around third-generation synchrotron sources[6,7] allows structural investigations of fast photoinduced processes. Other new X-ray sources, such as laser-produced plasma ones, generate ultra-short pulses down to 100 fs.[8,9] This opens the way to femtosecond X-ray crystallography, but with rather low X-ray intensities and more limited experimental possibilities at present. However this new ultra-fast science rapidly progresses around these sources and new large-scale projects exist.[10,11] It will contribute significantly to materials science in the near future, authors strongly believe.

It is the aim of this contribution to overview the state of art and the perspectives offered by emerging fast and ultra-fast X-ray scattering techniques to study photoinduced phase transitions (here, the word ultra-fast is used for sub-picosecond time resolution). In particular, we would like to mainly present the contribution of crystallographic methods in comparison with optical methods, such as pump-probe reflectivity measurements, the reader being not necessarily familiar with X-ray scattering. Thus we want to present which type of physical information can be obtained from the positions of the Bragg peaks, their intensity and their shape, as well as from the diffuse scattering out of Bragg peaks. Within this purpose, it is useful to mention photo-crystallography works which do not necessarily concern fast or ultra-fast dynamics, for instance, in relation with photo-stationary state or long living molecular excited state.[12,13] Moreover, one main objective of this new research field is to transpose the progress in the crystallography of excited state to extreme time resolution. In addition, we have also to stress the strong impact of fast and ultra-fast photo-crystallography that can give direct structural information on chemical reactions and biochemical functions.[14–18] An important physical feature is to enlighten the fundamental difference in nature between a photoinduced phase transition and conventional homogeneous photoinduced chemical or biochemical processes where molecules transform in an independent way each other. Actually the photoinduced phase transition with the establishment of the new electronic and structural macroscopic order is preceded by precursor co-operative phenomena due to the formation of nano-scale correlated objects. These are the counter-part of pre-transitional fluctuations which take place above the transition temperature at thermal equilibrium (short range order preceding long range one). Moreover ultra-fast X-ray scattering will play a central role within the field of manipulating coherence, for instance to directly observe coherent atomic motions induced by a light pulse, such as optical phonons.

In the first part of this contribution we present the experimental features accessible by X-ray scattering, that allow to give a physical picture of a photoinduced structural phase transition. The second part shows how a time-resolved X-ray scattering experiment can be performed with regard to the different pulsed X-ray sources. The first time-resolved X-ray diffraction experiments on photoinduced phase transitions are described in the third part and discussed, in particular in relation with recent time-resolved optical experiments. Finally some challenges for the future are briefly indicated in the conclusion.

2. What X-Ray Scattering Can See of a Photoinduced Phase Transition Accompanied by Structural Changes?

This part is devoted for a general presentation of X-ray scattering as a fundamental tool to investigate matter and its transformations, going up to crucial structural information on photoinduced phase transitions. If one looks at a scattering pattern recorded on a modern two-dimensional (2D) detector such as a CCD camera we observe a number of generally intense peaks and sometimes additional weaker diffused intensities outside of these peaks (Fig. 1). Which kind of information can be extracted from the positions of peaks, the intensities of peaks, the shapes of peaks and the more or less spreading out of additional diffuse intensities? The occurrence of sharp peaks, i.e. of Bragg reflections, is the manifestation of a periodic or quasi-periodic long range structural order. That means that the state (position, orientation, configuration, ...) of one molecule in a place determines the state of other over very long distances. This macroscopic ordering presents some intrinsic periodicity : simple for a crystal and more complex in the case of incommensurate structure or quasicrystal. This 3D long range ordering which is manifested by Bragg diffraction only concerns the average structure. In addition, the local fluctuations give rise to diffuse scattering more or less spread out within the reciprocal space, depending on the nature of correlation, i.e. of the short range order.

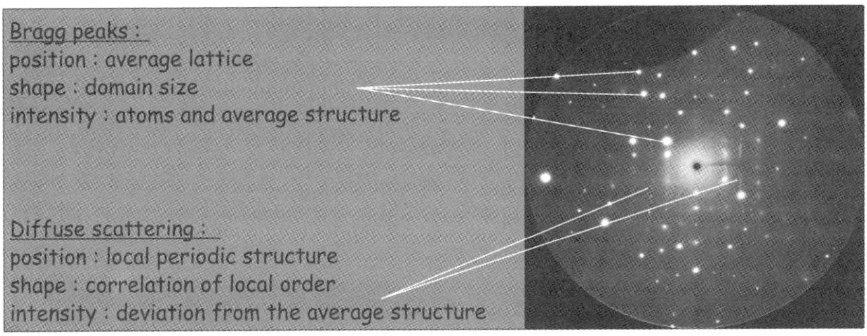

Fig. 1. X-ray scattering pattern recorded on a CCD camera in a quasi 1D system.

2.1. Bragg diffraction from periodic lattice

Now remember some standard aspects of X-ray diffraction by a crystal.[1,2,19] These will be discussed in the kinematical approximation where the magnitude of the incident X-ray wave amplitude is assumed to be the same at all points in the sample. In general, this approximation is valid for many crystalline compounds such as the ones discussed in this book. However for a perfect crystal this approximation is no longer valid and dynamical theory of diffraction has to be utilized. We will return to this delicate point only when it is necessary. Then, in the kinematical approximation, the intensity is proportional to the square modulus of the coherent addition of the amplitudes scattered by all the electrons in the diffracting crystal.

Thus in the simplest case of a crystal built with N × N × N identical unit cells on a 3D periodic lattice (Fig. 2) the scattered intensity in a direction defined by a scattering vector $\vec{Q} = \vec{k}_d - \vec{k}_i$, where $\vec{k}_d$ and $\vec{k}_i$ are respectively the diffracted and incident wave vector, takes on the form:

$$I(\vec{Q}) = \left| \sum_{u=1}^{N} \sum_{v=1}^{N} \sum_{w=1}^{N} F(\vec{Q}) e^{-i\vec{Q}\cdot\vec{R}_{uvw}} \right|^2, \tag{1}$$

where $F(\vec{Q})$ is the scattered amplitude along $\vec{Q}$ for one unit cell. The periodic summation leads to

$$I(\vec{Q}) = \frac{\sin^2(\pi N h)}{\sin^2(\pi h)} \times \frac{\sin^2(\pi N k)}{\sin^2(\pi k)} \times \frac{\sin^2(\pi N l)}{\sin^2(\pi l)} |F(\vec{Q})|^2, \tag{2}$$

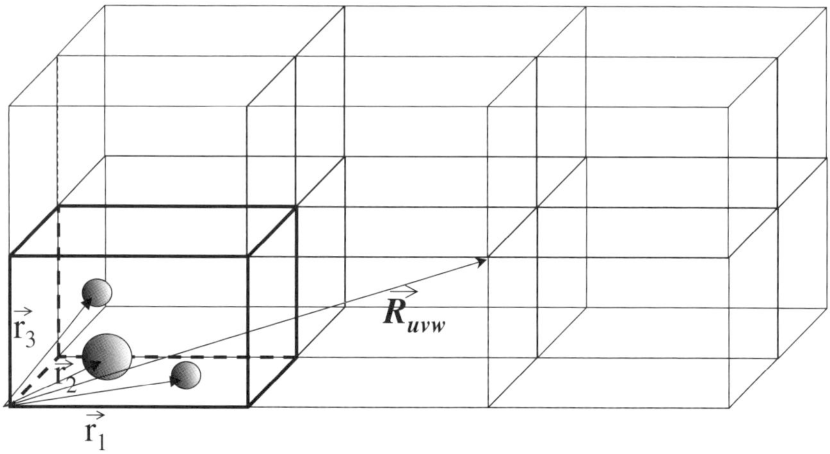

Fig. 2. Schematic representation of a three-dimensional crystal. Each unit-cell contains three atoms at positions $\vec{r}_1, \vec{r}_2, \vec{r}_3$.

where the scattering wave vector $\vec{Q}$ has been decomposed on the reciprocal space basis $\vec{a}^*$, $\vec{b}^*$ and $\vec{c}^*$:

$$\vec{Q} = h\vec{a}^* + k\vec{b}^* + l\vec{c}^* \tag{3}$$

Constructive interference is only observed around integer values of the *hkl* coordinates, then located on the nodes of the reciprocal lattice associated with the periodicity in the direct-space. This is the condition for Bragg scattering which manifests by sharp peaks in the diffraction pattern. Notice that the sharpness of Bragg peaks is governed by the inverse of N, then larger the crystal is sharper the peaks are. Actually this is governed by some imperfections in the crystal (mosaic), nevertheless sharp peaks in the diffraction pattern is always the indication of a macroscopic long range order. Moreover, the intensity of every Bragg peaks is proportional to the square modulus of the amplitude scattered by all the electrons of a unit cell. In the case of X-ray scattering, this last one is the Fourier Transform of the electron density in the unit cell. By considering a spherical distribution of electrons around atoms at rest, a structure factor of the unit cell can be introduced:

$$F(\vec{Q}) = F_{hkl} = \sum_j f_j e^{-\vec{Q}\cdot\vec{r}_i} = \sum_j f_j e^{-2i\pi(hx_j+ky_j+lz_j)}, \tag{4}$$

where f_j is the atomic scattering factor or form factor of spherical atom j, at position r_j. The integrated intensity of a Bragg peak, which is characterized by the three integer values *hkl*, is then determined by F_{hkl}. Thus the information about the distribution of atoms within the unit cell is contained in the structure factor associated with the various nodes of the reciprocal lattice. It should be remarked that accurate determination of the electron density is possible beyond the spherical approximation. To obtain the detailed information about the electron density in the unit cell, it is necessary to collect high resolution diffraction data, in particular for large Bragg angles to be sensitive to subatomic details.

To illustrate these different basic features of X-ray diffraction, Fig. 3 considers a very simple example where every unit cell only contains a small molecule. This shows how the position of Bragg peaks is related to the 3D periodic lattice, which allows to determine these characteristics (unit cell parameters), their intensity to the atomic structure, and their shape to the size of diffracting volume, in general controlled by the imperfections of crystal.

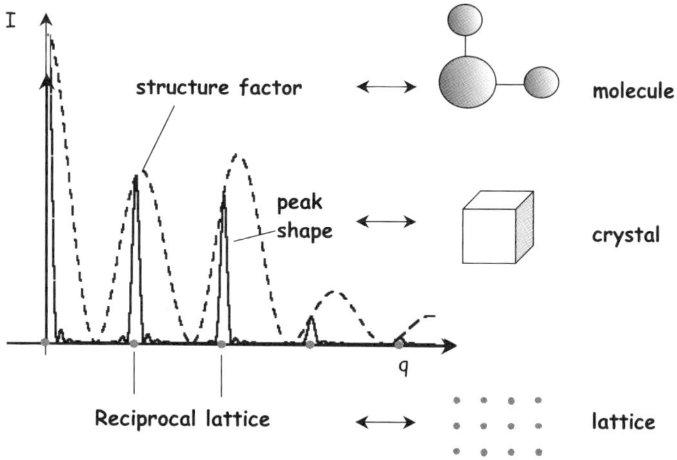

Fig. 3. Dotted lines represent the structure factor of the unit-cell $F(Q)$ in one Q direction; Periodicity of this function depends on the inter-atomic distances inside the unit-cell, i.e. intensity of Bragg peaks will directly depend on atomic positions inside the unit-cell and the presence of symmetry elements can make the intensity at some nodes of the reciprocal lattice to be zero. Three-dimensional lattice of the crystal defines the position of the nodes of the reciprocal lattice, i.e. position of the Bragg peaks allows to determine the unit-cell parameters. The peak shape is controlled by the size of the diffracting volume and related to the imperfections of the crystal.

2.2. *Average structural state*

Within the above description every unit cell are considered as identical. Actually this is the consequence of an average.[1,2] Indeed, an instantaneous picture of the crystal shows that unit cells are different from each other with respect to the atomic positions as a result of intrinsic thermal motions and/or other dynamic or static disordering processes. Thus the instantaneous structure factor of a unit cell "m" is composed of two terms, an average value of the structure factor over the crystal and a fluctuation which depends of m:

$$F_m = \langle F \rangle + \Delta F_m. \tag{5}$$

The mean structure factor $\langle F \rangle$ is associated with an average structure which is periodic, so that Bragg peaks remains sharp and well defined, only their intensity is affected with respect to atoms at rest. Thus the thermal vibration leads to an attenuation of the intensity of Bragg reflections due to the well-known Debye–Waller factor $\exp(-W)$. This is the consequence of incoherent thermal excitations of phonons which manifest by a gaussian distribution of the position of an atom within the harmonic approximation. In a conventional X-ray diffraction experiment the measurement of the integrated intensity of a Bragg reflection extends on a relative long time. Therefore a time averaging is performed which allows the investigation of the state at thermal equilibrium. In contrast a pump-probe fast or ultra-fast X-ray experiment investigates a quasi-instantaneous state. Indeed the averaging period for measurements is determined by the duration of the X-ray probe pulse. It is essential to bear in mind this aspect so that the effects for some degrees of freedom can be time averaged and other ones can not. This is especially important to consider in case of ultra-short X-ray pulses, such as for femtosecond crystallography.

It is important to stress out the difference in the averaging during the dynamical process of a photoinduced phase transition and an homogeneous process. Indeed, in general, there is coexistence of stable and photoinduced phases since the transformation is not often complete. This cannot be discussed as homogeneous random distribution of local photoinduced states (Fig. 4a), where the crystal structure is described by an average structure

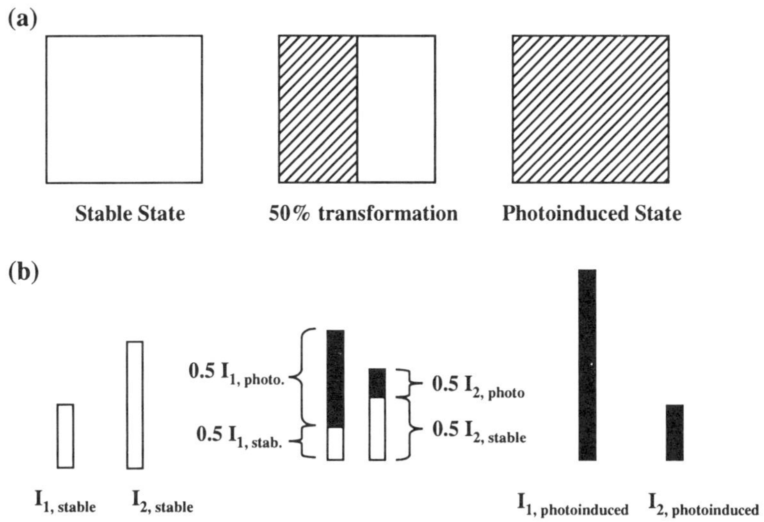

Fig. 4. (a) Schematic descriptions of the crystal in the stable state (left), the fully transformed photoinduced one (right), and a partial transformed state where the fraction volume of the photoinduced domains is 50% (middle). (b) Schematic description by stack columns of the intensity of two Bragg reflections, I_1 and I_2, in the stable (left) and photoinduced (right) states. When the transformation is not complete and occurs via a nucleation of domain process, each measured intensity I is the weighted incoherent contribution of stable and photoinduced domains with respective concentration $(1-x)$ and x in Eq. (7). This is illustrated in a case of a 50% transformation (middle).

factor $\langle F \rangle$, as in independent chemical molecular transformations. This factor is a weighted contribution of the photoinduced entity (F_{photo}, concentration x) and stable one (F_{stable}, concentration $1 - x$):

$$\langle F \rangle = xF_{\text{photo}} + (1 - x)F_{\text{stable}} \tag{6}$$

In a photoinduced phase transition a long range 3D ordering occurs, so that the crystal structure is modified at the macroscopic scale: a very large number of adjacent unit cells are transformed in the same way forming macroscopic 3D domains. Therefore, each domain, photoinduced or stable one, diffracts X-rays, with an associate intensity, I_{photo} and I_{stable}. Since the lattice parameters are different, the diffraction process is incoherent between domains. If the cell parameters are sufficiently different, the two different Bragg peaks, respectively associated with one of the two coexisting phases, are spatially separated on the 2D detector. However, in this case large strains occur and very often the crystal breaks when the phase transition takes place. With respect to this crucial experimental problem the situation is more favorable when the lattice parameters are close, but it becomes difficult to spatially separate the Bragg peaks originating from the two different coexisting phases. Therefore, for each observed Bragg spot the measured intensity I_{measured} is the weighted independent contribution of photoinduced and stable domains, with respective fraction x and $(1-x)$, as schematically shown in Fig. 4:

$$I_{\text{measured}} = xI_{\text{photo}} + (1 - x)I_{\text{stable}}. \tag{7}$$

Starting from the measured intensity, the refinement of both the structure of the photoinduced transformed phase and its fraction x can be difficult since they are highly correlated. However, if some hypothesis is used for the photoinduced structure, the refinement can be simplified.

2.3. *Symmetry change*

Another essential information which can be deduced from time-resolved X-ray diffraction is the symmetry change between the photoinduced phase and the stable one. It is well known that the positions of the nodes of the reciprocal lattice give the lattice periodicity in real space. Thus the crystalline system, with the unit cell parameters, can be extracted from the positions of Bragg peaks and in principle some ferroelastic distortions between the two phases can be detected, as for instance from cubic to tetragonal or from monoclinic to triclinic. In the case of the photo-irradiation of a paraelastic phase the formation of ferroelastic domains in the low symmetry phase has to be taken into account since it yields to a specific twinning. However, in relation with the problem discussed at the end of the last paragraph, if the elastic distortions are weak, it becomes difficult not only to spatially separate on the 2D detector the Bragg reflections between the two coexisting phases, but also between different ferroelastic domains. In addition, some reciprocal nodes are systematically absent in the diffraction pattern due to a structure factor set to zero by symmetry. These systematic absences of nodes, hence of reflections, can be related to the choice of a non-primitive unit cell, as in the textbook examples of centered and face-centered cubic lattices. Another origin is the presence of non-symmorphic symmetry elements. Both are most valuable in determining the space group, an essential part to establish the symmetry change, in particular in the case of symmetry breaking. An illustrated example is provided by the neutral-ionic (N-I) phase transition in tetrathiafulvalene-*p*-chloranil (TTF-CA) classified as a prototypical example of a photoinduced phase transition. It will be discussed in detail in Subsec. 4.1. This is a favorable situation where the symmetry change between a photoinduced phase and a stable one can be easily investigated. The growing of the photoinduced phase can be followed either by the appearance of the new Bragg reflections in the case of the establishment of a photoinduced long range order or with the decrease of the intensity of these specific reflections in the opposite case (disordering process). The situation is similar for a cell doubling ordering or disordering process manifested by the evolution of superstructure Bragg reflections. On the contrary, the appearance or disappearance of an inversion center is in general not easy to establish by X-ray diffraction. Some specific techniques exist but they are not so definitive. Finally an extreme situation is given by the crystalline melting where the whole of Bragg peaks disappears and is replaced by diffuse scattering.

It is well known from the Landau theory of phase transition, beside reconstructive phase transition (no direct symmetry relation), that every density of probability (electronic, atomic position, spin, ...) describing the crystalline structure in the two phases can be expressed as

$$\rho = \rho_0 + \Delta\rho, \tag{8}$$

where the first term ρ_0 is totally symmetric with respect to the high-symmetry phase and the second term $\Delta\rho$ describes the symmetry lowering arising in the low-symmetry phase.[20] This last term is non-zero only in this low-symmetry phase and is proportional to an order parameter associated with the symmetry breaking (for simplicity only a scalar order parameter is considered here). In general, structural phase transitions are associated with a change of space group manifesting symmetry breaking but without drastic change of ρ_0. A typical example is given by displacive ferroelectric phase transition where the order parameter is proportional to the relative static displacements between ionic sub-lattices. As another simple illustration, we may consider order-disorder phase transition with two possible states for a molecule on a crystalline state (orientation, conformation, composition,...). In this last case, the order parameter becomes proportional to the difference in probabilities between these two states. However, there are also fewer examples of isostructural phase transition without change of space group, i.e. without change of symmetry, and which manifests by a discontinuous change of ρ_0 (spin crossover transition, metal-insulator Mott transition,...). This totally symmetric quantity can serve to define an order parameter in the same sense as the density in the liquid-gas phase transition. This is non-zero in the two-phases. As an illustration, the concentration of high-spin molecules is often used to describe the spin crossover phase transition. Notice that the N-I phase transition involves on an equal footing the two types of order parameter, the concentration of I species and dimerization polar order, which can lead to a singular pressure-temperature phase diagram, analogous of gas-liquid-solid, and where the two related kinds of phase transition can occur successively or simultaneously. When symmetry breaking manifests by the appearance of new Bragg peaks, their intensities are proportional to the square of the related order parameter η. Then its temperature evolution can be easily determined from diffraction measurements.

2.4. *Dynamical physical photoinduced features*

This type of Landau approach can also be used to describe the photoinduced phase transition when it is established with respect to the stable one. It should be noted that an order parameter related to a symmetry breaking can appear or disappear depending on whether the photoinduced phase is the low symmetry phase or the one with high symmetry. A central aspect is to know how to describe the physical features in dynamics and then to understand how the photoinduced phase transformation proceeds. After the excitation laser pulse, both the totally symmetric and symmetry breaking density terms become time dependent. However it is essential to be very careful in the use of this description, for instance via time-dependent concentration of transformed species $c(t)$ and symmetry breaking order parameter $\eta(t)$. Indeed, as already discussed, when the phase separation between the photoinduced phase and the stable one takes place at some delay time one pertinent parameter becomes the respective fraction of each (original and photoinduced) phase due to incoherent diffraction process between them. In other words, due to this intrinsic heterogeneous process, an homogeneous description in terms of only $c(t)$ and $\eta(t)$ can not be used anymore. Then the intensities of Bragg reflections depend on both kinds of effects.

Furthermore, the situation can be varied for precursor phenomena leading to the metastable photoinduced phase. A fascinating one is the excitation of coherent phonons consequently to the laser pulse. Thus coherent zone center optical phonons give rise to a periodic oscillation of the intensity of Bragg peaks, since the structure factor can be expressed as:

$$\begin{aligned} F_{hkl}(t) &= \sum_j f_j e^{-\vec{Q}(\vec{r}_j+\Delta\vec{r}_j(t))} \\ &= \sum_j f_j e^{-2i\pi(h(x_j+\Delta x_j(t))+k(y_j+\Delta y_j(t))+l(z_j+\Delta z_j(t)))} \end{aligned} \quad (9)$$

This is basically different from the Debye–Waller effect manifesting the incoherent thermal excitation of phonons. In addition, the ultra-fast X-ray scattering is particularly well adapted to investigate coherent phonons since not only the period but also the polarization (normal coordinate) of the vibrational mode can be measured by the analysis of the oscillation amplitude of different Bragg peaks. Another situation is the formation of precursor nano-domains before the establishment of a new 3D structural order, for instance with the formation of lattice-relaxed CT strings, i.e. 1D segments of several adjacent donor-acceptor pairs in an I (N) state within a N (I) 3D matrix. This introduces characteristic spatial fluctuations around the average structure and so it manifests by related diffuse scattering. Indeed in presence of spatial fluctuations the total scattered X-ray intensity by M unit cells is composed of two terms:

$$I(\vec{Q}) = \sum_m \left|\langle F(\vec{Q})\rangle\right|^2 e^{-i\vec{Q}\cdot\vec{R}m} + M\sum_m \langle \Delta F_0(\vec{Q}) \cdot \Delta F_m(\vec{Q})\rangle e^{-i\vec{Q}\cdot\vec{R}m}, \quad (10)$$

where the first term corresponds to the Bragg diffraction by the average periodic structure and the second one the additional diffuse scattering more or less spread within the reciprocal space, depending of the nature of correlation.[1,2] This diffused scattering is governed by the Fourier transform of the correlation spatial extension, for instance diffuse plane in the case of the formation of decorrelated1D nano-objects. Thus the time evolution of the shape and the size of these precursor nano-objects can be experimentally determined. Moreover this diffuse scattering will condense in Bragg reflections when the 3D order will take place. Photoinduced phase transitions provide a richness of more or less coherent transformations and motions triggered by ultra-short laser pulse. However, these manifest on different intrinsic spatial scales and associated different time scales. Thus, impulsive coherent atomic or molecular precursor short-range transformations and motions (oscillatory and/or not)on sub-ps time scale take place first. On the other hand, the long range transformation in a new photoinduced phase in general leads to a phase separation and is mainly governed by the motion of phase fronts (often called domain walls) on the acoustic phonon time scale. These motions at a longer scale can also be coherent, and for instance acoustic phonons, domain wall oscillations and/or other coupled more or less oscillatory motions can be generated. A deep understanding of such phenomena at different scales requires the combined utilization of X-ray scattering at different time scales, fast and ultra-fast ones.

3. How Fast or Ultra-Fast Time-Resolved X-Ray Scattering Experiment Proceeds?

In this section, a short description of experimental facilities to perform fast or ultra-fast X-ray diffraction experiments with laser pump and X-ray probe technique is given. The pump is generally a classic femtosecond laser, such as a Ti:sapphire one, producing light pulses of about 100 fs, while the width of X-ray pulse probes depends of the nature of the source: about 100 ps for synchrotron source and about 100 fs for other ones such as laser-produced plasma sources.

3.1. *Fast and ultra-fast X-ray sources*

A new generation of X-ray tools is associated with the development of fast and ultra-fast X-ray sources, and this is of fundamental interest for many users coming from different scientific community such as biology, chemistry and physics. These new X-ray sources make it possible for an X-ray pulse duration to reach the 100 picosecond down to the 100 femtosecond time scales. This really opens a new field for the investigation of fast or ultra-fast processes in nature associated with molecular or atomic motions, from the impulsive atomic motions to the macroscopic transformation of a material. Time-resolved X-ray diffraction is the most appropriate technique to get key information on the structural rearrangements associated with such ultra-fast phenomena and the available time resolutions allow us to track very short-lived excited states. Different possibilities allow us to reach sub-nanosecond time resolution. Two main techniques are used in the generation of X-ray pulses, where the time resolution is limited by the width of the pulses.

The first possibility is to use the electromagnetic radiation emitted by the acceleration of charged particles.[1] This technique is used around synchrotron sources, taking advantage of the time structure of the electrons bunches in the ring, each bunch emitting its own X-ray pulse. As bunches of electrons are running inside the synchrotron ring with a speed close to the one of light, the time width of an emitted X-ray pulse is typically between 30 to 150 ps, depending on the specification of the machine, the current in the ring, ...[6] It is this technique that has been used to obtain the results presented hereafter on N-I photoinduced phase transition.

Laser-produced plasma sources are a second possibility to generate ultra-short X-ray pulses.[8,9] The technique is based on the X-ray radiation emitted by the relaxation of electrons in a plasma, excited by a 100 fs optical pulse. The X-ray pulse width is then limited by the width of the optical exciting pulse and typically 100 fs X-ray pulses can be generated in this way. At the present stage, the weak intensity of the X-ray beam produced by this technique limits its use to the measurement of few Bragg reflections. However, laser-plasma based X-ray sources are table-top facility in comparison with large facilities, such as synchrotron sources.

Other techniques, also based on the acceleration of charged particles, are now developed.[10,11] This is the case for Linac-based light sources, which uses the bunch length compression technique to allow us to generate 80 fs pulses, as the Sub-Picosecond Pulsed Source (SPPS) in Stanford. Another possibility to generate 100 fs X-rays has also been demonstrated for third generation synchrotron machines.[21] It consists of generating a temporal structure inside an electron bunch in a storage ring, by using the optical slicing of the bunch generated by its interaction with an 100 fs visible laser pulse. The 100 fs electron bunch generated in this way will then radiate into an undulator to produce synchrotron 100 fs duration X-ray pulses. All these developing techniques represent the state of the art for time-resolved X-ray diffraction experiments, but it is also an intermediate state before future sources based on X-ray free electron-laser sources (FEL). Of course the properties of the beam strongly depends on the source in terms of brillance, divergence, pulse width and the different sources are complementary. Future sources should combine the quality of the synchrotron beam and the time scale reached by 100 fs sources.

The use of ultra-fast X-ray detectors, such as streak camera, is another possibility for reaching sub-picosecond time resolution, allowing to access time information during an X-ray pulse.[22–24] However, the sensitivity strongly depends on the time resolution used and the delay dependence or the time resolution is somewhat limited.

3.2. *Optical laser pump and X-ray probe techniques*

Fast time-resolved X-ray diffraction measurements utilize the pulsed structure of the synchrotron radiation to probe photoinduced structural changes in condensed matter with a time resolution of about 100 ps. The pioneering work was performed on the ID09B beam-line at the European Synchrotron Radiation Facility (ESRF) in France.[6,7] It has made this technique operational and it is also developing now in other part of the world. The optical pump and X-ray probe techniques consist of exciting the sample with an optical pulse (typically of 100 fs width) and probing it with an X-ray pulse at a given delay *dt* during the relaxation of the metastable photoinduced state or, what is even more interesting, during the transformation as shown in Fig. 5. Such a pulsed laser excitation can trigger molecular switching or

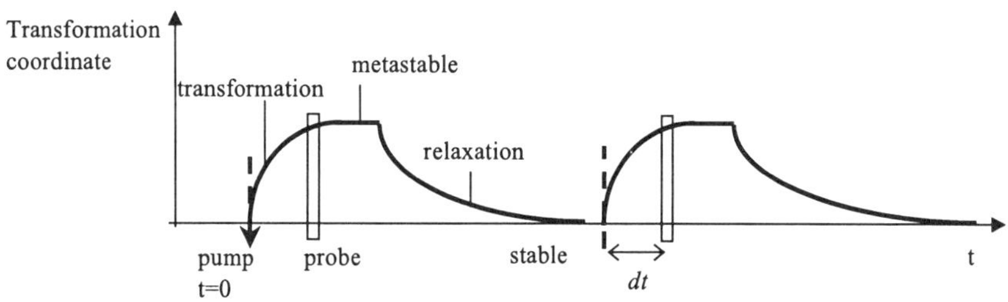

Fig. 5. Schematic drawing of the principle of laser pump and X-ray probe technique. At $t = 0$ the laser pump the sample which transforms to a metastable state with a finite life-time followed by the relaxation and the recovery of the thermal equilibrium. A synchronized X-ray pulse can probe the sample at a given delay dt. As single pulse measurements are not accurate enough, the experiment is performed in a stroboscopic way, X-ray probing the sample always at the same delay.

other photoinduced phenomena, such as photoinduced phase transitions. However, the number of excitation photon available in each laser pulse, which can reach one photon per molecule in a crystal of few 100 μm thick, is related to the repetition rate because of the finite power delivered by the pumping laser. In addition, as one wants to probe the system in the transient state in a stroboscopic way, one as to wait for the complete relaxation of the sample before exciting it again as shown in Fig. 5. That is the reason why the repetition rate is often set at about 1 kHz.

Let us take the example of a 100 ps experiment performed at the ESRF in the 16 bunch mode. As shown in Fig. 6, the different X-ray pulses of 100 ps, generated by the different electron bunches inside the ring, are separated only by about 176 ns. If one wants to probe the sample every ms, in relation with a laser repetition rate of 1 kHz, only one pulse every 6400 has to be selected and utilized for observation. That is the reason why a mechanical chopper is used (Fig. 6). The main part of technical difficulty for the pulse selection to control the delay time is the synchronization technique among the mechanical chopper, the synchrotron pulses and the laser ones, so that the time $t = 0$, corresponding to the interaction of the laser with the sample can be well defined. The diffracted X-ray, always probing the sample at a given delay *dt*, are then collected on an accumulation-type detector. Depending on the kind of diffraction experiment one wants to perform, monochromatic or polychromatic X-ray beam can be used. Thus such a technique has been successfully utilized to perform structural investigations of local chemical or biochemical photoinduced phenomena occurring on this 100 ps time scale: watching myoglobin in action (Lauë method),[17] iodine dissociation in solution (diffuse scattering),[15] transient structural change in a photosensitive organic solid (powder),[14] and other ones.[13,16]

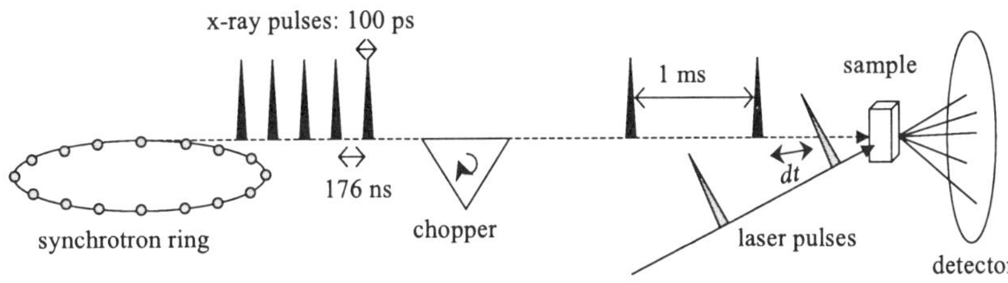

Fig. 6. Schematic drawing of the temporal structure of the X-ray pulses generated by the ESRF synchrotron in the 16 bunch mode. A chopper selects 1 pulse every ms probing the sample at the delay dt after the laser excitation. The diffraction pattern is collected on a 2D detector.

It is also very promising to study photoinduced structural phase transition as it will be illustrated in the following for the N-I transition. For this situation the investigation of small molecules systems requires monochromatic experiments. It can be performed on powder or single crystal, and in this last case many Bragg reflections are accurately measured (about one thousand). In addition, the use of a single crystal makes it possible to control the laser polarization effect.

The experimental facilities around laser-produced plasma top-table sources give the possibility to perform ultra-fast X-ray diffraction experiments.[8,9,25] At present their ultimate time resolution, of about 100 fs, is well suitable for the time scale of small atomic displacements. In particular this gives the possibility to directly observe the oscillations of an optical phonon, such as for instance a dimerization mode. Another interesting intrinsic property of this technique is that the time $t = 0$ is very well defined since it is the same optical laser that will pump the crystal to be studied and generate the plasma X-ray pulse. Thus difficult synchronization problems are avoided with respect to synchrotron sources and other electron based X-ray generation large facilities (LINAC, FEL, ...). However, the x-Ray pulse is emitted through a solid angle of 2π. That is the reason why such a X-ray source is highly divergent. In addition, only a small part

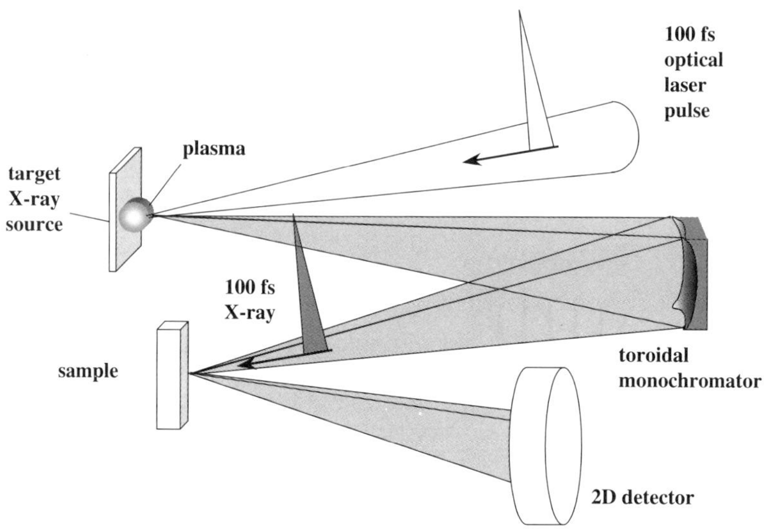

Fig. 7. Schematic drawing of a X-ray diffraction experiment around a laser-produced plasma source.

of the typically 10^9 emitted X-ray photons per pulse can be collected by a toroidal monochromator (Fig. 7) so that the number of photons on the sample is in the order of 10^3–10^4. Such a divergent technique makes it possible to collect Bragg reflections without requiring oscillation of the sample, as it is the case for well collimated sources where one needs to perform the so-called rocking-curve to integrate the Bragg reflection. Large improvement of this technique are in progress and an important increase of several order of magnitude of the X-ray intensity on the sample is expected.

3.3. *X-ray probed volume vs. photo-excited part*

An important problem concerns the comparison between the volume probed by X-ray and the part which is efficiently excited by the light pulse, as schematically shown in Fig. 8. When one excites a material with a laser pulse, the light excitation can only transform the material on a finite depth. This depends of the light penetration depth d_L and of the intensity of laser pulse with respect of the value of the intensity threshold since the cooperative photoinduced processes are highly nonlinear. Depending on the material, this may typically extend on a few penetration depth d_L, then from sub-μm to few 100 μm. In addition, X-rays are also absorbed by a material and can only probe a sample over few penetration depth d_x, which can be of the order of a few μm for compounds with heavy elements, such as conventional semi-conductors, to a few mm for molecular materials composed of light atoms. In addition, for a single crystal which is sufficiently perfect, the reflectivity is always less than predicted by the kinematical approximation

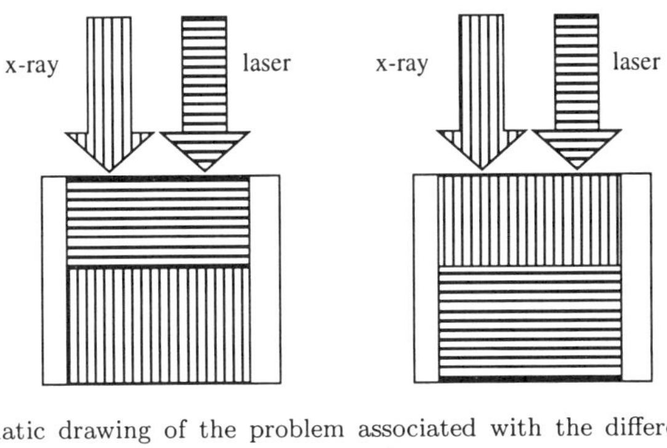

Fig. 8. Schematic drawing of the problem associated with the different penetration depths of the laser (d_L) and X-ray (d_x). If (d_L) is small, the laser excites the sample mainly around the surface (left). If (d_x) is small, the sample can be probed only round the surface.

and an extinction length is introduced.[1] Such a difference in the penetration depths d_L and d_x is a crucial problem which will be more deeply discussed in the following presentation of experimental results. From now let's indicate that in the case of low d_x the reflection diffraction geometry is used. In contrast, when d_x is large, all the sample is probed both in reflection and transmission geometry. This means that in the general case, the X-ray may not probe an homogeneous sample as described in Subsec. 2.2.

4. First Fast or Ultra-Fast Time-Resolved X-Ray Diffraction Experiments

This part describes some experimental results which have been recently obtained by fast or ultra-fast time-resolved X-ray diffraction on some photo-induced phase transitions in condensed matter. First, we present the 100 ps investigation of the N-to-I as well as I-to-N photoinduced phase transitions in the TTF-CA prototype compound. This illustrates how fast the X-ray facility around a synchrotron source allows, thanks to the measurement of diffraction patterns comprising a large number of Bragg spots, to determine the symmetry change and the nature of a photoinduced structural phase transition in relatively complex materials, such as those discussed in this book. Second, we describe three recent sub-ps works utilizing a laser produced plasma source: the non-thermal surface melting in conventional semi-conductors, the insulating-to-metal phase transition in VO_2 and coherent optical phonons. These high-temporal resolution experiments on more conventional hard solids have been limited to the measurement of a few Bragg peaks.

4.1. *Fast probing of N-I photoinduced phase transition*

Remember some characteristic properties of the N-I phase transition in the quasi-one-dimensional molecular mixed-stack charge-transfer material TTF-CA.[26,27] In this compound donor (D) and acceptor (A) molecules alternate along the same stack. Strong electronic and structural coupling gives rise to a chain multistability between a regular neutral (N) state, $\ldots D^\circ A^\circ D^\circ A^\circ D^\circ A^\circ \ldots$, and two degenerate and polar dimerized ionic (I) states, $\ldots(D^+A^-)(D^+A^-)(D^+A^-)\ldots$ or $\ldots(A^-D^+)(A^-D^+)(A^-D^+)\ldots$ At thermal equilibrium, this compound undergoes an electronic-structural phase transition, the so-called N-I transition, from a high symmetry N phase towards a low symmetry ferroelectric I phase (Fig. 9)[28] and this by lowering temperature[27] or increasing pressure.[29] In the N phase the

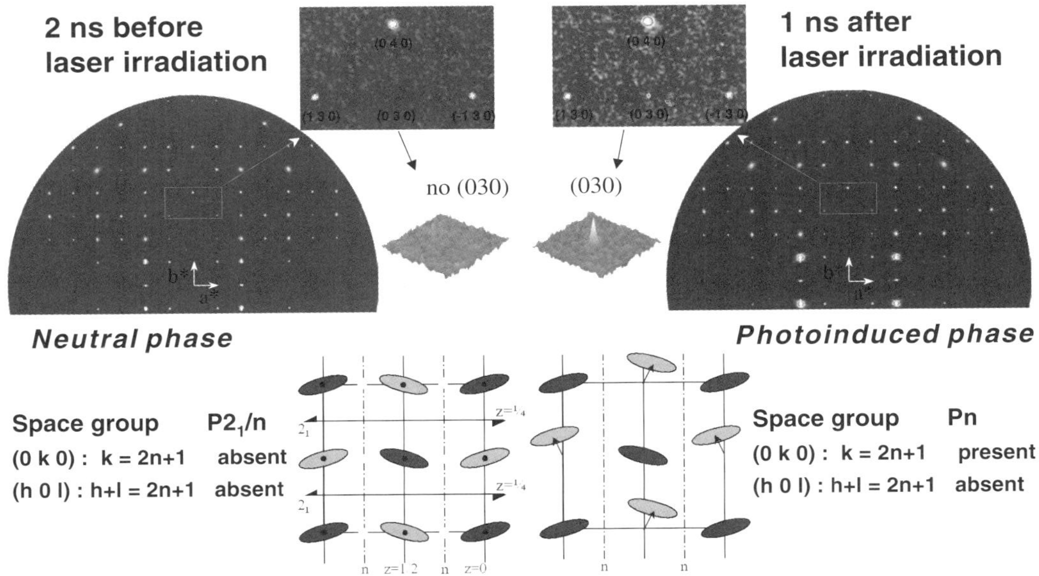

Fig. 9. Investigation of the N-to-I photoinduced transformation. Reconstructed Intensity in the reciprocal (a^*, b^*) planes before $(\Delta t = -2 - ns)$ and after $(\Delta t = +1 - ns)$ laser irradiation. The appearance of the (030) reflection, visible directly on the CCD image (inset), signs the ferroelectric nature of the three-dimensional ordered photoinduced state. Schematic drawing of the symmetry breaking associated with the phase transition from the N (left) and I (right) states. In the N states, donor (black) and acceptor (gray) molecules are located on inversion symmetry site (•). In the I state, the dimerization process gives rise to the loss of the inversion center and the loss of the screw axis corresponds to a ferroelectric ordering between the stacks.

space group is $P2_1/n$, manifested in the diffraction pattern by two types of systematic absences, $0k0$: $k = 2n + 1$ due to the presence of two-fold axis and $h0l$: $h + l = 2n + 1$ due to the presence of glide plane. The reason for such systematic absences can be easily described. For instance in the first case, the diffracted amplitude for 0k0 Bragg reflection, manifested by the F_{0k0} structure factor, is definitely sensitive to y co-ordinates, with the result that the two TTF (CA) molecules in the unit cell, interrelated by a screw axis, actually become equivalent by a translation of **b**/2. In the I phase, the screw axis symmetry is lost simultaneously with the center of inversion and new $0k0$: $k = 2n + 1$ appear characteristic of the symmetry breaking (space group Pn). At atmospheric pressure, the temperature of the strongly first-order transition is $T_c = 81\,K$. This phase transition has been analysed as a cascade of cooperative multi-scale phenomena: first the formation of nano CT-strings accompanied by their one-dimensional 1-D lattice relaxation,[30] (conventionally represented by $\ldots D^\circ A^\circ D^\circ A^\circ (D^+A^-)(D^+A^-)(D^+A^-)D^\circ A^\circ \ldots$), as recently evidenced by X-ray diffuse scattering,[31] and then their three-dimensional condensation and ordering.[32,33]

Pump and probe optical reflectivity spectroscopy has evidenced the photoinduced change of electronic state, both for the I-to-N and for the N-to-I, on the 100 ps time scale.[34–38] The photoinduced phenomena are highly cooperative and highly nonlinear because the efficiency, which can be as high as a few hundred of transformed DA pairs, is not simply proportional to the total absorbed energy and a threshold effect on the laser intensity exists. This is considered as one of the main characteristics of a photoinduced phase transition.[39] Time-resolved fast X-ray diffraction provides an outstanding opportunity for the direct observation with the appropriate time-resolution, in this prototypical example, of the photoinduced structural changes and symmetry breaking triggered by a pulsed laser irradiation.

The experiments[40,41] were performed on the beam-line ID09B at the European Synchrotron Radiation Facility (ESRF), using the optical pump & X-ray probe method. Monochromatic X-ray pulses of 100 ± 10 ps are used to probe a single crystal which is cooled by either a nitrogen stream for investigating the N-to-I photoinduced transformation or by a Helijet helium stream for studying the I-to-N. Because the system recovers the thermal equilibrium state on the μs-ms time scale, it is possible to probe the sample by recording the diffraction pattern stroboscopically using the ID09B set up. The use of cold stream allows an easy light irradiation of the crystal. It was pumped by ultra-short laser pulses at 1.55 eV (800 nm), with the polarization parallel to the stacking axis. This is an off resonant excitation because the CT band is centered on 0.65 eV. A sufficient photon density is needed in order to excite a large part of the crystal, i.e. to be above the threshold intensity, extending over a few penetration depths (estimated in a few μm range). Rotations of the crystal of small amplitude (2°) around the stacking axis were performed to collect diffraction data on a 2D detector (MAR-CCD). A well-adapted software (CrysAlis) was utilized for the data reduction and the reconstruction of the scattered intensity in the reciprocal space. Thus, more of 6000 peaks can be collected, corresponding to more than 800 or 1600 unique reflections, depending on whether this is the high-symmetry or the low-symmetry phase.

For both the N-to-I photoinduced transformation at 90 K and the I-to-N one at 70 K, drastic changes were observed in the intensity of some Bragg reflections (Figs. 10 and 11), with some decreasing and others increasing, excluding simple laser-heating effects that could only lower the intensities via an increase of the Debye–Waller factor. Such important intensity changes are a direct signature of a strong structural reorganization in the photoinduced state. For the N-to-I photoinduced transformation, complete data collections have been performed at 93 K for two time delays, −2 ns

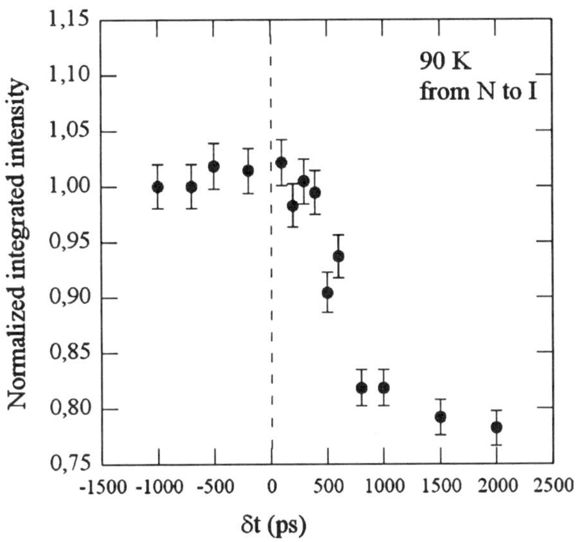

Fig. 10. Normalized intensity of the (0 6 7) Bragg reflection versus the delay between the laser pump and the X-ray probe pulses. The important change in the intensity signs a structural reorganization associated with neutral-to-ionic transformation as previously reported. A significant change occurs in the 500 ps range.

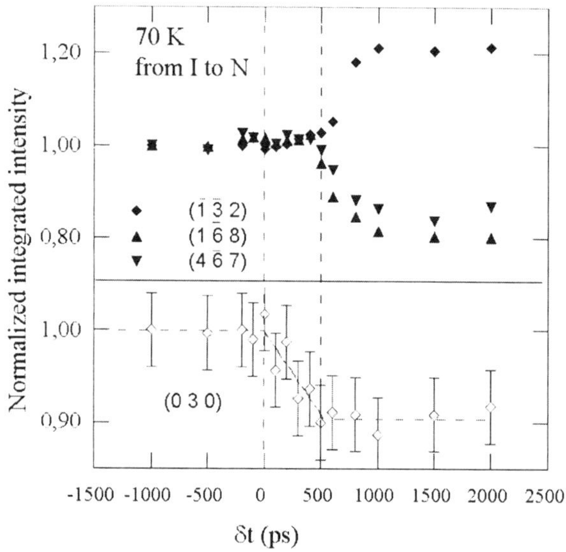

Fig. 11. Normalized intensity dependence of some Bragg reflections with the delay between the laser pump and the X-ray probe pulses. This is associated with the photoinduced transformation of the I phase to the N one. General reflections (top, symbols include error bars) are modified after an incubation time of about 500 ps, whereas the intensity of the (0 3 0) reflection starts to decrease just after the laser pulse excitation (bottom), indicating a two steps mechanism with an intermediate I_{para} disordered phase.

and +1 ns, between the laser excitation (1.7×10^{16} photon/cm^2) and the X-ray probe. At −2 ns, the extracted parameters and the resolved structure (Fig. 9, left) are in perfect agreement with the N structure ($\mathrm{P2_1/n, Z = 2}$) obtained at 90 K at thermal equilibrium, both for the intra-and intermolecular geometries. At +1 ns, the important feature is the observation of a symmetry lowering with the appearance of (0k0) reflections with k odd, forbidden by the symmetry of the neutral phase (Fig. 9, right). The space group of the photoinduced phase is then Pn, Z = 2, as for the I phase at thermal equilibrium, implying a ferroelectric long range order between dimerized ionic chains. Besides, we have also investigated the I-to-N photoinduced phase transformation at 70 K. As seen in Fig. 11, normal reflections (that are all Bragg reflections except the (0k0) ones with k odd) are modified after an incubation time estimated of about 500 ps, whereas intensities of (0k0) reflections with k odd are modified on a shorter time scale. Thus the intensity of the (0 3 0) reflection characteristic of the change of symmetry starts to decrease just after the laser irradiation to stabilize after the 500 ps. Such an observation is very similar to the results of time-resolved optical reflectivity study where a two step process has been evidenced.[33] Indeed, the signal of second harmonic generation decreases and disappears much rapidly than reflectivity changes for intra-molecular bands. This is characteristic of a faster restoring of the center of symmetry before the molecular transformation from I to N state. This dynamics feature is discussed as resulting from a first step disordering process of I strings after the photo-excitation in a similar way that the static high pressure behavior at thermal equilibrium with a succession of a ferroelectric I, paraelectric I and paraelectric N phases.[32] It is responsible for the decrease of (030) reflection intensity and nearly constant ones for normal reflections which are not affected by such a disordering process between I dimers. Furthermore, the intensity of the (0 3 0) reflection does not reach zero because the transformation of the sample is not complete. In the same way for the N-to-I photoinduced transformation, in the case of a lower excitation density, the (0 3 0) reflection which is the most intense new one at low Bragg angles, was not observed, nor were there significant changes in any other Bragg peak intensities. This supports the idea that these cooperative photoinduced processes are highly nonlinear. When the number of transformed molecular species is below a critical threshold, the photo-switching cannot take place at the macroscopic scale. Thus with a crystal width in the 100-μm range, the transformation may not extend over all the volume. As the laser light penetrates the sample, the excitation density decreases and may become lower than the critical threshold after a given depth. We have also to stress that since the TTF-CA is composed of relatively light elements X-ray really probes the bulk. In contrast, optical reflectivity measurements probe a region close to the surface. Consequently the probe dynamics may be different. In particular nucleation processes may be strongly influenced by the vicinity of the surface.

An important problem to debate is the coexistence of stable and photoinduced phases. We have to stress out that the N-to-I transformation does not proceed via an homogeneous random distribution of local photoinduced states, but via a phase separation between very large 3D domains of the stable and the photoinduced phases (macroscopic ordering). Therefore, each domain incoherently diffracts X-rays with an associated intensity. The spatial resolution used, limited in particular by the pixel size of the MAR CCD camera, did not allow to observe a significant shift of the lattice parameters or of the spot position between the stable and photoinduced phases. Therefore, for each Bragg spot the measured intensity $I_{measured}$ is the weighted incoherent contribution of stable (I_{stable}) and photoinduced (I_{photo}) domains, with respective concentration

$(1-x)$ and x:

$$I_{\text{measured}} = x\, I_{\text{photo}} + (1-x) I_{\text{stable}}. \tag{11}$$

As already mentioned, it is very difficult to refine both the structure of the photoinduced phase and the photoinduced fraction x from the measured intensities, since they are highly correlated. Assuming that the structure of the photoinduced ferroelectric phase is the same as the one of the ferroelectric I phase stable at low temperature and therefore with identical diffracted intensities, it is possible to estimate the transformation rate from the variation of the intensities. In such a case, there is a linear dependence of the variation of the intensities with x:

$$I_{\text{measured}} - I_{\text{neutral}} = x \cdot (I_{\text{ionic}} - I_{\text{neutral}}). \tag{12}$$

The comparison of more than 6000 reflections with $F^2/\sigma(F^2) > 5$ was performed for the N-to-I photoinduced transition. The result is reported in Fig. 12 giving an estimate of the transformation rate in the 35% range. This means that the crystal is transformed over about 20–40 μm, then over a few penetration depths. One major problem originates from the dependence of the effective penetration depth of the laser light with regards to the orientation of the sample. Therefore, during the data collection where the rotation of the crystal extends over 180°, different transformation rates

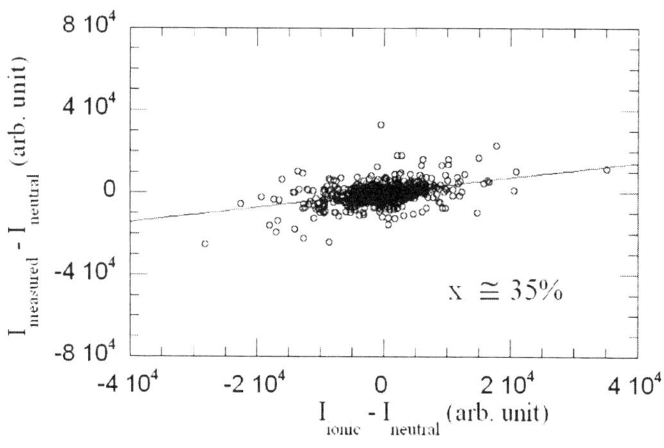

Fig. 12. Dependence of the variation of the intensity of more than 6000 Bragg reflections between the measurements in the N-to-I photoinduced experiment and the neutral state ($I_{\text{measured}} - I_{\text{neutral}}$) with the variation of the intensity between the neutral and ionic phases ($I_{\text{ionic}} - I_{\text{neutral}}$). The slope makes it possible to estimate the transformation rate around 35%.

should be generated for the 91 collected frames and our result has to be considered as an average. The order of magnitude of the transformation rate is a few tenth of percent, in agreement with the decrease of the (0 3 0) reflection observed for the I-to-N transition (Fig. 11). As this intensity is zero in the photoinduced N phase, a decrease of 10% corresponds to 10% of transformation.

4.2. *Ultra-fast probing in hard condensed matter*

Now we will describe some new experimental results with a sub-ps time resolution, down to 100 fs, which have been obtained by utilizing a laser produced plasma source. These results concern some more conventional and more simple hard solids, i.e. inorganic solids with only a few number of atoms by unit cell. In these compounds, contrary to many systems in this book, the electronic excitations are delocalized. Thus, when there are a very large fraction of valence electrons which are photo-excited in the conduction band this leads to an electronic-lattice destabilization which can generate a new periodic lattice structure or even melting, often related to an insulating-to-metal instability. In other words, intense light pulses may trigger long range lattice instabilities which can be ultra-fast, since only small coherent atomic motions are involved. This differs from the situation of more complex systems such as molecular crystals. In addition, these samples were perfect single crystals with the result that only a region in the vicinity of the surface can be probed by X-ray diffraction. Indeed the kinematical approximation is no more valid and dynamical theory of diffraction has to be utilized. In this situation the probed region is not determined by the absorption penetration depth but by the extinction length, in general 100 nm–1 μm range. This may be of the same order that the laser light penetration depth, but different. Since laser produced plasma sources provide rather low X-ray intensities at present, it is important have for signal/noise optimum conditions by matching of the optical excitation depth with the X-ray probing one. This can be performed by the use of single crystal thin film with appropriate thickness; but also by the use of special geometry such as discussed in the following paragraph. However if the present studied processes are ultra-fast near the surface, the extension to the bulk may be longer.

The first example is given by the non thermal photoinduced surface melting in semi-conductors.[42–45] The term "non thermal" means that this surface melting takes place on the sub-ps time scale before the typical electron-phonon thermalization one. This is observed in different materials and it is characterized by the disappearance of the Bragg reflections associated with the loss of the periodic lattice. Such processes are really ultra-fast since the highly excitation of electronic degrees of freedom triggers impulsive coherent atomic motions on a time scale which can reach few hundred fs only. The material transforms in an disordered quasi-liquid state before the thermal relaxation occurs. Let us take the case of InSb semi-conductor studied by A. Rousse *et al.*[42] The generation of a dense electron-plasma was confirmed by optical measurements, which can not allow to evidence melting since the optical signal associated with the liquid state is very difficult to separate from other contributions. Therefore, in order to investigate such an ultra-fast melting, the 100 fs diffraction technique is appropriate. The pump-and-probe technique was used to excite the sample with 120-fs pulses of laser light and to detect the resulting structural changes. 100 fs x-ray pulses were generated by focusing a second beam from the laser on a silicon target and the pulses are then directed at the sample. X-rays were detected with a cooled CCD camera. The excitation pulse causes disorder in the sample, eliminating diffraction peaks so that the x-ray signal drops to zero. The results are presented in Fig. 13. As discussed above, an important point concerns the matching of the optical excitation depth (around 1000 Å) with the X-ray probing one. Playing on

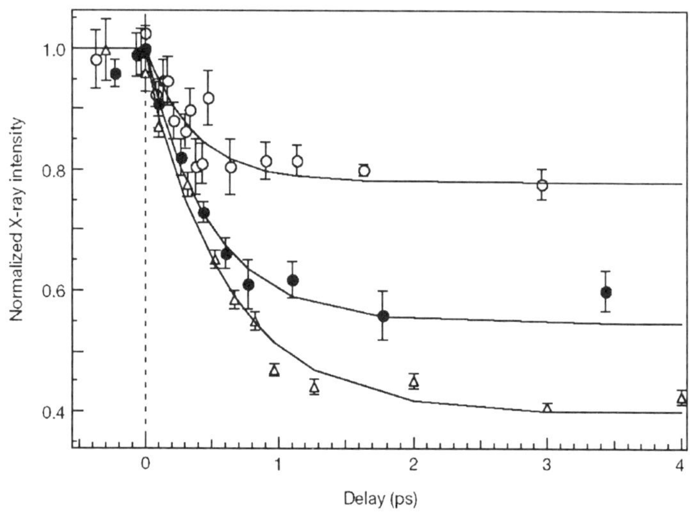

Fig. 13. From Ref. 43. Delay dependence of the normalized X-ray diffracted intensity of the (111) Bragg reflection for different X-ray probing thicknesses of 3500 Å (open circles) 1700 Å (filled circles) and 1200 Å (triangles). The exciting laser pulse penetration depth is around 1000 Å. Bragg reflection disappears as the surface melts. Nature's copyright permission.

the diffraction geometry, it was possible to play on the penetration depth of the X-ray from about 3500 to 1200 Å as shown in Fig. 13. The lower the penetration depth is, the higher the decrease of the Bragg reflection is, but a non melted part is always probed. This is a direct signature of the formation of a liquid film at the surface of the sample. Similar results have also been obtained by other groups on other semi-conductor materials and by utilizing thin films. It will be interesting in future, when higher X-ray intensity will be available, to investigate in which way the quasi-instantaneous disordered state obtained just after the disappearance of Bragg peaks, is similar or different to the liquid one at thermal equilibrium.

The second example we want to discuss concerns a fs photoinduced solid-solid phase transition in the oxyde of vanadium VO_2.[46] It is known to undergo an insulator-to-metal transition at 340 K and changes in the electronic band structure are associated with structural changes between a low-T monoclinic and a high-T rutile phase. Structural probing was achieved using a laser-produced plasma x-ray source after a 50 fs, 800 nm, p-polarized optical pump pulse with a fluence equal to $15\,\mathrm{mJ/cm^2}$, where no cumulative damage was observed. Figure 14 shows the formation in two steps of the metallic rutile phase consecutive to the laser excitation. The first step, occurring at the very surface (depth estimate: 40–60 nm and thus very high degree of electronic excitation, about $5.10^{21}\,\mathrm{cm^{-3}}$), corresponds to a macroscopic transition within less than 500 fs (Fig. 14 left inset). Such a time scale, shorter than the typical internal thermalization time of a non equilibrium phonon distribution, means that this excitation regime can not correspond to the conventional pathway for a first-order phase transition (nucleation and then growth of the new phase). A. Cavalleri *et al.* suggest that excitation of a dense carrier population across the 600 meV-band gap may significantly perturb the potentiel energy surface of the electronic ground state and depress the barrier separating the two phases. Further experiments are necessary to answer the fundamental questions : which comes first in the sample, the structural change or the electronic one, what microscopic process is responsible for initiating the structural distortion and what causes the disappearance of the gap after optical excitation.

The third example corresponds to the first observation of coherent optical phonons in a material by ultra-fast X-ray diffraction.[47] Bismuth is one of the prototype materials in which coherent optical phonons are generated upon femtosecond laser excitation. The optical phonons at the center of the Brillouin zone should affect the X-ray diffraction intensity because of the changes in the structure factor through the oscillatory atomic motion which occurs coherently. From the polarization of the A1g phonon involved

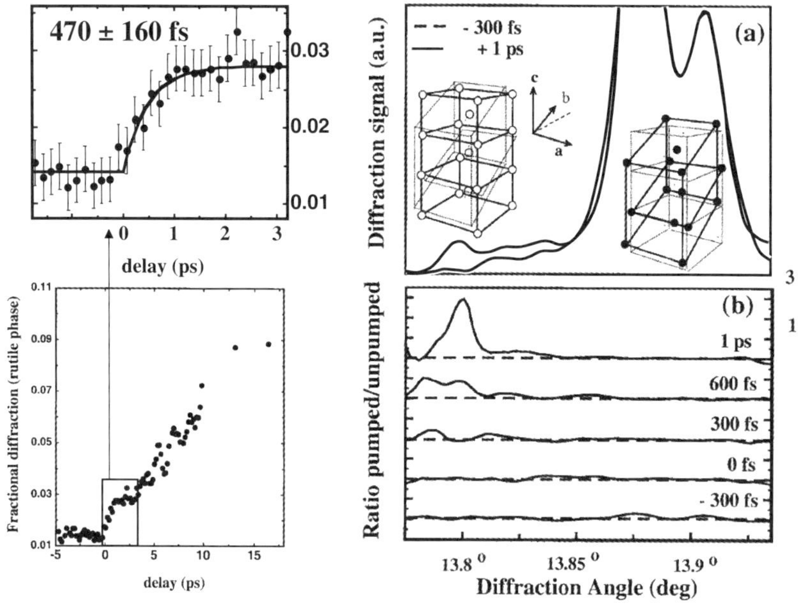

Fig. 14. From Ref. 46. Left: Time-dependent, integrated diffraction signal from the metallic rutile phase normalized to the integrated diffraction from the monoclinic phase. Inset: region near zero time delay. Right: *Upper plot*: Angle dependent x-ray diffraction signal of the (110) Bragg reflection measured for negative and positive delay times. *Lower plot*: Ratio between the X-ray reflectivity of the excited and unpumped crystal. Copyright credit: A. Cavalleri *et al.*, *Phys. Rev. Lett.* **87** 237401 (2001).

(giving the displacements $\Delta xj, \Delta yj$ and Δzj discussed in Subsec. 2.4), it is possible to calculate the effect of the amplitude of atomic oscillation around the stable position on the structure factor. These changes affect in an opposite way the (111) and the (222) Bragg reflections. Following the effect from a short laser pulse, K. Sokolowski–Titen *et al.* directly observed from the changes in the Bragg peaks intensity the large-amplitude coherent atomic vibrations. As shown in Fig. 15 for low laser fluence, clear oscillations of the (111) and (222) reflections are observed and the atomic displacements affect in an opposite way the two reflections mentioned here. From the measurements, it is possible to extract the phonon frequency, and, as the variation of the intensity occurs with a different sign for different Bragg reflections, it is possible to separate such coherent atomic motion inside the unit cell from the signal associated with incoherent dynamics, for example melting as discussed above. The frequency of the observed

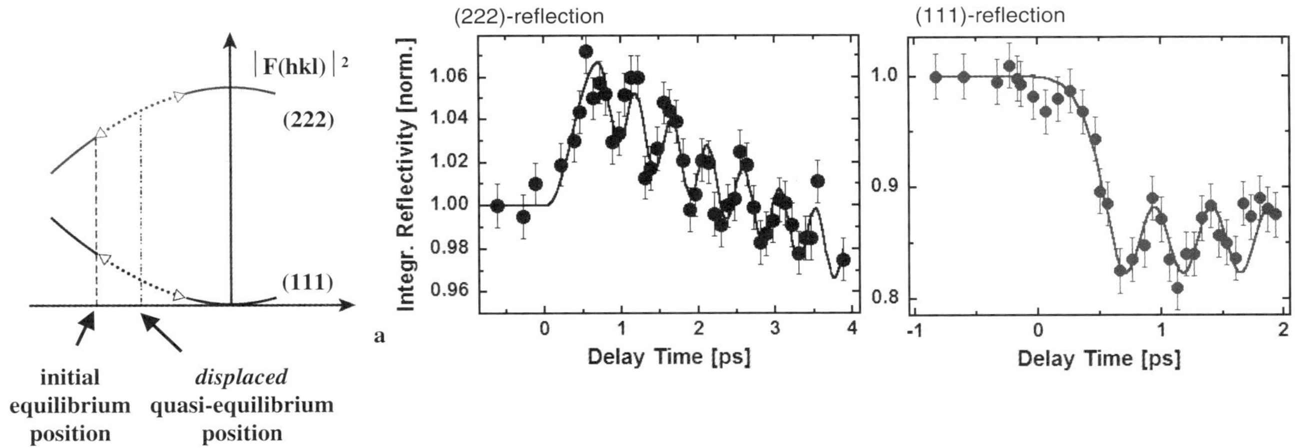

Fig. 15. From Ref 47. Variation of the intensity of (111) and (222) Bragg reflections with the distance between the two basis atoms of bismuth (left). The optical phonon generated by the optical laser pulse is associated with the coherent oscillation of the atoms, giving rise to the oscillation of the intensity of the reflections. (111) and (222) reflections are affected in an opposite way (center and left) as expected from the calculation (left). The decrease of the mean value of this last one is associated with Debye–Waller effect. Nature's copyright permission.

phonon shifts from 2,94 THz down to 2.1 THz and this is attributed to lattice softening and strong anharmonicity due to the large phonon amplitude (approx. 0.1–0.2 Å). For fluences below the melting threshold the diffraction signal exhibits a pronounced oscillatory behavior. For higher fluences (>10 mJ/cm^2) the initial stage of large amplitude optical phonon motion is followed by disordering/melting of the material as evidenced by a diffraction order independent drop of the diffraction efficiency. From the intensity dependence it is estimated that the initial coherent atomic displacement exceeds 10% of the nearest neighbor distance at these high excitation levels. It may thus be speculated whether this can lead to a destabilization of the lattice which triggers the transition to the disordered liquid state. Finally, let us indicate that the observation of lattice strain wave, then of impulsively generated acoustic phonons, had been reported before.[48,49]

5. Conclusion

Fast and ultra-fast X-ray diffraction is still a young field. Advances in present pulsed X-ray sources and facilities, as well as the carrying out of next-generation X-ray light sources, promise a new level in structural science and in investigation of dynamics in condensed matter. The reviewed results presented here give an illustration of what can be done and open exciting perspectives. Some present scientific challenges exist, for instance, to extend the utilization of ultra-fast X-ray diffraction in more complex materials such as those discussed in this book. Among the large contributions that this new advanced method may provide an essential one is the light pulse control of phase transitions. A deep understanding of such phenomena at different scales requires the combined use of X-ray diffraction over different time scales (fast and ultra-fast), but also of temporal X-ray absorption and optical spectroscopies. This is the key for elucidating this new kind of coherent manipulation of matter by light and for controlling ultra-fast macroscopic switching of materials.

Acknowledgments

The fast X-ray diffraction project is an international fruitful collaboration with Michael Wulff (ESRF), Tadeusz Luty (Technical University, Wroclaw), Simone Techert (Max Planck Institut, Gôttingen) and A. Plech (University of Konstanz). This work was partly supported by the New Energy and Industrial Technology Development Organisation and the Ministry of

Education, Science, Culture and Sports of Japan. We acknowledge the helpful contribution during this work of Mathias Meyer (Oxford Diffraction, Wroclaw) and Loïc Toupet and Philippe Rabiller at GMCM (Rennes). In addition we gratefully acknowledge fruitful discussions and exchanges of information with Antoine Rousse (LOA, Palaiseau), Andrea Cavalleri (LBL, Berkeley) and Klaus Sokolowski–Tinten (University of Essen).

References

1. *Neutron and Synchrotron Radiation for Condensed Matter Studies*, Vol. 1: *Theory, Instruments and Methods*, eds. J. Baruchel, J. L. Hodeau, M. S. Lehmann, J. R. Regnard and C. Schlenker (Springer-Verlag, Berlin, Editions de Physique, Les Ulis, 1993).
2. *Neutron and Synchrotron Radiation for Condensed Matter Studies*, Vol. 2: *Applications to Solid state Physics and Chemistry*, eds. J. Baruchel, J. L. Hodeau, M. S. Lehmann, J. R. Regnard and C. Schlenker (Springer-Verlag, Berlin, Editions de Physique, Les Ulis, 1993).
3. K. A. Nelson, *Science* **286**, 1310 (1999).
4. J. Wark, *Nature* **398**, 284 (1999).
5. C. W. Siders and A. Cavalleri, *Science* **300**, 591 (2003).
6. M. Wulff, F. Schotte, G. Naylorand and D. Bourgeois, *Nucl. Instrum. Methods Phys.* **A398**, 69 (1997).
7. M. Wulff, A. Plech, L. Eybert, *et al.*, *Faraday Discuss.* **122**, 13 (2002).
8. A. Rousse, C. Rischel and J. C. Gauthier, *Rev. Mod. Phys.* **73**, 17 (2001).
9. A. Cavalleri, C. W. Siders, K. Sokolowski-Tinten, *et al.*, *Opt. Photonics News* **12**, 28 (2001).
10. R. F. Service, *Science* **298**, 1356 (2002).
11. H. C. Kapteyen and T. Ditmire, *Nature* **420**, 467 (2002).
12. *Time-Resolved Diffraction*, ed. J. R. Helliwell and P.M. Rentzepis (Clarendon Press, Oxford, 1997).
13. P. Coppens and I. Novozhilova, *Faraday Discuss.* **122**, 1 (2002).
14. B. Perman, V. Srajer, Z. Ren, *et al.*, *Science* **279**, 1946 (1998).
15. S. Techert, F. Schotte and M. Wulff, *Phys. Rev. Lett.* **86**, 2030 (2001).
16. R. Neutze, R. Wouts, S. Techert, *et al.*, *Phys. Rev. Lett.* **87**, 195508 (2001).
17. A. Geis, M. Bouriau, A. Plech, *et al.*, *J. Lum.* **94–95**, 493 (2001).
18. F. Schotte, M. Lim, T. A. Jackson, *et al.*, *Science* **300**, 1944 (2003).
19. C. Kittel, *Introduction to Solid State Physics* (John Wiley & Sons, New York, 1986).
20. E. M. Lifshitz and L. P. Pitaevskii, *Landau and Lifshitz Course of Theoretical Physics*, Vol. 5 *Statistical Physics*, Part 1 (Pergamon Press, Oxford, 1980).
21. R. W. Schoenlein, S. Chattopadhyay, H. H. W. Chong, *et al.*, *Science* **287**, 2237 (2000).
22. M. Wulff, D. Bourgeois, T. Ursby, *et al.*, *Time-Resolved Diffraction*, eds. J. R. Helliwell and P. M. Rentzepis (Clarendon Press, Oxford, 1997), p. 195.
23. J. Larsson, Z. Chang, E. Judd, *et al.*, *Opt. Lett.* **22**, 1012 (1997).

24. J. Larsson, P. A. Heimann, A. M. Lindenberg, *et al.*, *Appl. Phys. A: Mater. Sci. Process* **66**, 587 (1998).
25. C. Rischel, A. Rousse, I. Uschmann, *et al.*, *Nature* **390**, 490 (1997).
26. J. B. Torrance, J. E. Vasquez, J. J. Mayerle and V. Y. Lee, *Phys. Rev. Lett.* **46**, 253 (1981).
27. J. B. Torrance, A. Girlando, J. J. Mayerle, *et al.*, *Phys. Rev. Lett.* **47**, 1747 (1981).
28. M. Le Cointe, M. H. Lemée-Cailleau, H. Cailleau, *et al.*, *Phys. Rev.* **B51**, 3374 (1995).
29. H. Okamoto, T. Koda, Y. Tokura, *et al.*, *Phys. Rev.* **B39**, 10693 (1989).
30. N. Nagaosa, *Solid State Comm.* **57**, 3488 (1986).
31. E. Collet, M. H. Lemée-Cailleau, M. Buron-Le Cointe, *et al.*, *Europhys. Lett.* **57**, 67 (2002).
32. M. H. Lemée-Cailleau, M. LeCointe, H. Cailleau, *et al.*, *Phys. Rev. Lett.* **79**, 1690 (1997).
33. T. Luty, H. Cailleau, S. Koshihara, *et al.*, *Europhys. Lett.* **59**, 619 (2002).
34. S. Koshihara, Y. Tokura, T. Mitani, *et al.*, *Phys. Rev.* **B42**, 6853 (1990).
35. S. Koshihara, Y. Takahashi, H. Sakai, *et al.*, *J. Phys. Chem.* **B103**, 2529 (1999).
36. T. Suzuki, T. Sakamaki, K. Tanimura, *et al.*, *Phys. Rev.* **B60**, 6191 (1999).
37. K. Tanimura and I. Akimoto, *J. Lumin.* **94–95**, 483 (2001).
38. S. Iwai, S. Tanaka, K. Fujinuma, *et al.*, *Phys. Rev. Lett.* **88**, 057402 (2002).
39. K. Nasu, H. Ping and H. Mizouchi, *J. Phys. : Condens. Matter* **13**, 693 (2001).
40. E. Collet, M. H. Lemée-Cailleau, M. Buron-Le Cointe, *et al.*, *Science* **300**, 612 (2003).
41. L. Guérin, E. Collet, M. H. Lemée-Cailleau, *et al.*, *Chem. Phys.* in press.
42. C. W. Siders, A. Cavalleri, K. Sokolowski-Tinten, *et al.*, *Science* **286**, 1340 (1999).
43. A. Rousse, C. Rischel, S. Fourmaux, *et al.*, *Nature* **410**, 65 (2001).
44. K. Sokolowski-Tinten, C. Blome, C. Dietrich, *et al.*, *Phys. Rev. Lett.* **87**, 225701 (2001).
45. S. K. Sundaram and E. Mazur, *Nature Materials* **1**, 217 (2002).
46. A. Cavalleri, C. Toth, C. W. Siders, *et al.*, *Phys. Rev. Lett.* **87**, 237401 (2001).
47. K. Sokolowski-Tinten, C. Blome, J. Blums, *et al.*, *Nature* **422**, 287 (2003).
48. A. H. Chin, R. W. Schoelein, T. E. Glover, *et al.*, *Phys. Rev. Lett.* **83**, 336 (1999).
49. A. Cavalleri, C. W. Siders, F. L. H. Brown, *et al.*, *Phys. Rev. Lett.* **85**, 586 (2000).

INDEX